T0137311

Springer Proceedings in Complexity

Springer Complexity

Springer Complexity is an interdisciplinary program publishing the best research and academic-level teaching on both fundamental and applied aspects of complex systems—cutting across all traditional disciplines of the natural and life sciences, engineering, economics, medicine, neuroscience, social, and computer science.

Complex Systems are systems that comprise many interacting parts with the ability to generate a new quality of macroscopic collective behavior the manifestations of which are the spontaneous formation of distinctive temporal, spatial, or functional structures. Models of such systems can be successfully mapped onto quite diverse "real-life" situations like the climate, the coherent emission of light from lasers, chemical reaction–diffusion systems, biological cellular networks, the dynamics of stock markets and of the Internet, earthquake statistics and prediction, freeway traffic, the human brain, or the formation of opinions in social systems, to name just some of the popular applications.

Although their scope and methodologies overlap somewhat, one can distinguish the following main concepts and tools: self-organization, nonlinear dynamics, synergetics, turbulence, dynamical systems, catastrophes, instabilities, stochastic processes, chaos, graphs and networks, cellular automata, adaptive systems, genetic algorithms, and computational intelligence.

The three major book publication platforms of the Springer Complexity program are the monograph series "Understanding Complex Systems" focusing on the various applications of complexity, the "Springer Series in Synergetics", which is devoted to the quantitative theoretical and methodological foundations, and the "SpringerBriefs in Complexity" which are concise and topical working reports, case-studies, surveys, essays, and lecture notes of relevance to the field. In addition to the books in these two core series, the program also incorporates individual titles ranging from textbooks to major reference works.

More information about this series at http://www.springer.com/series/11637

Sean Cornelius · Kate Coronges
Bruno Gonçalves · Roberta Sinatra
Alessandro Vespignani
Editors

Complex Networks IX

Proceedings of the 9th Conference
on Complex Networks CompleNet 2018

 Springer

Editors
Sean Cornelius
Center for Complex Network Research
Northeastern University
Boston, MA
USA

Kate Coronges
Network Science Institute
Northeastern University
Boston, MA
USA

Bruno Gonçalves
Center for Data Science
New York University
New York, NY
USA

Roberta Sinatra
Center for Network Science, and Department
  of Mathematics and its Applications
Central European University
Budapest
Hungary

Alessandro Vespignani
Bouvé College of Health Sciences
Northeastern University
Boston, MA
USA

ISSN 2213-8684   ISSN 2213-8692   (electronic)
Springer Proceedings in Complexity
ISBN 978-3-319-89240-5   ISBN 978-3-319-73198-8   (eBook)
https://doi.org/10.1007/978-3-319-73198-8

Printed on acid-free paper

This Springer imprint is published by Springer Nature
The registered company is Springer International Publishing AG
The registered company address is: Gewerbestrasse 11, 6330 Cham, Switzerland

Preface

The International Workshop on Complex Networks CompleNet (www.complenet. org) was initially proposed in 2008, and the first workshop took place in 2009 in Catania. The initiative was the result of efforts from researchers from the (i) BioComplex Laboratory in the Department of Computer Sciences at Florida Institute of Technology, USA, and the (ii) Dipartimento di Ingegneria Informatica e delle Telecomunicazioni, Universit di Catania, Italy. CompleNet aims at bringing together researchers and practitioners working on complex networks or related areas. In the past two decades, we have indeed witnessed an exponential increase in the number of publications in this field. From Biology to Computer Science, from Economics to Social Systems, Complex Networks are becoming pervasive in many fields of science. It is this interdisciplinary nature of complex networks that CompleNet aims at addressing. CompleNet 2018 was the ninth event in the series and was hosted at Northeastern University in Boston, MA, from March 5–8, 2018.

This book includes the peer-reviewed list of works presented at CompleNet 2018. We received an unprecedented 222 submissions from 36 countries around the world. Each submission was reviewed by at least three members of the Program Committee. Acceptance was judged based on the relevance to the symposium themes, clarity of presentation, originality and accuracy of results, and proposed solutions. After the review process, 13 full papers and 15 short papers were selected to be included in this book. The 26 contributions in this book address many topics related to complex networks and have been organized in five major groups: (1) Theory of complex networks, (2) Graph Embeddings, (3) Network Dynamics, (4) Network Science Applications, (5) Human Behavior and Social Networks. We would like to thank the Program Committee members for their work in promoting the event and refereeing submissions. We are grateful to our speakers: Aaron

Clauset, Kayle De La Haye, James Evans, Jon Kleinberg, David Lazer, Yelena Mejova, J. P. Onnela, Sam Scarpino, Olaf Sporns, Jessika Trancik, Milena Tsvetkova, Fernada B. Viégas, and Martin Wattenberg; their presentation is one of the reasons CompleNet 2018 was such a success.

Boston, MA, USA   Sean Cornelius
                                                                    Kate Coronges
                                                                  Bruno Gonçalves
                                                                  Roberta Sinatra
                                                             Alessandro Vespignani

Contents

Contributors

Naveed Afzal Department of Computer Science, School of Computing Tokyo Institute of Technology, Meguro, Tokyo, Japan

Edoardo M. Airoldi Department of Statistics, Harvard University, Cambridge, USA

Younis Al Rozz BioComplex Laboratory, Computer Science, Florida Institute of Technology, Melbourne, FL, USA

Hend Alrasheed Department of Computer Science, Kent State University, Kent, OH, USA

Alessia Amelio DIMES, University of Calabria, Rende (CS), Italy

Mary Jean Amon Department of Psychological and Brain Sciences, Indiana University, Bloomington, IN, USA

Gonzalo Bacigalupe CIGIDEN, National Research Center for Disaster Risk Management, Edificio Hernan Briones, Campus San Joaquin, Universidad Catolica de Chile, Santiago, Chile; CSP, CEHD, University of Massachusetts Boston, Boston, MA, USA

Khalid Bakhshaliyev Computer Science and Engineering Department, University of Nevada, Reno, NV, USA

Amotz Bar-Noy City University of New York, New York City, NY, USA

Niloy Biswas Department of Statistics, Harvard University, Cambridge, USA

Glencora Borradaile School of EECS Oregon State University, Corvallis, OR, USA

Torsten Braun University of Bern, Bern, Switzerland

Muhammed Abdullah Canbaz Computer Science and Engineering Department, University of Nevada, Reno, NV, USA

V. Carchiolo Dip. di Ingegneria Elettrica Elettronica Informatica (DIEEI), UniversitÃ degli Studi di Catania, Catania, Italy

Dvir Cohen Software and Information Systems Engineering, Ben-Gurion University of the Negev, Beersheba, Israel

Ariadne A. Costa Department of Psychological and Brain Sciences, Indiana University, Bloomington, IN, USA

Catherine Cramer New York Hall of Science, Queens, NY, USA

Farnaz Zamani Esfahlani Department of Systems Science and Industrial Engineering, Center for Collective Dynamics of Complex Systems, Binghamton University, Binghamton, NY, USA

Daniele Fadda University of Pisa, 2 Pisa, Italy

Luis H. Favela Department of Philosophy and Cognitive Sciences Program, University of Central Florida, Orlando, FL, USA

Raphaël Fournier-S'niehotta CEDRIC CNAM, Paris, France

Takayasu Fushimi School of Computer Science, Tokyo University of Technology, Hachioji-city, Tokyo, Japan

Lorenzo Gabrielli University of Pisa, 2 Pisa, Italy

Ralucca Gera Naval Postgraduate School, Monterey, CA, USA

Fosca Giannotti KDD Lab. ISTI-CNR, 1 Pisa, Italy; University of Pisa, 2 Pisa, Italy

Patrick Gildersleve Oxford Internet Institute, University of Oxford, Oxford, UK

Silvia Giordano University of Applied Sciences and Arts of Southern Switzerland (SUPSI), Manno, Switzerland

Mehmet Hadi Gunes Computer Science and Engineering Department, University of Nevada, Reno, NV, USA

Harith Hamoodat BioComplex Laboratory, Computer Science, Florida Institute of Technology, Melbourne, FL, USA

Ryohei Hisano Social ICT center, University of Tokyo, Tokyo, Japan

Matan Hugi Software and Information Systems Engineering, Ben-Gurion University of the Negev, Beersheba, Israel

Guo-Ping Jiang Nanjing University of Posts and Telecommunications, Nanjing, China

Noriko Kando Information and Society Research Division, National Institute of Informatics, Chiyoda-ku, Tokyo, Japan

Yui Kazawa Graduate School of Systems and Information Engineering, University of Tsukuba, Tsukuba, Ibaraki, Japan

Vesa Kuikka Finnish Defence Research Agency, Riihimäki, Finland

Michaela Labriole New York Hall of Science, Queens, NY, USA

Matthieu Latapy CNRS, UMR 7606, LIP6, Sorbonne Universités, UPMC Univ Paris 06, Paris, France

Patrizia Lattarulo IRPET, Firenze, Italy

A. Longheu Dip. di Ingegneria Elettrica Elettronica Informatica (DIEEI), UniversitÃ degli Studi di Catania, Catania, Italy

Luca Luceri University of Applied Sciences and Arts of Southern Switzerland (SUPSI), Manno, Switzerland; University of Bern, Bern, Switzerland

Clémence Magnien CNRS, UMR 7606, LIP6, Sorbonne Universités, UPMC Univ Paris 06, Paris, France

M. Malgeri Dip. di Ingegneria Elettrica Elettronica Informatica (DIEEI), UniversitÃ degli Studi di Catania, Catania, Italy

G. Mangioni Dip. di Ingegneria Elettrica Elettronica Informatica (DIEEI), UniversitÃ degli Studi di Catania, Catania, Italy

Ronaldo Menezes BioComplex Laboratory, Computer Science, Florida Institute of Technology, Melbourne, FL, USA

Theresa Migler Cal Poly, San Luis Obispo, CA, USA

Letizia Milli University of Pisa, Pisa, Italy; KDD Lab. ISTI-CNR, Pisa, Italy

Terrence J. Moore U.S. Army Research Lab, adelphi, MD, USA

Tsuyoshi Murata Department of Computer Science, School of Computing Tokyo Institute of Technology, Meguro, Tokyo, Japan

Mirco Nanni University of Pisa, 2 Pisa, Italy

Siddharth Pal Raytheon BBN Technologies, Cambridge, MA, USA

Pedro Paredes CRACS & INESC-TEC, DCC-FCUP, Universidade do Porto, Porto, Portugal

Dino Pedreschi University of Pisa, 2 Pisa, Italy

Leonardo Piccinini IRPET, Firenze, Italy

M. Previti Dip. di Ingegneria Elettrica Elettronica Informatica (DIEEI), UniversitÃ degli Studi di Catania, Catania, Italy

Rami Puzis Software and Information Systems Engineering, Ben-Gurion University of the Negev, Beersheba, Israel

Ram Ramanathan Gotenna Inc, New York City, NY, USA

Pedro Ribeiro CRACS & INESC-TEC, DCC-FCUP, Universidade do Porto, Porto, Portugal

Giulio Rossetti KDD Lab. ISTI-CNR, 1 Pisa, Italy

Benedek Rozemberczki School of Informatics, University of Edinburgh, Edinburgh, U.K.

Rik Sarkar School of Informatics, University of Edinburgh, Edinburgh, U.K.

Tetsuji Satoh Faculty of Library, Information and Media Science, University of Tsukuba, Tsukuba-city, Ibaraki, Japan

Hiroki Sayama Department of Systems Science and Industrial Engineering, Center for Collective Dynamics of Complex Systems, Binghamton University, Binghamton, NY, USA

Lori Sheetz Center for Leadership and Diversity in STEM, U.S. Military Academy at West Point, West Point, NY, USA

Zion Sofer Software and Information Systems Engineering, Ben-Gurion University of the Negev, Beersheba, Israel

Olaf Sporns Department of Psychological and Brain Sciences, Indiana University, Bloomington, IN, USA

Ananthram Swami U.S. Army Research Lab, adelphi, MD, USA

Andrea Tagarelli DIMES, University of Calabria, Rende (CS), Italy

Emma Towlson Center for Complex Network Research, Northeastern University, Boston, MA, USA

Sho Tsugawa Graduate School of Systems and Information Engineering, University of Tsukuba, Tsukuba, Ibaraki, Japan

Stephen Uzzo New York Hall of Science, Queens, NY, USA

Javier Velasco-Martin CIGIDEN, National Research Center for Disaster Risk Management, Edificio Hernan Briones, Campus San Joaquin, Universidad Catolica de Chile, Santiago, Chile

Tiphaine Viard CEDRIC CNAM, Paris, France

Xinwei Wang Nanjing University of Posts and Telecommunications, Nanjing, China

Gordon Wilfong Bell Labs, Murray Hill, NJ, USA

Xu Wu Nanjing University of Posts and Telecommunications, Nanjing, China

Taha Yasseri Oxford Internet Institute, University of Oxford, Oxford, UK; Alan Turing Institute, The British Library, London, UK

Feng Yu City University of New York, New York City, NY, USA

Part I
Theory of Complex Networks

On the Eccentricity Function in Graphs

Hend Alrasheed

Abstract Given a graph $G = (V, E)$, the eccentricity of a vertex u is the distance from u to a vertex farthest from u. The set of vertices that minimizes the maximum distance to every other vertex (has minimum eccentricity) constitutes the center of the graph. The minimum eccentricity value represents the graph's radius. The eccentricity function of a graph can be unimodal or non-unimodal. A graph with unimodal eccentricity function has the property that the eccentricity of every vertex equals its distance to the center plus the radius. A graph with non-unimodal eccentricity function lacks this property. In this work, we characterize each type of eccentricity function and study the impact of each type on the intersection of shortest paths among distant vertex pairs with the center. A shortest path intersects the center if it includes at least one vertex that belongs to the center. In particular, we show that if the eccentricity function is unimodal, all shortest paths among distant vertex pairs intersect the graph's center. We also discuss when those paths do not intersect the center in graphs with non-unimodal eccentricity functions.

1 Introduction

Research shows that in many real-world networks, traffic tends to concentrate on a subset of densely connected vertices (known as the core) that is also central to the network [1, 5, 12]. While the definition of which vertices are core varies, there are multiple centrality measures that are commonly used to assist core identification. First, one or more centrality measures are used to rank vertices according to their importance. Then, vertices of higher ranks are included in the core [2, 5]. Some of the centrality measures are the closeness, the betweenness, and the eccentricity centrality. The eccentricity of a vertex u in a given unweighted graph is the maximum distance (number of edges) from u to any other vertex in the graph. The eccentricity centrality determines the center as the vertex (or vertices) that minimizes the maximum distance. The importance of those central vertices in applications is due to their

H. Alrasheed (✉)
Department of Computer Science, Kent State University, Kent, OH 44242, USA
e-mail: halrashe@kent.edu

© Springer International Publishing AG 2018
S. Cornelius et al. (eds.), *Complex Networks IX*, Springer Proceedings
in Complexity, https://doi.org/10.1007/978-3-319-73198-8_1

role as organizational hubs. Generally, distant vertices communicate faster through such intermediate vertices [2]. This work is motivated by systems in which information is being sent between distant vertices (people, locations, etc.) and must pass through a central vertex (where the information is processed) before the transmission is completed.

Let $d(u, v)$ denote the distance between vertices u and v in a connected, unweighted, and undirected graph $G = (V, E)$. For a vertex u, we define $F(u)$ as the set of vertices at maximum distance from u. The *eccentricity* $ecc(u)$ is the distance from u to a vertex $v \in F(u)$. The minimum and maximum eccentricities represent the graph's *radius* $rad(G)$ and *diameter* $diam(G)$. The vertices with minimum eccentricity constitute the *center* $C(G)$. A *path* of length $k - 1$ from vertex u_0 to vertex u_k is a sequence of adjacent vertices $u_0, u_1, \ldots, u_k$ and a *shortest path* from u_0 to u_k, denoted by $\rho(u_0, u_k)$, is a path that minimizes this length.

By definition, the values of the eccentricities represent the distances among vertices (vertices with smaller eccentricities are closer to other vertices). However, eccentricity values also provide some insight into the distances between vertices and the center of the graph. Based on this observation, the eccentricity function can be described as either *unimodal* or *non-unimodal*. A graph with unimodal eccentricity function has the property that $ecc(u) = d(u, C(G)) + rad(G)$ for every vertex u. A graph with non-unimodal eccentricity function lacks this property. In this work, we characterize the unimodality of the eccentricity function in graphs based on the monotonicity of the shortest paths that connect noncentral vertices in a graph to its center. We also study the impact of unimodality on the intersection of shortest paths among distant vertex pairs with the center (a shortest path intersects the center if it includes at least one vertex that belongs to the center). A vertex pair (u, v) is a *distant vertex pair* if $v \in F(u)$ and in this case, we say that $\rho(u, v)$ is a long shortest path. In particular, we show that if the eccentricity function is unimodal, all shortest paths among distant pairs intersect the center. We also discuss when those paths do not intersect the center in graphs with non-unimodal eccentricity functions. Finally, we examine the eccentricity functions of a set of real-world networks.

2 Eccentricity, Locality, and Monotonicity

Let $G = (V, E)$ be a connected, unweighted, and undirected graph with eccentricity function $ecc(u)$ defined for every vertex and let $C(G)$ be its center. In this work, we use two kinds of vertex layering (both related to vertex eccentricities).

Graph eccentricity layering. The *eccentricity layering* $\mathcal{EL}(G)$ partitions the vertex set V into layers $\ell_r(G)$, $r = 0, 1, \ldots$ based on their eccentricities. Each layer r includes all vertices with eccentricities $r = ecc(G) - rad(G)$, i.e., $\ell_r(G) = \{u \in V : ecc(u) - rad(G) = r\}$. Here, r represents the index of the layer. Note that vertices located at the outermost layer have eccentricities equal to the graph's diameter.

Distance-to-center layering. The *distance-to-center layering* $\mathcal{DC}(G)$ partitions the vertex set V into layers $\zeta_r(G)$, $r = 0, 1, \ldots$ based on their distances to $C(G)$. Each layer r includes all vertices at distance r from $C(G)$, i.e., $\zeta_r(G) = \{u \in V : d(u, C(G)) = r\}$. Here, r represents the index of the layer.

The possible difference between the two layerings of a noncentral vertex is provided in the following proposition.

Proposition 1 *For a vertex $u \in V \backslash C(G)$, where $u \in \ell_r(G)$ of the graph eccentricity layering, the distance between u and the center $d(u, C(G))$ is bounded as follows.*
$r \leq d(u, C(G)) \leq rad(G), r \geq 1$.

The upper bound follows from the fact that the maximum distance between u and the center cannot exceed the graph's radius. The importance of this observation is in the lower bound. If vertex $u \in \ell_1(G)$, then $1 \leq d(u, C(G)), \leq rad(G)$, if vertex $u \in \ell_2(G)$, then $2 \leq d(u, C(G)), \leq rad(G)$, and so on. Note that the lower bound increases with the layer which results in a decrease of the distance between the vertex and the center. That is, the smaller the eccentricity of a vertex, the more distant it can be from the center.

Let $N(u)$ be the *neighborhood* of a vertex u $(N(u) = \{v \in V : uv \in E\})$ and $degree(u) = |N(u)|$ be the degree of u. We define the *locality* of a noncentral vertex u, denoted as $loc(u)$, according to the eccentricities of its neighbors as $loc(u) = \min\{d(u, v) : v \in V \text{ and } ecc(u) = ecc(v) + 1\}$.

The difference between the eccentricities of any pair of adjacent vertices is always ≤ 1. The locality decides the number of hops between u and a closest vertex v such that the eccentricity of v is strictly smaller than the eccentricity of u. We assign the locality of any $u \in C(G)$ a value of one since a vertex with less eccentricity does not exist. In Fig. 1a, $loc(a) = 1$ since a has a neighbor vertex (c) with smaller eccentricity. $loc(u) = 2$ and $loc(w) = 3$ because a vertex with smaller eccentricity is at distance two and three, respectively.

Remark 1 $loc(u) \leq rad(G)$ for any vertex $u \in V$.

The next remark, which follows Proposition 1, sets bounds on the value of the locality of a vertex based on its eccentricity.

Remark 2 For a vertex $u \in V \backslash C(G)$, where $u \in \ell_r(G)$ of the graph eccentricity layering, the locality of u is bounded as follows. $1 \leq loc(u) \leq rad(G) - r + 1$ for any integer r, $1 \leq r \leq rad(G)$.

That is, the smaller the eccentricity of a given vertex, the higher can be its locality. For example, consider a graph G with $rad(G) \neq diam(G)$ and two vertices u with $ecc(u) = rad(G) + 1$ and v with $ecc(v) = diam(G)$. It may be the case that u and v have equal distances to the center. Next, we describe the monotonicity of shortest paths that connect noncentral vertices to the center with respect to the eccentricities of their vertices. A shortest path $\rho(u, v) = (w_0, w_1, \ldots, w_i)$, $w_0 = u$ and $w_i = v$, that connect a noncentral vertex u to a vertex $v \in C(G)$ $(ecc(u) > ecc(v))$, is *monotonically decreasing* if $ecc(w_i) > ecc(w_{i+1})$ $\forall$ $w_i \in$

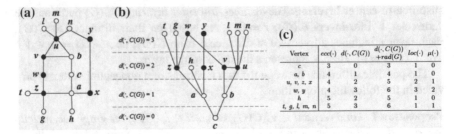

Fig. 1 a A graph with non-unimodal eccentricity function. **b** Distance-to-center layering of the graph in **a**. **c** List of vertices, their eccentricities, distances to the center, localities, and error values

$\rho(u, v)$, $1 \le i \le d(u, v) - 1$, *monotonically nonincreasing* if $ecc(w_i) \ge ecc(w_{i+1})$ $\forall$ $w_i \in \rho(u, v)$, $1 \le i \le d(u, v) - 1$, or *non-monotonic* if $w_i \in \rho(u, v)$ such that $ecc(w_i) < ecc(w_{i+1})$ where $1 \le i \le d(u, v) - 1$. A vertex has a monotonically nonincreasing path if no other monotonically decreasing path exists. Similarly, a vertex has a non-monotonic path if no other monotonically decreasing or monotonically nonincreasing paths exist.

3 Unimodality of the Eccentricity Function and the Center

The vertices of a graph $G = (V, E)$ can be partitioned into two subsets: $C(G)$ and $G \backslash C(G)$ corresponding to the subsets of central and noncentral vertices, respectively. According to the eccentricities of the vertices on the shortest paths that connect noncentral vertices to the center, the eccentricity function of a graph can be described as unimodal or non-unimodal. The eccentricity function is *unimodal* if the eccentricities along the shortest paths that connect noncentral vertices to the center are monotonically decreasing. Because of the unimodality, the local minimum (a vertex with local minimum eccentricity) and the global minimum (the graph's center) coincide, i.e., no local centers exist in the graph. Some standard classes of graphs have unimodal eccentricity functions. Those include trees, complete graphs, and block graphs.

Definition 1 Given a graph $G = (V, E)$, its eccentricity function is unimodal if for every vertex $u \in V \backslash C(G)$, there exists at least one vertex $v \in N(u)$ such that $ecc(u) = ecc(v) + 1$.

The eccentricity function is *non-unimodal* if at least one of the shortest paths that connect noncentral vertices to the center is monotonically nonincreasing or non-monotonic. This causes the existence of local centers. A vertex can be considered a local center if no other vertex with less eccentricity exists within its neighborhood. That is, if it has locality greater than one. See Fig. 1.

A. Unimodality and distances to the graph's center. Here, we analyze the impact of the unimodality of the eccentricity function on the distances between a vertex and the center and the eccentricity of that vertex. Mainly, we show that *if the eccentricity function is unimodal, then all vertices with equal eccentricities have equal distances to the center of the graph*. This is not necessarily true for graphs with non-unimodal eccentricity functions. Given a graph $G = (V, E)$ with unimodal eccentricity function, the relationship between the eccentricity of a vertex u, its distance to the center, and the graph radius satisfies:

Lemma 1 ([4]) *If the eccentricity function of a given graph $G = (V, E)$ is unimodal, then $ecc(u) = d(u, C(G)) + rad(G)$ for every vertex $u \in V$.*

It follows from Lemma 1 that for a graph with unimodal eccentricity function, its eccentricity layering and distance-to-center layering return equal partitioning for the vertex set. That is, all vertices with equal eccentricities have the same distance to the center of the graph. The relationship in Lemma 1 can be generalized (to include graphs with non-unimodal eccentricity functions) as follows.

Proposition 2 *Let $G = (V, E)$ be a graph. For every vertex $u \in V$, $ecc(u) \leq d(u, C(G)) + rad(G)$.*

In a graph with non-unimodal eccentricity function, the eccentricity layering and the distance-to-center layering provide different partitions for the vertex set. Consider a set of vertices with equal eccentricities. Vertices with locality $= 1$ are closer to the center compared to vertices with locality > 1. For example, in Fig. 1a, both vertices a and z have eccentricity 4; however, $d(a, C(G)) = 1$ and $d(z, C(G)) = 2$. Note that $loc(a) = 1$ and $loc(z) = 2$. Moreover, The difference between the eccentricity layering and the distance-to-center layering of a vertex u is either due to the locality of u or the locality of some vertex on a shortest path from u to $C(G)$. For example, for vertex m in Fig. 1a, $ecc(m) \neq d(m, C(G)) + rad(G)$ even though $loc(m) = 1$. This is because vertex $u \in \rho(m, C(G))$ has $loc(u) > 1$.

Since vertex locality is a local measure (within the vertex neighborhood), we extend it to include the localities of all vertices along a shortest path from a vertex to the center. Let $u \in G \backslash C(G)$ be a vertex and $c \in C(G)$ be a closest vertex in the graph's center to u and $\rho(u, c) = w_1, w_2, \ldots, w_j$, where $w_1 = u$ and $w_j = c$ be a path between vertices u and c. The *error* between the distance-to-center layering and the eccentricity layering for a vertex u denoted by $\mu(u)$ is defined as $\mu(u) = d(u, C(G)) + rad(G) - ecc(u)$.

In Fig. 1a, $\mu(h) = 0$, $\mu(g) = 1$, and $\mu(y) = 2$. Every vertex u in a graph with unimodal eccentricity function has error $\mu(u) = 0$. The value of the error has the following bound.

Remark 3 For any graph $G = (V, E)$, $\mu(u) \leq rad(G) - 1$.

Proposition 3 *Let $G = (V, E)$ be a graph and $\mathcal{EL}(G)$ be its eccentricity layering. For a vertex $u \in V \backslash C(G)$ and $u \in \ell_r(G)$, $\mu(u) \leq rad(G) - r$ for any integer r, $1 \leq r \leq rad(G)$.*

That is, if $\mu(u) = rad(G) - k$, where $1 \le k \le rad(G)$, then $rad(G) < ecc(u) \le rad + k$. That is, the larger the eccentricity of a vertex, the smaller error it can have.

B. Unimodality, long shortest paths, and the graph's center. Let the *interval* $I(u, v)$ includes all vertices on the shortest paths between u and v, i.e., $I(u, v) = \{w \in V : d(u, w) + d(w, v) = d(u, v)\}$. A shortest path $\rho(u, v)$ intersects (passes) a subset W if there is at least one vertex $w \in W$ such that $w \in I(u, v)$. Here, we analyze the intersection of shortest paths with the center. Mainly, we show that *if the eccentricity function of a graph is unimodal, then there is at least one shortest path between a vertex u and a vertex v at most distant from u such that $\rho(u, v)$ intersects the graph's center (Lemma 2).* Moreover, we show that *if the eccentricity function is non-unimodal, a shortest path between a vertex u and its most distant vertex v does not intersect the center if (1) Vertex u has error $\mu(u) > 0$ and $ecc(v) > ecc(u)$ (Lemma 3) or (2) Vertex u has error $\mu(u) > 0$, $ecc(v) \ge ecc(u)$, and $\mu(v) \ge 1$ (Remark 5). That is, both u and v are far from center.*

Lemma 2 *Let $G = (V, E)$ be a graph with unimodal eccentricity function. For any vertex $u \in V \backslash C(G)$ and a vertex $v \in F(u)$, $I(u, v) \cap C(G) \ne \emptyset$.*

Proof Let $c \in C(G)$. Because of the unimodality, $ecc(u) = d(u, c)) + rad(G)$. By triangle inequality $d(u, v) \le d(u, c) + d(c, v)$. Since $v \in F(u)$, $ecc(u) = d(u, v)$, and $rad(G) \le d(v, c) \le rad(G)$. Thus, $c \in \rho(u, v)$.

Remark 4 Let $G = (V, E)$ be a graph with non-unimodal eccentricity function. For any vertex $u \in V \backslash C(G)$ with $\mu(u) = 0$ and a vertex $v \in F(u)$, $I(u, v) \cap C(G) \ne \emptyset$.

In Fig. 1a, vertices with error (μ) of zero have a long shortest path that intersects the center such as $\rho(h, l)$.

Lemma 3 *Let $G = (V, E)$ be a graph with non-unimodal eccentricity function, $u \in V \backslash C(G)$ be a vertex with $\mu(u) = rad(G) - k$, $1 \le k < rad(G)$, and let v be a vertex $v \in F(u)$ with $ecc(v) > rad(G) + k$. Then $I(u, v) \cap C(G) = \emptyset$.*

Proof Let c be a central vertex. Using the error, the distance $d(u, c)$ is $d(u, c) = ecc(u) + \mu(u) - rad(G)$. Similarly, $d(v, c) = ecc(v) + \mu(v) - rad(G)$. Since $v \in F(u)$, $d(u, v) = ecc(u)$. Assume by contradiction that $c \in I(u, v)$. Then $d(u, v) = d(u, c) + d(c, v) = ecc(u) + \mu(u) - rad(G) + d(c, v) = ecc(u) - k + ecc(v) + \mu(v) - rad(G) > ecc(u)$.

Remark 5 From Lemma 3, we conclude that if $\mu(u) = rad(G) - k$, $1 \le k < rad(G)$, then for a vertex $v \in F(u)$ with $ecc(v) \ge rad(G) + k$ and $\mu(v) \ge 1$, $I(u, v) \cap C(G) = \emptyset$.

Lemma 3 considers vertices that are far from the center because of the error of one vertex and the eccentricity of the other. For example, in Fig. 1a, consider the interval $I(z, m)$. $I(z, m) \cap C(G) = \emptyset$ since $\mu(z) = 1$ and $ecc(m) = 5$.

4 Experiments on Real-World Networks

We examine the eccentricity functions of a broad set of real-world networks (see Table 1). All networks are undirected and unweighted. In Table 1, we report the size of the center of each network and the number of vertices of locality 1, 2, and ≥ 3. It is clear from Table 1 that the majority of vertices have small localities in general and a locality of 1 in particular. Also, the US-AIRLINES network has unimodal eccentricity function.

In Table 2, we show the locations of vertices with higher localities (locality > 1) with respect to the graph eccentricity layering. Note that vertices in layer 0 are central vertices, vertices in layer 1 are vertices with eccentricities equal to $rad(G) + 1$, and so on. Table 2 shows that the vertices of locality > 1 generally concentrate in lower layers (vertices with small eccentricities) and mostly in layer 1. A closer look at those vertices reveals that the higher their localities, the smaller their eccentricities are. For example, in the DUTCH-ELITE network, the maximum locality is 5, and all vertices with locality of 5 belong to layer 1.

To see the impact of vertices of higher locality on the shortest paths, we compare the number of vertices of locality = 1 and locality > 1 with the values of the error (μ) of each vertex. Vertices of locality > 1 have an error value $\mu \geq 1$. A vertex u with $loc(u) = 1$ will have $\mu(u) = 0$ if no other vertex of higher locality exists in a shortest path from u to the center. Generally, if a graph has x number of vertices of locality > 1 and x number of vertices with error > 1, then all vertices of locality = 1 have their eccentricities equal to their layers (they have the same index in both the eccentricity and the distance-to-center layerings). In other words, no vertices of locality > 1 exist in a shortest path from those vertices to the center. For our real-world networks, we compare the number of vertices of locality = 1, locality > 1, and error > 0 in Table 3. For the POWER-GRID network, even though only about 1% of vertices have locality > 1, 54% of the vertices have error > 1. This indicates that the majority of vertices include other vertices of locality > 1 in their shortest paths to the center. The last column in Table 3 shows how many of the shortest paths that connect distant vertex pairs intersect the center in each network. All long shortest paths intersect the center of the network with unimodal eccentricity function. Networks with high maximum locality (such as the DUTCH-ELITE and Facebook networks) have generally fewer long shortest paths that pass the graph's center.

We also differentiate between two types of vertices with locality > 1: (A) vertices with locality > 1 and with at least one neighbor with equal eccentricity and (B) vertices with locality > 1 and with all neighbors with greater eccentricities. This is important for investigating the monotonicity of the shortest paths that connect noncentral vertices to the center. Vertices of type B will naturally have non-monotonic paths to the center. The question is what effect do they have on the monotonicity of the shortest paths of other vertices? In our set of networks, the networks with vertices of type B are the PPI, DUTCH-ELITE, POWER-GRID, and GNUTELLA networks with 0.05, 3, 0.3, and 0.1% of their vertices are of this type. Close inspection of the monotonicity of the shortest paths that connect noncentral vertices to the center shows

Table 1 Statistics of the analyzed networks: $|V|$: number of vertices, $|E|$: number of edges; $|C(G)|$: size of the center (% to $|V|$); % of vertices with locality 1, 2, and ≥ 3

Network	Type	$|V|$	$|E|$	$rad(G)$	$diam(G)$	$|C(G)|$ (%)	ver w $loc = 1$ (%)	ver w $loc = 2$ (%)	ver w $loc \geq 3$ (%)
US-Airways [3]	Transportation	332	2126	3	6	0.3	99.7	–	–
PPI [8]	Biological	1458	1948	11	19	3	91	5	1
Dutch-Elite [3]	Social	3621	4310	12	22	1	96	–	3
Facebook [9]	Social	4039	88234	4	8	0.01	98	0.09	1
Power-Grid [15]	Infrastructure	4941	6594	23	46	0.2	99	0.4	0.4
AS-Graph-3 [7]	Internet graph	5357	10328	5	9	0.3	97	2.7	–
ROUTEVIEW [14]	Internet graph	10515	21455	5	10	0.02	99.3	0.7	0
HOMO-PI [13]	Biological	16635	115364	5	10	0.8	98.2	≈ 1	0
GNUTELLA [10]	Peer to peer	26498	65359	6	11	0.004	88.6	8	3.3
EMAIL-ENRON [11]	Social	33696	180811	7	13	0.7	98.9	0.4	0
SLASHDOT [11]	Social	77360	905468	6	12	0.001	89.2	9.5	1.3
ITDK [6]	Internet graph	190914	607610	14	26	0.08	96.5	3.4	0.01

Table 2 Distribution of vertices with loc > 1 over the layers of the eccentricity layering. For each layer, number of vertices with loc > 1/total number of vertices

Network	$|V|$	No. of vertices with $loc > 1$	Layer		
			1	2	≥3
US- AIRLINES	332	0	0/24	0/250	0/57
PPI	1458	48	65/366	8/533	5/511
DUTCH- ELITE	3621	119	60/74	49/687	10/2857
FACEBOOK	4039	49	49/112	0/2579	0/1347
POWER- GRID	4941	39	0/2	0/3	16/4935
AS- GRAPH- 3	5357	145	122/1816	23/2766	0/765
ROUTEVIEW	10515	76	54/1102	21/7212	1/2199
HOMO-PI	16635	143	141/7002	2/8843	0/655
GNUTELLA	26498	3027	1579/1934	1448/13021	0/11542
EMAIL-ENRON	33696	136	135/12210	1/17051	0/4187
SLASHDOT	77360	8369	7978/9694	391/53665	0/14000
ITDK	190914	6521	3312/9967	2624/51939	528/128853

Table 3 Max locality and error in each network. Last column: % of distant pairs with at least one shortest path that passes $C(G)$

Network	Max locality	Max error μ	% of ver w $loc = 1$	% of ver w $loc > 1$	% of ver w $\mu \geq 1$	% of sh ρ pass $C(G)$
US- AIRLINES	1	0	100	0	0	100
PPI	4	3	94	≈5	25	92
DUTCH- ELITE	5	4	97	3	64	36
FACEBOOK	3	2	98	≈2	48	53
POWER- GRID	5	4	99	0.7	54	58
AS- GRAPH- 3	2	1	97	2.7	6	100
ROUTEVIEW	2	2	99.3	0.7	8	96
HOMO-PI	2	1	99	1	1.3	99
GNUTELLA	4	3	88.6	11.4	70	62
EMAIL- ENRON	2	1	99.6	0.4	1.14	99
SLASHDOT	3	2	89.2	10.8	67.7	33
ITDK	3	2	96.6	3.41	21.3	91.6

that only those four networks have non-monotonic paths to the center. Moreover, those vertices affect the monotonicity of the shortest paths of other vertices in the DUTCH-ELITE, POWER-GRID, and GNUTELLA networks. See Table 4.

Table 4 Effect of vertex locality on the monotonicity of shortest paths to $C(G)$. % *loc* > $1_=$: % of ver with loc > 1 and with at least one neighbor of equal ecc; % *loc* > $1_>$: % of ver with loc > 1 and with all neighbors of greater ecc; % dec, % n-inc, and % non: percent of ver with monotonically decreasing, monotonically nonincreasing, and non-monotonic shortest paths to $C(G)$ respectively

Network	% *loc* = 1	% *loc* > $1_=$	% *loc* > $1_>$	% dec	% n-inc	% non
PPI	94	4.7	0.3	75	24.6	0.4
DUTCH- ELITE	97	0	3	36	0	64
POWER-GRID	99	0.7	0.3	46	38	16
GNUTELLA	88.6	11.3	0.1	30	64	6

5 Concluding Remarks

In systems with situations where information is exchanged between distant vertices and is expected to pass a central vertex, the center can be considered as its core set (the set through which all traffic passes). This set is sufficient if the graph has a unimodal eccentricity function. However, when the eccentricity function is non-unimodal, some long shortest paths may not pass the center. Therefore, more vertices need to be added to the core set. We believe the implications of our results are as follows. First, the unimodality of the eccentricity function can be used to identify the core vertices. Second, it can aid in designing networks with some desired properties by engineering the eccentricity function.

References

1. Adcock, A., Sullivan, B., Mahoney, M.: Tree-Like structure in large social and information networks. In: ICDM (2013)
2. Alrasheed, H., Dragan, F.F.: Core-periphery models for graphs based on their δ-hyperbolicity: an example using biological networks. Complex Networks VI, pp. 65–77. Springer International Publishing, Berlin (2015)
3. Batagelj, V., Mrvar, A.: Pajek datasets. http://vlado.fmf.uni-lj.si/pub/networks/data/ (2006)
4. Dragan, F.F.: Conditions for coincidence of local and global minima for the eccentricity function on graphs and the Helly property. Studies in Applied Mathematics and Information Science, pp. 49–56 (1990)
5. Holme, P.: Core-periphery organization of complex networks. Phys. Rev. E **72**(4), 046111 (2005)
6. http://www.caida.org/data/internet-topology-data-kit
7. http://web.archive.org/web/20060506132945/
8. Jeong, H., et al.: Lethality and centrality in protein networks. Nature **411**, 41–42 (2001)
9. Leskovec, J., Mcauley, J.: Learning to discover social circles in ego networks. Advances in Neural Information Processing Systems, pp. 539–547 (2012)
10. Leskovec, J., Kleinberg, J., Faloutsos, C.: Graph evolution: densification and shrinking diameters. ACM TKDD **1**(1) (2007)
11. Leskovec, J., et al.: Community structure in large networks: natural cluster sizes and the absence of large well-defined clusters. Internet Math. **6**(1), 29–123 (2009)

12. Narayan, O., Saniee, I.: The large scale curvature of networks. Phys. Rev. E **84**(6), 066108 (2011)
13. Stark, C., et al.: BioGRID: a general repository for interaction datasets. Nucleic Acids Res. (2006)
14. University of Oregon route-views project. http://www.routeviews.org/
15. Watts, D., Strogatz, S.: Collective dynamics of small-world networks. Nature **393**(6684), 440–442 (1998)

Density Decompositions of Networks

Glencora Borradaile, Theresa Migler and Gordon Wilfong

Abstract We introduce a new topological descriptor of a network called the density decomposition which is a partition of the nodes of a network into regions of uniform density. The decomposition we define is unique in the sense that a given network has exactly one density decomposition. The number of nodes in each partition defines a density distribution which, we find, is measurably similar to the degree distribution of given *real* networks (social, internet, etc.) and measurably dissimilar in synthetic networks (preferential attachment, small world, etc.).

1 Introduction

A better understanding of the topological properties of real networks can be advantageous for two major reasons. First, knowing that a network has certain properties, e.g., bounded degree or planarity, can sometimes allow for the design of more efficient algorithms for extracting information about the network or for the design of more efficient distributed protocols to run on the network. Second, it can lead to methods for synthesizing artificial networks that more correctly match the properties of real networks thus allowing for more accurate predictions of future growth of the network and more accurate simulations of distributed protocols running on such a network.

We show that networks decompose naturally into regions of uniform density, a *density decomposition*. The decomposition we define is unique in the sense that a given network has exactly one density decomposition. The number of nodes in each

G. Borradaile
School of EECS Oregon State University, Corvallis, OR, USA
e-mail: glencora@eecs.oregonstate.edu

T. Migler (✉)
Cal Poly, San Luis Obispo, CA, USA
e-mail: tmigler@calpoly.edu

G. Wilfong
Bell Labs, Murray Hill, NJ, USA
e-mail: gordon.wilfong@nokia-bell-labs.com

© Springer International Publishing AG 2018
S. Cornelius et al. (eds.), *Complex Networks IX*, Springer Proceedings
in Complexity, https://doi.org/10.1007/978-3-319-73198-8_2

region defines a distribution of the nodes according to the density of the region to which they belong, that is, a *density distribution* (Sect. 2). Although density is closely related to degree, we find that the density distribution of a particular network is not necessarily similar to the degree distribution of that network. For example, in many synthetic networks, such as those generated by popular network models (e.g., preferential attachment and small worlds), the density distribution is very different from the degree distribution (Sect. 3.1). On the other hand, for *all* of the real networks (social, internet, etc.) in our data set, the density and degree distributions are measurably similar (Sect. 3). Similar conclusions can be drawn using the notion of *k-cores* [31], but this suffers from some drawbacks which we discuss in Sect. 2.3.

1.1 Related Work

We obtain the density decomposition of a given undirected network by first orienting the edges of this network in an *egalitarian*[1] manner. Then, we partition the nodes based on their indegree and connectivity in this orientation.

Fair orientations have been studied frequently in the past. These orientations are motivated by many problems. One such motivating problem is the following telecommunications network problem: Source–sink pairs (s_i, t_i) are linked by a directed s_i-to-t_i path c_i (called a *circuit*). When an edge of the network fails, all circuits using that edge fail and must be rerouted. For each failed circuit, the responsibility for finding an alternate path is assigned to either the source or sink corresponding to that circuit. To limit the rerouting load of any vertex, it is desirable to minimize the maximum number of circuits for which any vertex is responsible. Venkateswaran models this problem with an undirected graph whose vertices are the sources and sinks and whose edges are the circuits. He assigns the responsibility of a circuit's potential failure by orienting the edge to either the source or the sink of this circuit. Minimizing the maximum number of circuits for which any vertex is responsible can thus be achieved by finding an orientation that minimizes the maximum indegree of any vertex. Venkateswaran shows how to find such an orientation [33]. Asahiro, Miyano, Ono, and Zenmyo consider the edge-weighted version of this problem [2]. They give a combinatorial $\{\frac{w_{max}}{w_{min}}, (2 - \epsilon)\}$-approximation algorithm where w_{max} and w_{min} are the maximum and minimum weights of edges, respectively, and ϵ is a constant which depends on the input [2]. Klostermeyer considers the problem of reorienting edges (rather than whole paths) so as to create graphs with given properties, such as strongly connected graphs and acyclic graphs [20]. De Fraysseix and de Mendez show that they can find an indegree assignment of the vertices, given particular properties [11]. Biedl, Chan, Ganjali, Hajiaghayi, and Wood give a $\frac{13}{8}$-approximation algorithm for finding an ordering of the vertices such that for each vertex v, the neighbors of v are as evenly distributed to the right and left of v as possible [8]. For the purpose of

---

[1]An egalitarian orientation is one in which the indegrees of the nodes are as balanced as possible as allowed by the topology of the network.

deadlock prevention [35], Wittorff describes a heuristic for finding an acyclic orientation that minimizes the sum over all vertices of the function $\delta(v)$ choose 2, where $\delta(v)$ is the indegree of vertex v. This objective function is motivated by a problem concerned with resolving deadlocks in communications networks [36].

In our work, we show that the density decomposition can isolate the densest subgraph. The densest subgraph problem has been studied a great deal. Goldberg gives an algorithm to find the densest subgraph in polynomial time using network flow techniques [17]. There is a 2-approximation for this problem that runs in linear time [10]. As a consequence of our decomposition, we find a subgraph that has density no less than the density of the densest subgraph less one. There are algorithms to find dense subgraphs in the streaming model [4, 16]. There are algorithms that find all densest subgraphs in a graph (there could be many such subgraphs) [30].

We consider many varied real networks in our study of the density decomposition. We find our results to be consistent across biological, technical, and social networks.

2 The Density Decomposition

In order to obtain the density decomposition of a given undirected network, we first orient the edges of this network in an egalitarian manner. Then, we partition the nodes based on their indegree and connectivity in this orientation.

The following procedure, the PATH- REVERSAL algorithm, finds an egalitarian orientation [9]. A *reversible path* is a directed path from a node v to a node u such that the indegree of v, $\delta(v)$, is at least greater than the indegree of u plus one: $\delta(v) > \delta(u) + 1$

> Arbitrarily orient the edges of the network.
> While there is a reversible path
>   reverse this path.

Since we are only reversing paths between nodes with differences in indegree of at least 2, this procedure converges; the running time of this algorithm is quadratic [9]. The orientation resulting from this procedure suggests a hierarchical decomposition of its nodes:

Let k be the maximum indegree in an egalitarian orientation.
Ring k (R_k) contains all nodes of indegree k and all nodes that reach them.
Iteratively, given $R_k, R_{k-1}, \ldots$, and R_{i+1}, R_i contains all the remaining nodes
    with indegree i along with all the remaining nodes that reach them.

By the termination condition of the above procedure, only nodes of indegree k or $k - 1$ are in R_k. Further, nodes in R_i must have indegree i or $i - 1$. By this definition, an edge between a node in R_i and a node in R_j is directed from R_i to R_j when $i > j$ and all the isolated nodes are in R_0. The running time to give this decomposition is bounded by the running time to find an egalitarian orientation, $O(|E|^2)$.

Density can be defined in two ways: either as the ratio of number of edges to number of nodes ($\frac{|E|}{|V|}$) or as the ratio of number of edges to total number of possible edges ($\frac{2|E|}{|V|(|V|-1)}$). In this discussion, we use the former definition. This definition of density is closely related to node degree (the number of edges adjacent to a given node): the density of a network is equal to half the average total degree.

We *identify* a set S of nodes in a graph by merging all the nodes in S into a single node s and removing any self-loops (corresponding to edges of the graph both of whose endpoints were in S). Our partition $R_k, R_{k-1}, \ldots, R_0$ induces regions of uniform density in the following sense:

Density Property For any $i = 0, \ldots, k$, identifying the nodes in $\cup_{j>i} R_j$ and deleting the nodes in $\cup_{j<i} R_j$ leave a network G whose density is in
the range $(i - 1, i]$ (for $|R_i|$ sufficiently large).

In particular, R_k isolates a *densest* region in the network. Consider the network G_i formed by identifying the nodes $\cup_{j>i} R_j$ and deleting the nodes in $\cup_{j<i} R_j$; this network has one node (resulting from identifying the nodes $\cup_{j>i} R_j$) of indegree 0 and $|R_i|$ nodes of indegree i of $i - 1$, at least one of which must have indegree i. Therefore, for any i, the density of G_i is at most i and density at least

$$\frac{(|R_i| - 1)(i - 1) + i}{|R_i| + 1} \xrightarrow{|R_i| \gg i} i - 1.$$

In Sect. 2.1, we observe that this relationship between density and this decomposition is much stronger.

2.1 Density and the Density Decomposition

In this section, we discuss the following three properties:

Property D1 The density of a densest subnetwork is at most k. That is, there is no
denser region R_j for $j > k$.
Property D2 The density decomposition of a network is unique and does not depend
on the starting orientation.
Property D3 Every densest subnetwork contains only nodes of R_k.

These properties allow us to unequivocally describe the density structure of a network. We summarize the density decomposition by the *density distribution*: $(|R_0|, |R_1|, \ldots |R_{k-1}|, |R_k|)$, i.e., the number of nodes in each region of uniform density. We will refer to a node in R_i as having *density rank i*.

The subnetwork of a network G *induced* by a subset S of the nodes of G is defined as the set of nodes S and the subset of edges of G whose endpoints are both in S; we denote this by $G[S]$. First, we will note that both the densest subnetwork and the subnetwork induced by the nodes of highest rank have density between $k - 1$ and k.

Recall that k is the maximum indegree of a node in an egalitarian orientation of G and that R_i is the set of nodes in the ith ring of the density decomposition. We will refer to R_k as the densest ring.

We use the following two lemmas to prove Property D1.

Lemma 1 *The density of the subnetwork induced by the nodes in R_k is in the range $(k-1, k]$.*

We could prove Lemma 1 directly with a simple counting argument on the indegrees of nodes in R_k or by using a network flow construction similar to Goldberg's and the max flow-min cut theorem [17].

Lemma 2 *The density of a densest subnetwork is in the range $(k-1, k]$.*

The upper bound given in Lemma 2 may be proven directly by using a counting argument for the indegrees of vertices in an egalitarian orientation of the densest subnetwork or by using the relationship between the density of the maximum density subgraph and the psuedoarboricity [21].

This upper bound proves Property D1 of the density decomposition. Property D1 has been proven in another context. It follows from a theorem of Frank and Gyárfás [13] that if ℓ is the maximum outdegree in an orientation that minimizes the maximum outdegree then the density of the network, d, is such that $\lceil d \rceil \leq \ell$.

Corollary 1 *The subgraph induced by the nodes of R_k is at least as dense as the density of the densest subgraph less one.*

Note that the partition of the rings does not rely on the initial orientation, or, more strongly, nodes are uniquely partitioned into rings, giving Property D2.

Theorem 1 *The density decomposition is unique.*

We can prove this by noting that the maximum indegree of two egalitarian orientations for a given network is the same [2, 9, 33]. For a contradiction, we consider two different egalitarian orientations of the same graph that yield two distinct density decompositions. We then compare corresponding rings in each orientation and find that they are, in fact, the same.

The following theorem relies on the fact that the density decomposition is unique and proves Property D3.

Theorem 2 *The densest subnetwork of a network G is induced by a subset of the nodes in the densest ring of G.*

We could prove Theorem 2 directly by comparing the density of the subgraph induced by the vertices in the densest subgraph intersected with the vertices in R_k and the density of the densest subgraph. Or we could use integer parameterized max flow techniques [2, 14].

Note that there are indeed cases where the densest subgraph is induced by a strict subset of nodes in the top ring. For example, consider the graph, G, consisting of K_3 and K_4 with a single edge connecting the two cliques. K_4 is the densest subgraph in G, however all of G is contained in the top ring (R_2).

2.2 Interpretation of Density Rank

We can interpret orientations as assigning responsibility: if an edge is oriented from node a to node b, we can view node b as being *responsible* for that connection. Indeed several allocation problems are modeled this way [2, 3, 9, 18, 33]. Put another way, we can view a node as wishing to shirk as many of its duties (modeled by incident edges) by assigning these duties to its neighbors (by orienting the linking edge away from itself). Of course, every node wishes to shirk as many of its duties as possible. However, the topology of the network may prevent a node from shirking too many of its duties. In fact, the egalitarian orientation is the assignment in which every node is allowed to simultaneously shirk as many duties as allowed by the topology of the network. For example, consider two situations in which a node has degree 7, in the first situation, a is the center of the star network with eight nodes, in the second situation, b is a node in the clique on eight nodes. Although nodes a and b both have degree 7, in the star network a can shirk all of its duties, but in the clique network b can only shirk half of its duties. There is a clear difference between these two cases that is captured by the density rank of a and b that is not captured by the degree of a and b. For example, if these were coauthorship networks, the star network may represent a network in which author a only coauthors papers with authors who never work with anyone else whereas the clique network shows that author b coauthors with authors who also collaborate with others. One may surmise that the work of author b is more reliable or respected than the work of author a.

Theorem 3 *For a clique on n nodes, there is an orientation where each node has indegree either* $\lfloor n/2 \rfloor$ *or* $\lfloor n/2 \rfloor - 1$.

A proof for Theorem 3 can be given by construction of such an egalitarian orientation or by using a nonlinear programming approach [27].

2.3 Relationship to k-Cores

A k-core of a network is the maximal subnetwork whose nodes all have degree at least k [31]. A k-core is found by repeatedly deleting nodes of degree less than k while possible. For increasing values of k, the k-cores form a nesting hierarchy (akin to our density decomposition) of subnetworks $H_0, H_1, \ldots, H_p$ where H_i is an i-core and p is the smallest integer such that G has an empty $(p + 1)$-core. For networks generated by the $G_{n,p}$ model, most nodes are in the p-core [24, 28] For the preferential attachment model, all nodes except the initial nodes belong to the c-core, where c is the number of edges connecting to each new node [1].

These observations are similar to those we find for the density distribution (Sect. 3) and many of the observations we make regarding the similarity of the degree and density distributions of real networks also hold for k-core decompositions [25]. However, the local definition of cores (depending only on the degree of a node) provides

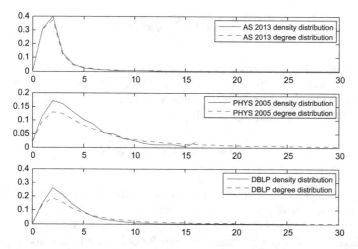

Fig. 1 In the AS network nodes represent autonomous systems and two autonomous systems are connected if there is a routing agreement between them [38] (44,729 nodes and 170,735 edges). In the PHYS network, nodes represent condensed matter physicists and two physicists are connected if they have at least one coauthored paper [26] (40,421 nodes and 175,692 edges). In the DBLP network, nodes represent computer scientists and two computer scientists are connected if they have at least one coauthored paper [37] (317,080 nodes and 1,049,866 edges). The (truncated) normalized density and degree distributions are displayed. The degree distributions have long diminishing tails. AS 2013 has 67 non-empty rings, but rings 31 through 66 contain less than 1.5% of the nodes; ring 67 contains 0.75% of the nodes. DBLP has 4 non-empty rings denser than ring 30 that are disconnected; rings 32, 40, 52, and 58 contain 0.02, 0.01, 0.03, and 0.04% of the nodes, respectively

a much looser connection to density than the density decomposition, as we make formal in Lemma 3.

The density of the top core may be less then the density of the top ring. Also, there are graphs for which the densest subgraph is not contained in the top core.

Lemma 3 *Given a core decomposition $H_0, H_1, \ldots, H_k$ of a network, the subnetwork formed by identifying the nodes in $\cup_{j>i} H_j$ and deleting the nodes in $\cup_{j<i} H_j$ has density in the range $[\frac{i}{2}, i)$ for $|H_i|$ sufficiently large.*

3 The Similarity of Degree and Density Distributions

The normalized density ρ and degree δ distributions for three networks (AS 2013, PHYS 2005, and DBLP) are given in Fig. 1, illustrating the similarity of the distributions. We quantify the similarity between the density and degree distributions of these networks using the Bhattacharyya coefficient, β [7]. For two normalized **p** and **q**, the Bhattacharyya coefficient is:

$$\beta(\mathbf{p}, \mathbf{q}) = \sum_i \sqrt{p_i \cdot q_i}.$$

$\beta(\mathbf{p}, \mathbf{q}) \in [0, 1]$ for normalized, positive distributions; $\beta(\mathbf{p}, \mathbf{q}) = 0$ if and only if $\mathbf{p}$ and $\mathbf{q}$ are disjoint; $\beta(\mathbf{p}, \mathbf{q}) = 1$ if and only if $\mathbf{p} = \mathbf{q}$. We denote the Bhattacharyya coefficient comparing the normalized density ρ and degree δ distributions, $\beta(\rho, \delta)$ for a network G by $\beta_{\rho\delta}(G)$. Specifically,

$$\beta_{\rho\delta}(G) = \beta(\rho, \delta) = \sum_i \sqrt{\rho_i \cdot \delta_i},$$

where ρ_i is the fraction of nodes in the ith ring of the density decomposition of G and δ_i is the fraction of nodes of *total* degree i in G; we take $\rho_i = 0$ for $i > k$ where k is the maximum ring index. Refer to Fig. 2. For all the networks in our data set, $\beta_{\rho\delta} > 0.78$. Note that if we exclude the Gnutella and Amazon networks, $\beta_{\rho\delta} > 0.9$. We point out that the other networks are self-determining in that each relationship is determined by at least one of the parties involved. On the other hand, the Gnutella network is highly structured and designed and the Amazon network is a is a one-mode projection of the buyer-product network (which is in turn self-determining).

Perhaps this is not surprising, given the close relationship between density and degree; one may posit that the density distribution ρ simply bins the degree distribution δ. However, note that a node's degree is its *total* degree in the undirected graph, whereas a node's rank is within one of its *indegree* in an egalitarian orientation. Since the total indegree to be shared amongst all the nodes is half the total degree of the network, we might assume that, if the density distribution is a binning of the degree distribution, the density rank of a node of degree d would be roughly $d/2$. That is, we may expect that the density distribution is halved in range and doubled in magnitude ($\rho_i \approx 2\delta_{2i}$). If this is the case, then

$$\beta(\rho, \delta) \approx \sum_d \sqrt{\rho_i \delta_i} \approx \sum_d \sqrt{2\delta_d \delta_{2d}}.$$

If we additionally assume that our network has a power-law degree distribution such as $\delta_x \propto 1/x^3$,

$$\beta(\rho, \delta) \approx \int_1^\infty \sqrt{\frac{2}{x^3}\left(2\frac{2}{(2x)^3}\right)} dx = 0.5$$

(after normalizing the distributions and using a continuous approximation of β). Even with these idealized assumptions, this does not come close to explaining $\beta_{\rho\delta}$ being in excess of 0.78 for the networks in our data set. Further, for many synthetic networks $\beta_{\rho\delta}$ is small, as we discuss in the next section. We note that this separation between similarities of density and degree distributions for the empirical networks and synthetic networks can be illustrated with almost any divergence or similarity measure for a pair of distributions.

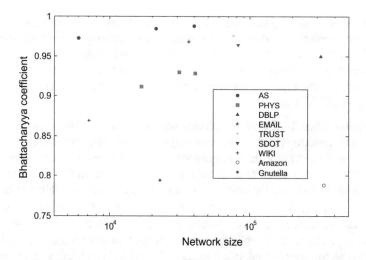

Fig. 2 Similarity $(\beta_{\rho\delta})$ of density and degree distributions for nine diverse networks. We introduced AS, PHYS, and DBLP in Fig. 1. In the EMAIL network, nodes represent Enron email addresses and two addresses are connected if there has been at least one email exchanged between them [19] (36,692 nodes and 183,831 edges). In the TRUST network, nodes represent www.epinions.com members and two members are connected if one trusts the other [32] (75,879 nodes and 405,740 edges). In the SDOT network, nodes represent www.slashdot.org members and two members are connected if they are friends or foes [23] (82,168 nodes and 504, 230 edges). In the WIKI network, nodes represent www.wikipedia.org users and two users are connected if one has voted for the other to be in an administrative role [22] (7,115 nodes and 103,689 edges). In the Amazon network, nodes represent products and two products are connected if they are frequently purchased together [37] (334,863 nodes and 925,872 edges). In the Gnutella network, nodes represent network hosts and two hosts are connected if they share files [29] (22,687 nodes and 54,704 edges). EMAIL, TRUST, and WIKI are naturally directed networks. For these networks, we ignore direction and study the underlying undirected networks. Notice that both the Amazon and Gnutella networks are highly structured. It is not surprising that these networks would have a weaker connection between the density and degree distributions

3.1 The Dissimilarity of Degree and Density Distributions of Random Networks

In contrast to the measurably similar degree and density distributions of real networks, the degree and density distributions are measurably *dis*similar for networks produced by many common random network models; including the preferential attachment (PA) model of Barabasi and Albert [6] and the small world (SW) model of Watts and Strogatz [34]. We use $\tilde{\beta}_{\rho\delta}(M)$ to denote the Bhattacharyya coefficient comparing the expected degree and density distributions of a network generated by a model M.

Preferential attachment networks. In the PA model, a small number, n_0, of nodes seed the network and nodes are added iteratively, each attaching to a fixed number, c, of existing nodes. Consider the orientation where each added edge is directed toward the newly added node; in the resulting orientation, all but the n_0 seed nodes have indegree c and the maximum indegree is c. At most cn_0 path reversals will make this orientation egalitarian, and, since cn_0 is typically very small compared to n (the total number of nodes), most of the nodes will remain in the densest ring R_c. Therefore, PA networks have nearly trivial density distributions: $\rho_c \approx 1$. On the other hand, the expected fraction of degree c nodes is $2/(c+2)$ [5]. Therefore $\tilde{\beta}_{\rho\delta}(\text{PA}) \approx \sqrt{2/(c+2)}$.

Small-world networks. A small-world network is one generated from a d-regular network by reconnecting (uniformly at random) at least one endpoint of every edge with some probability. For probabilities close to 0, a network generated in this way is close to d-regular; for probabilities close to 1, a network generated this way approaches one generated by the random-network model ($G_{n,p}$) of Erdös and Rényi [12]. In the first extreme, $\tilde{\beta}_{\rho\delta}(SW) = 0$ (Lemma 4 below) because all the nodes have the same degree and the same rank. As the reconnection probability increases, nodes are not very likely to change rank while the degree distribution spreads slightly. In the second extreme, the highest rank of a node is $\lfloor c/2 \rfloor + 1$ [15] and, using an observation of the expected size of the densest subnetwork,[2] with high probability nearly all the nodes have this rank. It follows that

$$\tilde{\beta}_{\rho\delta}(G_{n,p}) \approx \sqrt{\frac{c^{c/2}}{e^{-c}(c/2)!}},$$

which approaches 0 very quickly as c grows. We verified this experimentally finding that $\tilde{\beta}_{\rho\delta}(G_{n,p}) < 0.5$ for $c \geq 5$.

Lemma 4 *For $d \geq 3$, $\beta_{\rho\delta}(G) = 0$ for any d-regular network G with $d \geq 3$.*

We can prove Lemma 4 by showing that $\rho_d = 0$ since $\delta_d = 1$ for regular networks.

4   Conclusion

We have introduced the density decomposition and summarized the decomposition with the density distribution. We found that the hierarchy of vertices within this decomposition is partitioned according to the density of the induced subgraphs. We found that the density and degree distributions are remarkably similar in real graphs and dissimilar in synthetic networks. In further work, we plan to use the density distribution to build more realistic random graph models. For more details, a full version of the paper may be found on arxiv.

---

[2]E-mail exchange between Glencora Borradaile and Abbas Mehrabian.

References

1. Alvarez-Hamelin, J., DallAsta, L., Barrat, A., Vespignani, A.: K-core decomposition of Internet graphs: hierarchies, self-similarity and measurement biases. Netw. Heterog. Media 3(2), 371 (2008)
2. Asahiro, Y., Miyano, E., Ono, H., Zenmyo, K.: Graph orientation algorithms to minimize the maximum outdegree. Int. J. Found. Comput. Sci. 18(2), 197–215 (2007)
3. Asahiro, Y., Jansson, J., Miyano, E., Ono, H.: Upper and lower degree bounded graph orientation with minimum penalty. In: Proceedings in Computing: The Australasian Theory Symposium, pp. 139–146 (2012)
4. Bahmani, B., Kumar, R., Vassilvitskii, S.: Densest subgraph in streaming and mapreduce. Proc. VLDB Endow. 5(5), 454–465 (2012)
5. Barabási, A.-L.: Network Science. Cambridge University Press, Cambridge (2016)
6. Barabási, A.-L., Albert, R.: Emergence of scaling in random networks. Science 286(5439), 509–512 (1999)
7. Bhattacharyya, A.K.: On a measure of divergence between two statistical populations defined by their probability distributions. Bull. Calcutta Math. Soc 35, 99–109 (1943)
8. Biedl, T., Chan, T., Ganjali, Y., Hajiaghayi, M.T., Wood, D.R.: Balanced vertex-orderings of graphs. Discret. Appl. Math. 148, 27–48 (2005)
9. Borradaile, G., Iglesias, J., Migler, T., Ochoa, A., Wilfong, G., Zhang, L.: Egalitarian graph orientations. J. Graph Algorithms Appl. 21, 687–708 (2017)
10. Charikar, M.: Greedy approximation algorithms for finding dense components in a graph. In: Proceedings of the Third International Workshop on Approximation Algorithms for Combinatorial Optimization, pp. 84–95. Springer, London, UK (2000)
11. de Fraysseix, H., de Mendez, P.O.: Regular orientations, arboricity, and augmentation. In: Proceedings of the DIMACS International Workshop on Graph Drawing, pp. 111–118. Springer, London, UK (1995)
12. Erdös, P., Rényi, A.: On random graphs I. Publicationes Mathematicae (Debrecen) 6, 290–297 (1959)
13. Frank, A., Gyárfás, A.: How to orient the edges of a graph? Colloquia Mathematica Societatis János Bolyai 1, 353–364 (1976)
14. Gallo, G., Grigoriadis, M., Tarjan, R.: A fast parametric maximum flow algorithm and applications. SIAM J. Comput. 18(1), 30–55 (1989)
15. Gao, P., Pérez-Giménez, X., Sato, C.: Arboricity and spanning-tree packing in random graphs with an application to load balancing. In: Proceedings of the Twenty-Fifth Annual ACM-SIAM Symposium on Discrete Algorithms, SODA '14, pp. 317–326. SIAM (2014)
16. Gibson, D., Kumar, R., Tomkins, A.: Discovering large dense subgraphs in massive graphs. In: Proceedings of the 31st international conference on very large data bases, pp. 721–732. VLDB Endowment (2005)
17. Goldberg, A.: Finding a maximum density subgraph. Technical report, University of California at Berkeley, Berkeley, CA, USA (1984)
18. Harvey, N.J.A., Ladner, R.E., Lovász, L., Tamir, T.: Semi-matchings for bipartite graphs and load balancing. J Algorithms 59, 53–78 (2006)
19. Klimt, B., Yang, Y.: Introducing the Enron Corpus. In: First Conference on Email and Anti-Spam (2004)
20. Klostermeyer, W.F.: Pushing vertices and orienting edges. Ars Comb. 51, 65–75 (1999)
21. Kowalik, Ł.: Approximation scheme for lowest outdegree orientation and graph density measures. In: Proceedings of the 17th International Conference on Algorithms and Computation, ISAAC'06, pp. 557–566. Springer, Berlin, Heidelberg (2006)
22. Leskovec, J., Huttenlocher, D., Kleinberg, J.: Signed networks in social media. In: Proceedings of the SIGCHI Conference on Human Factors in Computing Systems, CHI '10, pp. 1361–1370. ACM, New York, NY, USA (2010). http://snap.stanford.edu/data/

23. Leskovec, J., Lang, K., Dasgupta, A., Mahoney, M.: Community structure in large networks: natural cluster sizes and the absence of large well-defined clusters. CoRR (2008). http://snap.stanford.edu/data/. arXiv:0810.1355
24. Łuczak, T.: Size and connectivity of the k-core of a random graph. Discret. Math. **91**(1), 61–68 (1991)
25. Migler, T.: The Density Signature. Ph.D. thesis, Oregon State University (2014)
26. Newman, M.: Fast algorithm for detecting community structure in networks. Phys. Rev. E **69**:066133 (2004). http://www-personal.umich.edu/mejn/netdata/
27. Picard, J.-C., Queyranne, M.: A network flow solution to some nonlinear 0–1 programming problems, with applications to graph theory. Networks **12**, 141–159 (1982)
28. Pittel, B., Spencer, J., Wormald, N.: Sudden emergence of a giant k-core in a random graph. J. Comb. Theory Ser. B **67**(1), 111–151 (1996)
29. Ripeanu, M., Foster, I., Iamnitchi, A.: Mapping the Gnutella network: Properties of large-scale peer-to-peer systems and implications for system design. IEEE Internet Comput. J **6**:2002 (2002). http://snap.stanford.edu/data/
30. Saha, B., Hoch, A., Khuller, S., Raschid, L., Zhang, X-N.: Dense subgraphs with restrictions and applications to gene annotation graphs. In: Proceedings of the 14th Annual international conference on Research in Computational Molecular Biology, RECOMB'10, pp. 456–472. Springer, Berlin, Heidelberg (2010)
31. Seidman, S.: Network structure and minimum degree. Soc. Netw. **5**(3), 269–287 (1983)
32. Tahajod, M., Iranmehr, A., Khozooyi, N.: Trust management for semantic web. In: Computer and Electrical Engineering, 2009. ICCEE '09. Second International Conference, vol. 2, pp. 3–6 (2009). http://snap.stanford.edu/data/
33. Venkateswaran, V.: Minimizing maximum indegree. Discret. Appl. Math. **143**, 374–378 (2004)
34. Watts, D., Strogatz, S.: Collective dynamics of 'small-world' networks. Nature **393**(6684), 409–10 (1998)
35. Wimmer, W.: Ein Verfahren zur Verhinderung von Verklemmungen in Vermittlernetzen, October 1978. http://www.worldcat.org/title/verfahren-zur-verhinderung-von-verklemmungen-in-vermittlernetzen/
36. Wittorff, V.: Implementation of constraints to ensure deadlock avoidance in networks (2009). US Patent # 7,532,584 B2
37. Yang, J., Leskovec, J.: Defining and evaluating network communities based on ground-truth. In: Proceedings of the ACM SIGKDD Workshop on Mining Data Semantics, MDS '12, pp. 3:1–3:8. ACM, New York, NY, USA (2012). http://snap.stanford.edu/data/
38. Zhang, Y.: Internet AS-level Topology Archive. http://irl.cs.ucla.edu/topology/

Fast Streaming Small Graph Canonization

Pedro Paredes and Pedro Ribeiro

Abstract In this paper, we introduce the streaming graph canonization problem. Its goal is finding a canonical representation of a sequence of graphs in a stream. Our model of a stream fixes the graph's vertices and allows for fully dynamic edge changes, meaning it permits both addition and removal of edges. Our focus is on small graphs, since small graph isomorphism is an important primitive of many subgraph-based metrics, like motif analysis or frequent subgraph mining. We present an efficient data structure to approach this problem, namely a graph isomorphism discrete finite automaton and showcase its efficiency when compared to a non-streaming-aware method that simply recomputes the isomorphism information from scratch in each iteration.

1 Introduction

The *Graph Isomorphism* problem (GI) consists in finding a bijection between the vertex sets of two graphs that preserve the vertex adjacency or state that one does not exist. It is a widely studied problem in several domains. Its theoretical interest arises from the fact that GI is trivially in NP but is still unknown whether it is NP-COMPLETE or in P, even though it is considered unlikely that GI is NP-COMPLETE [5]. Recently, the upper bound on the complexity was improved to quasipolynomial time [2].

From a practical point of view, it is used as a primitive for several methods that tackle different problems, like frequent subgraph discovery [11], network motif analysis [16], and graph matching [6]. As such, efficient practical methods that compute isomorphism information were developed [9, 14] based on several heuristics. One of the most well-known algorithms is called `nauty`, an exponential algorithm that performs exceptionally well in most inputs.

P. Paredes (✉) · P. Ribeiro
CRACS & INESC-TEC, DCC-FCUP, Universidade do Porto, Porto, Portugal
e-mail: pparedes@dcc.fc.up.pt

P. Ribeiro
e-mail: pribeiro@dcc.fc.up.pt

© Springer International Publishing AG 2018
S. Cornelius et al. (eds.), *Complex Networks IX*, Springer Proceedings in Complexity, https://doi.org/10.1007/978-3-319-73198-8_3

27

The *Graph Canonization* problem (GC) is a variant of GI that consists of finding a canonical labeling (also called a *canon*) for a graph such that all isomorphic graphs have the same canon and that if two graphs are not isomorphic they have different canons. Solving GC implies solving GI, since after knowing the canonical labels of two graphs determining if they are isomorphic is simply checking if the two labels are equal. However, in general, GI is not known to be equivalent to GC [1]. The most common practical approach to GI is by solving GC [14], since it is better suited for most applications where a set of graphs needs to be partitioned into isomorphic classes.

The previous discussion focuses on algorithms and problems which are static, meaning the input graph or structure is fixed. However, there is an interest in studying graph problems on a dynamic or streaming environment, that is, where the input graph is changing. There are multiple models of streaming graphs [13], that allow for either edge addition, deletion, or both. Particularly in the graph mining realm, there has been an increasing interest in studying dynamic graphs problems, namely, by introducing or altering known metrics to suit temporal graphs (graphs where edges have timestamps that represent intervals of time where they are active) [7, 10, 15].

In this paper, we present a new problem that approaches the graph isomorphism problem in a dynamic environment. This formulation considers a streamed graph as a set of operations that add or remove edges in each iteration and it is required to calculate a canonical representation for each intermediate graph. Additionally, we focus on solving this problem for small graphs, that is, undirected graphs that have around 10 vertices or directed graphs that have around 6 vertices. Even though this apparently reduces the applicability of the introduced problem, it is important to note that many graph mining techniques focus on small graphs, like network motif analysis [16] or frequent subgraph mining [11]. In Sect. 4.3, we present a small case study that shows the applicability of the problem and apply it to practical complex networks.

Our main contribution is an algorithm that solves this problem in an efficient way, when compared to a simpler non-streaming-aware approach that fully recomputes isomorphism in each iteration. This algorithm is based on a data structure that resembles a discrete finite automaton that represents the full isomorphism class space. The method is agnostic in terms of the type of graph, meaning it is generic to work with multiple graph types (undirected, directed, vertex and edge colored, multigraphs, and more), however, in this paper, we only focus on simple undirected and directed graphs.

2 Preliminaries

A *network* or *graph* G is a pair $(V(G), E(G))$, where $V(G)$ is a set of *vertices* and $E(G)$ a set of *edges*, represented by pairs (a, b) where $a, b \in V(G)$. We assume that the graph is simple (no multiple edges or self-loops) an *labeled* so that every vertex of a graph G is assigned a distinct integer from 1 to $|V(G)|$. We denote the label of a vertex v by $L(v)$. For a given graph G, we write $V(G) = \{v_1, v_2, \ldots, v_{|V(G)|}\}$ to denote a vertex set where $L(v_i) = i$. *Graph equality* between two graphs G and H is observed if, assuming both $L(g_i) = i$, $g_i \in V(G)$ and $L(h_i) = i$, $h_i \in V(H)$, we have $(g_i, g_j) \in E(G) \Leftrightarrow (h_i, h_j) \in E(H)$.

A *permutation* π is an element of the symmetric group S_n, with its usual composition operation $\circ$. We denote the *image* of an integer x under the permutation π by π^x. For a permutation π, we denote by $\overline{\pi}$ the *inverse* of π, such that $\pi \circ \overline{\pi} = \mathbf{1}$, where $\mathbf{1}$ is the identity permutation. A *transposition* is a permutation that only swaps two elements and fixes all others. Given a graph G with vertex set $V(G) = \{v_1, v_2, \ldots\}$ and a permutation π, we denote by G^π the graph with vertex set $V(G^\pi) = \{v_{\pi^1}, v_{\pi^2}, \ldots\}$, meaning we permute the labels. To simplify notation, for a vertex v of a graph G with label i and a permutation π, we write π^v to denote the vertex in G^π with label π^i.

Two graphs G_1 and G_2 are said *isomorphic* if there is a permutation π such that $G_1^\pi = G_2$, we denote this by $G_1 \cong G_2$. The *isomorphism graph class* of a graph G is the equivalence class of G in the relation of isomorphism of graphs. An *automorphism* of a graph G is a permutation π such that $G^\pi = G$. We define Aut(G) as the set of automorphisms of G. The *orbits* of a graph G are the equivalence classes of vertices of G under the action of automorphisms, this means two vertices u, v have the same orbit if there is $\pi \in$ Aut(G) such that $\pi^u = v$ or $\pi^v = u$. A *canonical function* is a function C that, given a graph G, $C(G) \cong G$ and for any $\pi \in$ Aut(G) we have $C(G^\pi) = C(G)$.

A *graph changing operation* of cardinality n is a pair (x_1, x_2), where x_1 and x_2 are integers between 1 and n. The application of a graph changing operation $\Delta = (x_1, x_2)$ of cardinality n over a graph G with $|V(G)| = n$ is the graph $G' = G\Delta$ with the same vertex set of G, where, if $v, u \in V(G)$ are such that $L(v) = x_1$ and $L(u) = x_2$: if $(v, u) \notin E(G)$ then $E(G') = E(G) \cup \{(v, u)\}$; if $(v, u) \in E(G)$ then $E(G') = E(G) \setminus \{(v, u)\}$. Thus, the application of a graph changing operation (x_1, x_2) is equivalent to toggling on or off the edge between the two vertices with labels x_1 and x_2. A *graph stream* S of cardinality n is a sequence of graph changing operations with the same cardinality. We call the *size* of a stream $|S|$ to the number of elements in S. The application of a graph stream $S = [\Delta_1, \Delta_2, \ldots]$ with cardinality n over a graph G with $|V(G)| = n$ is a sequence of graphs $[G, G\Delta_1, G\Delta_1\Delta_2, \ldots]$, denoted by $S(G)$. For a given stream S over a graph G, if we are only interested in every other k graph, meaning $S(G)_1, S(G)_{1+k}, S(G)_{1+2k}, \ldots$, we say the stream S has step k.

2.1 Problem Definition

Now that we are armed with the appropriate set of tools, we can define the problem we aim to solve in this paper. We first define the static version of our problem in Definition 1. This problem is essentially providing a graph canonization function C.

Definition 1 In the *static canonization problem*, we are given a graph G and are asked to provide a canonical representation of G, such that for any $\pi \in \text{Aut}(G)$, G^π has the same representation.

This problem is a known problem and will be used as a primitive in this paper. We use `nauty` [14] in our method as the solver of this problem. However, note that any method that returns the canon of a graph could be used instead.

We now give the dynamic version of the above problem, which is the focus of this paper, and is included in Definition 2.

Definition 2 In the *dynamic canonization problem*, we are given a graph G with n vertices and a graph stream S of cardinality n, and we are asked to provide a canonical representation for each graph in $S(G)$.

Note that, with this formulation, we fix the number of vertices and only vary the edge set. It is also important to note that we focus on small graphs, as stated in the introduction.

3   Proposed Method

Our method explores the dimension of the total number of graphs of a certain size to build a data structure that compresses the relationship between their topologies. This data structure is analogous to a deterministic finite automaton (a finite-state machine), where each node represents a different graph and transitions represent additions or deletions of edges. The result is an algorithm which solves the dynamic canonization problem in an online fashion. We will first describe how the automaton works and how to use it, then we follow up with how to build the automaton efficiently. To avoid ambiguities, we use "node" and "transition" to refer to properties of the automaton and "vertex" and "edge" to refer to properties of the graphs represented by the automaton.

3.1   The Automaton

As mentioned above, we use a data structure that is analogous to an automaton to support our algorithm. This will be used as we iterate through each graph in $S(G)$ to follow the isomorphism graph class.

The node set of the automaton represents the different isomorphism graph classes of a fixed number of vertices n. For each different class, we fix one label function and associate to it a single node of the automaton. This equates to fixing a permutation per isomorphism class and using it as a canonical labeling. For each node, there is one transition coming out of it per possible pair of two vertices of the underlying graph. Each one of this transitions represents an edge toggle, meaning an addition

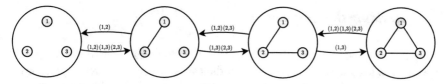

Fig. 1 An automaton representing undirected size 3 graphs

or removal of an edge to the represented graph, which depend on whether the two vertices of this transition are connected or not on the represented graph. Thus, the destination of each transition is the node whose isomorphism graph class is the one of the altered graph. We portray a pictorial representation of this object in Fig. 1.

Since every change between two consequent graphs in $S(G)$ is described by a single pair of vertices, it is natural to use the described automaton to follow the isomorphism graph class of each graph by walking through the automaton. On each step, we use the transition which is associated with the pair of vertices on the current graph changing operation. Initially, the automaton starts on the node that represents the empty graph with n vertices. To find the node that represents G, we build G by following all transitions that represent the pairs of vertices on each edge of the graph, in any order. Subsequently, each graph changing operation results in following one transition.

However, this is not enough to actually apply the automaton, since the order of vertices that was fixed on a certain node may not be the same as the one the current graph we are considering from $S(G)$. Thus, we keep a permutation π_p that tells us how to change the order of vertices of the current graph in order to have the same graph as the one the current node represents. If we think about labels, let G_c be the current graph and G_n be the graph represented by the current node (by definition we have $G_c \cong G_n$), π_p has the following property: $L(G_c) \circ \pi_p = L(G_n)$, since the label function works like a permutation from vertices to indices and taking $\circ$ as regular permutation composition.

To accommodate this change, we also need to update how the transitions work, since after following a transition the relation between the current graph and the graph represented by the current node may change. Thus, we associate a permutation with each transition that informs on how to update π_p. If the permutation for a certain transition is P, then the new π'_p is obtained by $\pi'_p = \pi_p \circ P$. Initially, π_p is set to the identity permutation, since the initial node represents the empty graph (where every permutation is valid). Note that in Fig. 1, the permutations were omitted for brevity.

The resulting automaton represents all different graphs of size n and can be used to keep track of the canonical representation of a graph after a vertex pair change by following a transition and composing a permutation. If we are applying a change of vertex pair (a, b), we follow the transition related to (π_p^a, π_p^b), since we always work on top of the representation the automaton gives.

3.2 Building the Automaton

Now that we have described how the automaton works and how to use it, we will specify how to build it. There are two important aspects here that heavily influence the complexity of the building process, but also the complexity of using the automaton. The first is how to fix the graph each node represents. The second is when to build the automaton, since we can pre-build it or build it as we process the graph stream. We will answer the first question through the following explanation and also point out why it is a relevant question. Regarding the second, we will first describe an on-the-fly method and then a method that pre-builds the automaton but leads to a more efficient automaton.

On-the-fly Method In order to fix a canonical order for each node, we use the representation `nauty` provides. Our method to dynamically build the automaton is based on following the supposed transitions as the stream is processed. Whenever we find ourselves on a nonexisting node, we run `nauty` to know where we should be and either create a new node or point the transition to the correct destination. Additionally, we fill the transition permutations accordingly.

The only node we pre-build is the node that represents the empty graph. Afterwards, we will process each new vertex pair (a, b). Let $a_p = \pi_p^a$ and $b_p = \pi_p^b$. On processing a new pair, we first check if the transition of (a_p, b_p) was already created. If not, we first run `nauty` on the transformed graph, that is, if G is the current graph after adding or removing the edge induced by (a, b), we do so on G' where $L(G') = L(G) \circ \pi_p$, meaning the graph from the current node altered by the pair (a_p, b_p). We do so because `nauty` not only returns the canonical adjacency matrix that we will use to represent the automaton node, but also a permutation P that transforms the graph represented by the canonical adjacency matrix into G'. We can then create a new transition by (a_p, b_p) from the current node C to the new node N (found with `nauty`) with permutation $\overline{P}$, since this permutation transforms the graph on C with added vertex pair (a_p, b_p) into the graph on N, which is the same that `nauty` returns.

This was implicit in the previous paragraph, but we also need a bookkeeping mechanism to store the node representations, so as to avert having a duplicated node representing the same graph class. This can be done using a dictionary data structure that maps canonical representations, as obtained through `nauty`, to automaton nodes (if they exist). Since the graph representation is fixed by the `nauty` canonical representation, the method described in the previous paragraph is exactly the same whether the destination node (N, in the previous paragraph's notation) has to be created or not. If the node is missing, we simply create a new node and feed it to the bookkeeping dictionary.

When processing a change (a, b), let P be the permutation `nauty` returns, C be the initial automaton node and N the destination node. Since $\overline{P}$ transforms graphs in the C representation to the N representation, the converse is also true, that is, P transforms graphs in the N representation to C. Thus, we can use this information to right away fill another transition, P, from N to C. However, since the representation changed, the vertex edge associated with this transition is not (a_p, b_p) but rather $(\overline{P}^{a_p}, \overline{P}^{b_p})$, since this is the corresponding edge pair in N.

It is important to note that the real temporal bottleneck of using this automaton lies on the application step rather than the building step, as we will observe in Sect. 4. This means that the advantage of using a dynamic building method is only observable if the full automaton is impossible to be generated. For example, if we are applying the method in an instance graph with a high number of vertices, and a low total number of different graph types in the stream, using the dynamic building method we only build a partial automaton.

Consequently, it is useful to optimize the automaton underlying representation and methodology if this improves the runtime of applying, even if it worsens the building procedure. With this in mind, we can compress the permutations associated with each transition in order to avoid iterating n integers. By observing the different canonical representations given by nauty, one can observe that they are fairly regular, meaning that often if two graphs differ by a single edge, their adjacency matrix only differs in one (or two, in the undirected case) entries. This implies that the permutation associated with the transition between the two is simple, often either the identity or a single transposition. Thus, we can compress these cases to a special representation that instead of composing a permutation with π_p, either does nothing or simply swaps two entries of π_p. We will see a detailed analysis of the effect of this in Sect. 4, but theoretically this would lower the complexity of following a transition.

Pre-building Method There are not many points to improve related to the on-the-fly building process, since this method does the bare minimum to know where each transition leads to. Consequently, our pre-building method works very similarly, but it does a depth-first search on the automaton in the beginning, generating all possible nodes and transitions. However, the advantage of doing a method that precomputes the automaton is that it is easier to fix a different representation of graphs per node, since there is no need to follow the canonical representation given by nauty (or to have one that works regardless of the order with which we build the automaton). This is important since changing the underlying representation changes the permutations associated with each edge and this has a direct effect on their compressibility and thusly on the runtime.

It is easy to prove that composing a permutation to the graph of each node does not change the correctness of the algorithm, as long as we update the transitions accordingly, since we are simply projecting the automaton to a different space. Hence, it is easy to change the underlying representation of each node by composing a permutation to the permutation nauty returns during the "create new transition" procedure, as long as we compose the same permutation to each transition coming into the same node. In practice, we are simply changing the representation to one that better suits our goals.

All that is left is to choose which permutations to compose with. Instead of focusing on individual permutations, one can determine the underlying representation and choose the permutation that generates this representation. To choose a representation, we can choose the order under which we initially traverse the automaton to

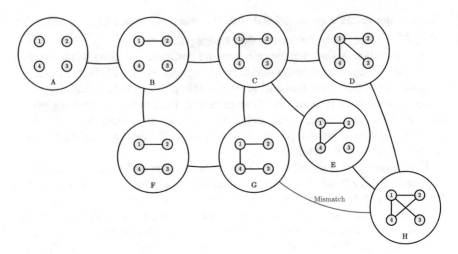

Fig. 2 A partial automaton representing undirected size 4 graphs

pre-build it and use the first graph to touch a each node as its representation. To implement this, the permutation we compose with each node is simply $\overline{P}$ (borrowing from the previous subsection's notation), where we fix the permutation P obtained on the first time we visit that node (which is when we actually create the node). This results in choosing the identity permutation as the permutation from C to N on the first visit to the node.

Different orders were tested, with the goal of increasing the percentage of transitions whose permutation was either the identity permutation or a single transposition. It would be possible to implement an optimization algorithm here, like a local search algorithm, that would repeatedly perturb the traversal order. Although, this would be computationally heavy and would probably not yield much better results than a simply greedy approach. Consequently, we chose an altered edge lexicographical order, that is, we first follow all pairs that create edges before any pair that removes edges and we break ties choosing the lexicographical first transition vertex pairs. We tested different approaches, but this one yielded the better results.

Note that for graphs with four or more vertices, it is impossible to build an automaton where each transition permutation is either the identity permutation or a single transposition. This is equivalent to saying that the graphs in two adjacent automaton nodes differ by at most one edge. To prove the impossibility premise, we will assume that it is possible to build an automaton for four vertex undirected graphs. Consider Fig. 2, which represents a partially constructed automaton for four vertex graphs. We omit the multiple transitions between nodes and simply fix each node's graph representation and show the relationships between nodes (where one or more transitions would be present). In this example, there is a mismatch between node G and H, since their graph's adjacency matrices differ by more than an edge. We can show that this example is "canonical", meaning that all possible automata for four vertex graphs are equivalent to this example, but here we omit the formal proof because of space constraints.

4 Analysis

4.1 Theoretical Analysis

First of all, a note on the automaton's general behavior. Let $\mathcal{G}_n$ denote the set of different graphs with n vertices (note this is an agnostic analysis, since it works for both undirected and directed graphs). Let E_n be the maximum number of edges for a graph with n vertices, that is, $E_n = n^2$ for directed graphs and $E_n = n(n+1)\frac{1}{2}$ for undirected graphs. Since the automaton has one node per different isomorphic graph and each node has a transition per possible pair of vertices, it has $|\mathcal{G}_n|$ nodes and $|\mathcal{G}_n|E_n$ transitions. These pose as the main bottleneck of the automaton method, since they are directly related with memory usage, where each node holds a canonical label and each transition a permutation and destination node. Since $\mathcal{G}_n$ grows rapidly with n, this method is only appropriate to small graphs, depending on the available memory.

For the base building on-the-fly method, we run `nauty` once per transition pair (since we build a transition and its reverse per `nauty` call), thus we call it $|\mathcal{G}_n|E_n\frac{1}{2}$ times. To follow a transition of the automaton, if it exists, it is necessary to compose a permutation, which takes at most $\mathcal{O}(n)$ time for a graph with n vertices. This is true if we have the default representation, if the permutation to compose with can be compressed, then the time needed is only $\mathcal{O}(1)$.

4.2 Empirical Analysis

This analysis is based on our implementation of the described method in C++, which is publicly available.[1] Our C++ code is compiled with GCC 4.8.3, and runs on a single core of an AMD Opteron(tm) with 2.30 GHz under Fedora 20, with 4GB of RAM. Here, we focus on two main themes: namely the compressibility of the transition permutations and the runtime of using the automaton versus using a simpler base approach, namely recalculating the isomorphism class for every instance using `nauty`.

We define two notions of compressibility: C_0 is the *zero compressibility* of an automaton, meaning the percentage of transition permutations that are the identity permutation; C_1 is the *one compressibility* of an automaton, meaning the percentage of transition permutations that are either a single transposition or the identity permutation. In Table 1, we show C_0 and C_1 values for some automata of different

---

[1] https://github.com/ComplexNetworks-DCC-FCUP/streaming-small-isomorphism.

Table 1 Values of C_0 and C_1 for different automata and build methods

	On-the-fly				Pre-build			
	Undirected		Directed		Undirected		Directed	
	C_0 (%)	C_1 (%)	C_0 (%)	C_1 (%)	C_0 (%)	C_1 (%)	C_0 (%)	C_1 (%)
3	25	75	31	73	33	78	34	69
4	18	52	25	62	24	53	29	56
5	14	39	20	53	20	44	26	46
6	12	29	–	–	15	30	–	–
7	9	21	–	–	11	22	–	–
8	6	15	–	–	11	19	–	–

sizes, both undirected and directed, for the two building methods. We omit the results pertaining to automata that were too memory intensive to compute (directed sizes 6, 7 and 8).

It is clear that the pre-build method achieves better compressibilities, specially C_0 compressibilities, which are more critical in terms of runtime. If we discount the building time, which is slightly higher for the pre-build method (but constant), in general, this results in a speedup of up to 2 times, for most input graph streams. However, the increased building time means that for higher vertex numbers (from 8 up), the runtime advantage only becomes noticeable for larger stream sizes. This result was obtained empirically using the graph streams used in the analysis of the following paragraphs.

To compare the temporal behavior of our method with the base `nauty` recomputation method, we generated several synthetic networks, with different goals and variants. Here, we use the version using the on-the-fly building method. We selected 13 graph streams descriptions with different properties and, for each one, studied the runtime of our method and the base recomputation method for several stream sizes. We summarize them in Table 2, where the step k is the number of graph changing operations between each canonization request, that is, we are only interested on the canonization of every other k element of $S(G)$, as we defined previously.

The following list summarizes each model used to generate graph streams:

- **ER Model**, is based on the Erdos-Rényi [4] random graph model, where each graph changing operation is chosen uniformly at random from all the possible vertex pairs. Its directed version, the **D-ER Model** is analogous.
- **PR Model**, is based on a preferential attachment rule for networks [3] where each vertex pair is chosen as a graph changing operation depending on the degree of each of its vertices. Its directed version, the **D-PR Model** is analogous.
- **SWAP Model**, simulates edge swapping operations, with each 4 contiguous graph changing operation representing a swapping of two edges (chosen uniformly at random). It has a step of 4 because we are only interested in the graphs after each swap.

To study each one, we generated multiple streams with increasing sizes, from 10^4 to 10^7 and observed the runtime of both methods. We plot the results of that analysis in Fig. 3 (note that the X-axis is in logarithmic scale). The top left figure pertains to

Table 2 Graphs used for the experimental analysis

Designation	Direction	$\|V(G)\|$	Origin	Step
ER-6, ER-7, ER-8	Undirected	6, 7, 8	ER Model	1
PR-6, PR-7, PR-8	Undirected	6, 7, 8	PR Model	1
SW-5, SW-6, SW-7	Undirected	5, 6, 7	SWAP Model	4
dER-4, dER-5	Directed	4, 5	D-ER Model	1
dPR-4, dPR-5	Directed	4, 5	D-PR Model	1

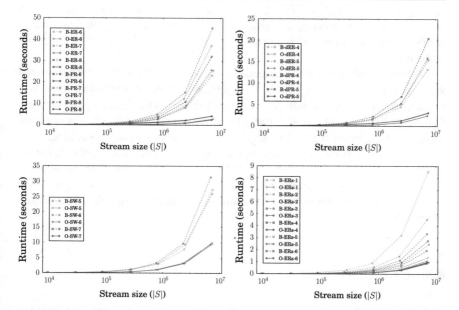

Fig. 3 Comparison of our method (solid lines and prefix O-) versus the base method (dashed lines and prefix B-) for multiple streams

the undirected models, the top right figure directed models, the bottom left figure contains all streams based on the SWAP model, and the bottom right figure represents a growing step experiment that will be further explained below.

It is noticeable that our method greatly outperforms the base method on all streams. Furthermore, the asymptotic behavior of our method suggests that for even greater stream sizes the benefit will increase. The same applies to the speedups obtained. For the unit step streams, the speedup grew approximately linearly from about 1 up to 15 times. For the SWAP model, the speedup was more stable, varying between 2.7 and 3.1. It is also interesting to note that our method had very similar results for different stream models with the same number of vertices, whereas the base method was much more input dependant, which shows that our method is agnostic to the input source.

In the bottom left figure, regarding the SWAP model, it is interesting to note that even though there is a step of 4, our method still maintains a good speedup when comparing to the base method. Note that the higher the step the worse is the benefit of our method, since the base method only performs computation when it is required to return a canonical label whereas our method has to update the automaton after each change.

There is a clear tipping point observable in the data, which represents the minimum stream size for which it is more beneficial to use our method instead of the base method. For the top-left figure, it appears to be around 10^5. This value is related to the automaton size and with the number of times that the method needs to run nauty in the building time. We can extrapolate from here and estimate for different streams sizes and different inputs (even with a number of vertices higher than memory restrictions would allow) and estimate how good our method is going to be in relation to the base method.

Building on this tipping point argument, the bottom-right figure shows a growing step experiment. We used the ER model to generate various networks with six vertices and artificially vary the step from 1 to 6 (each integer in the figure legend indicates the step of that measure). It is important to point out that for all different steps, our method outperformed the base method, with decreasing speedups. Additionally, as we increase the step, the mentioned tipping point of efficiency also increases. Further similar experiments indicate that there is always a tipping point when the step is of the order of $\mathcal{O}(n)$, which means our method is useful as long as the average number of edge modifications between required canonical labels is in the order of the number of vertices.

4.3   Case Study

To further show the usefulness of our method, we present a brief case study problem and present a solution based on our method. Recently, there have been many contributions to the study of network motif and subgraph counting analysis in temporal graphs, as stated in Sect. 1. Here, we present a problem formulation that is inserted in this trend.

Let a G be a temporal graph with edges changing. We want to analyze how patterns evolve in this network and for that we will focus on how a determined induced subgraph of G in a certain timestamp evolves through time. Thus, given two graph types H_1 and H_2 (with the same number of vertices, and possibly the same), we want to know the percentage of times that a set of nodes in a certain timestamp in G is isomorphic to H_1 and in a future timestamp isomorphic to H_2. If we do this for all possible graphs H_1 and H_2 of a certain size n, then we get a Markov chain of temporal subgraph transitions that can be used as a fingerprint of the network and be further used for multiple graph mining tasks. This technique is similar to what was done in [7], but here only patterns of at most three vertices were studied, and to

Table 3 Graphs used for the case study

Designation	Name	Direction	$|V(G)|$	$|S(G)|$	Origin
email	email-eu-core	Directed	986	332,334	Communication [12]
college	college-msg	Directed	1,899	20,296	Communication [12]
infectious	infectious	Undirected	410	17,298	Social [8]
arxiv	arxiv-hep-th	Undirected	22,908	2,673,133	Coauthorship [12]

Table 4 Runtimes, in seconds, for the case study analysis

	Using the base method				Using our method			
	email	college	infectious	arxiv	email	college	infectious	arxiv
3	10.86	8.81	8.33	32.81	6.62	5.08	3.19	31.52
4	22.55	17.12	17.81	66.72	10.73	8.62	4.98	60.48
5	34.45	30.01	34.36	113.95	16.24	13.34	8.56	98.74

what was done in [15], but here this was done in a edge-oriented fashion and with a slightly different formulation.

Doing a complete search of all possible patterns and transitions is possible, but very heavy, even for a relatively small network. Because of that, we only consider connected induced subgraphs and we propose an approximated approach to this problem. We will first sample a single-connected induced subgraph H from G in any timestamp. We then follow the vertex set of H through time in G. To do so, we use our automaton to first represent H and then follow the edge changes. We fix a time step δ, such that whenever δ units of time have passed, we record the current isomorphism class and add a transition on the Markov chain table from the previous class to the current one. By doing so, we can follow the isomorphism information of that particular vertex set throughout the whole lifetime of G. If we repeat this procedure enough times, we have effectively sampled a portion of the temporal transition space.

Since our goal is not to provide a graph mining method to the stated problem but to showcase a possible usage of our automaton, here we will not discuss this method much further. We implemented a basic version of this approach and ran it using both the base method and our method as the underlying isomorphism tool. To compare their runtimes, we ran them on a small set of complex networks with 1,000,000 samples, which we list in Table 3. The runtimes obtained for multiple subgraph size n are shown in Table 4. These runtimes include the time for sampling and performing other supporting computation, which lowers the speedup in relation to what was seen in Sect. 4.

5 Conclusion

In this paper, we introduced a new problem consisting of computing graph isomorphism on a fully dynamic streaming environment, supporting edge insertion and deletion. We presented an efficient algorithm that tackles this problem using a data structure similar to a discrete finite automaton to represent the full space of different isomorphism classes. Compared to a simple non-streaming-aware approach of recomputing the solution for each iteration of the stream, the automaton method and its variations obtained a much better performance, with speedups increasing with the stream size. We also briefly studied the applicability of our method, studying how the stream parameters (ex: the stream size and the stream step) vary while keeping the usefulness of our method in relation to the simpler approach, and we have shown a possible application.

Acknowledgements This work is partly financed by ERDF within project "POCI-01-0145-FEDER-006961", by FCT as part of project "UID/EEA/50014/2013", and by FourEyes, a research line within "TEC4Growth/NORTE-01-0145-FEDER-000020" financed by NORTE2020 through ERDF.

References

1. Arvind, V., Das, B., Köbler, J.: The space complexity of k-tree isomorphism. In: International Symposium on Algorithms and Computation, pp. 822–833. Springer, Berlin (2007)
2. Babai, L.: Graph isomorphism in quasipolynomial time [extended abstract]. In: 48th Annual ACM SIGACT Symposium on Theory of Computing, pp. 684–697. ACM (2016)
3. Barabási, A.L., Albert, R.: Emergence of scaling in random networks. Science **286**(5439), 509–512 (1999)
4. Erdos, P., Rényi, A.: On the evolution of random graphs. Publ. Math. Inst. Hung. Acad. Sci **5**(1), 17–60 (1960)
5. Goldreich, O., Micali, S., Wigderson, A.: Proofs that yield nothing but their validity or all languages in NP have zero-knowledge proof systems. J. ACM (JACM) **38**(3), 690–728 (1991)
6. Gori, M., Maggini, M., Sarti, L.: Exact and approximate graph matching using random walks. IEEE Trans. Pattern Anal. Mach. Intell. **27**(7), 1100–1111 (2005)
7. Huang, H., Tang, J., Liu, L., Luo, J., Fu, X.: Triadic closure pattern analysis and prediction in social networks. IEEE Trans. Knowl. Data Eng. **27**(12), 3374–3389 (2015)
8. Isella, L., Stehlé, J., Barrat, A., Cattuto, C., Pinton, J.F., Van den Broeck, W.: What's in a crowd? analysis of face-to-face behavioral networks. J. Theor. Biol. **271**(1), 166–180 (2011)
9. Junttila, T., Kaski, P.: Engineering an efficient canonical labeling tool for large and sparse graphs. In: 9th Workshop on Algorithm Engineering and Experiments, pp. 135–149 (2007)
10. Kovanen, L., Karsai, M., Kaski, K., Kertész, J., Saramäki, J.: Temporal motifs in time-dependent networks. J. Stat. Mech. Theory Exp. **2011**(11), P11005 (2011)
11. Kuramochi, M., Karypis, G.: An efficient algorithm for discovering frequent subgraphs. IEEE Trans. Knowl. Data Eng. **16**(9), 1038–1051 (2004)
12. Leskovec, J., Krevl, A.: SNAP Datasets: stanford large network dataset collection. http://snap.stanford.edu/data (2014)
13. McGregor, A.: Graph stream algorithms: a survey. ACM SIGMOD Rec. **43**(1), 9–20 (2014)
14. McKay, B.D., Piperno, A.: Practical graph isomorphism, ii. J. Symb. Comput. **60**, 94–112 (2014)
15. Paranjape, A., Benson, A.R., Leskovec, J.: Motifs in temporal networks. In: 10th ACM International Conference on Web Search and Data Mining, pp. 601–610. ACM (2017)
16. Wernicke, S.: Efficient detection of network motifs. IEEE/ACM Trans. Comput. Biol. Bioinform. **3**(4) (2006)

Silhouette for the Evaluation of Community Structures in Multiplex Networks

Alessia Amelio and Andrea Tagarelli

Abstract This paper focuses on the silhouette as validity criterion for community structures in networks, with emphasis on multiplex networks. We propose a versatile definition of the silhouette, by generalizing it to encompass different scenarios of proximity between entities in a network, where the distance notion can be geodesic-based or homophily-oriented. To the best of our knowledge, we are the first to propose this twofold perspective on the silhouette and its extension to deal with multiplex networks. We also define an approximate variant of the multiplex silhouette to speed up its computation on large networks, based on the exploitation of central nodes to be regarded as community representatives. Experimental results performed on benchmark real-world network datasets have revealed that the proposed multiplex silhouette is positively correlated with its approximate version, while the latter proved to be much faster in terms of execution time.

1 Introduction

Internal validity criteria for assessing a community structure in a network assume that the community structure should satisfy certain inner requirements of quality based on topological properties of the network graph [23]. Such requirements typically involve the use of notions defined on the connectivity internal to a community and on the connectivity external to a community, or according to a configuration model, such as for modularity [16].

It should be noted that internal validity criteria for community structures are often designed to focus on quality requirements based on the degree distribution of nodes according to their assignments to communities. One aspect that is scarcely considered in our opinion is the *affinity* between nodes. Nevertheless, this is particularly important in many network scenarios, especially those related to social environments,

A. Amelio · A. Tagarelli (✉)
DIMES, University of Calabria, Via Pietro Bucci 44, 87036 Rende (CS), Italy
e-mail: tagarelli@dimes.unical.it

A. Amelio
e-mail: aamelio@dimes.unical.it

© Springer International Publishing AG 2018
S. Cornelius et al. (eds.), *Complex Networks IX*, Springer Proceedings
in Complexity, https://doi.org/10.1007/978-3-319-73198-8_4

in order to explain the tendency of individuals to associate and bond with similar others; this is known as *homophily*, which is central in the modeling of social influence phenomena.

Research on data cluster analysis has provided a number of clustering quality criteria that depend on the use of a distance or similarity measure to capture affinity among the data objects w.r.t. their assignments to clusters. In this regard, one particularly meaningful of such measures is the *silhouette* coefficient [18]. Given a set of clusters, for every object the silhouette determines how similar it is to its own cluster in relation to how similar the object is to other clusters. In network analysis, the silhouette can also be applied to assess the node memberships to communities, where the notion of distance typically corresponds to the length of (shortest) paths between any two nodes [12, 13].

In this paper, we focus on the silhouette coefficient for the evaluation of community structures. We propose a versatile definition of the silhouette by generalizing it to encompass different scenarios of proximity between entities in a network, where the distance notion can be geodesic-based or homophily-oriented. This twofold perspective on silhouette has not been considered in the study of complex network systems. In this regard, our main contribution in this paper is the definition of *silhouette for multiplex networks*, for which we also propose an approximation to deal with well-known efficiency issues in silhouette. Indeed, to avoid the quadratic computation in the number of nodes, our proposed approximation of multiplex silhouette considers only distances between one node inside a community and the representatives of the other communities.

Experimental results conducted on real-world multiplex networks have provided interesting remarks on the use of multiplex silhouette and its comparison with the approximate variant, as well as with other quality criteria such as local clustering coefficient and modularity. In the rest of this paper, Sect. 2 discusses internal validity criteria for community detection. Section 3 introduces the proposed silhouette for multiplex networks. Sections 4 and 5 present evaluation methodology and results. Finally, Sect. 6 concludes the paper.

2   Related Work

One convenient way to organize measures falling into the category of internal validity criteria is by distinguishing them at (i) *graph-level*, (ii) *community-level*, and (iii) *vertex-level* [8].

Graph-level measures compute a score directly for the whole network. Two measures belonging to this category are *surprise* [1] and *significance* [21]. Community-level measures apply to each community of the network. They include *conductance* and *cut-ratio*, which account for the normalized external connectivity of a community. Other measures consider the ratio between the internal community degree and external community degree (*separability*), all possible edges between nodes in the community (*density*), or the diameter of the community (*compactness*) [7]. A remarkable mention needs to be made for the *modularity* measure, which captures the deviation between the actual connectivity of a community and the expected one based on a configuration model [16]. Several extensions of the modularity have been defined, including modularity for directed and weighted networks [5] and for overlapping community structures [17]. Finally, vertex-level measures apply to each node of the network. They include the local *clustering coefficient* of a node [22], based on the node neighbors' connections, and the *permanence*, based on the number of internal node community edges vs. the number of maximum edges to a single external community [6]. A few measures are also available for the evaluation of community structures in multiplex networks. These include the *multislice modularity* [15], the *multilayer modularity* [2], the *cross-layer edge clustering coefficient* [4], and the *redundancy* [3]. Note that all of the aforementioned measures do not explicitly consider any concept of affinity between nodes according to their memberships to communities.

3 Multiplex Silhouette

Let $G_{\mathcal{L}} = (V_{\mathcal{L}}, E_{\mathcal{L}}, V, \mathcal{L})$ be a *multilayer network* graph, where V is a set of entities and $\mathcal{L} = \{L_1, \ldots, L_\ell\}$ is a set of layers. Each layer represents a specific type of relation between entity nodes. Let $V_{\mathcal{L}} \subseteq V \times \mathcal{L}$ be the set containing the entity-layer combinations, i.e., the occurrences of each entity in the layers. $E_{\mathcal{L}} \subseteq V_{\mathcal{L}} \times V_{\mathcal{L}}$ is the set of undirected links connecting the entity-layer elements. For every $L_i \in \mathcal{L}$, we define $V_i = \{v \in V \mid (v, L_i) \in V_{\mathcal{L}}\} \subseteq V$ as the set of nodes in the graph of L_i, and $E_i \subseteq V_i \times V_i$ as the set of edges in L_i. Each entity must be present in at least one layer, but each layer is not required to contain all elements of V. In this work, we consider the special case of *multiplex* networks, such that the interlayer links only connect the same entity in different layers.

We are given a community structure $C = \{C_1, \ldots, C_k\}$ over $G_{\mathcal{L}}$. If we denote with $dist_L(\cdot, \cdot)$ the distance between any two nodes in V that are connected in layer L, the average distance between any node $v \in V$ and all objects belonging to its community C is defined as:

$$D(v|C, \mathcal{L}) = \frac{1}{|C|} \sum_{u \in C} \sum_{L \in \mathcal{L}} dist_L(u, v) \tag{1}$$

whereas the minimum over the values of average distance between v and all nodes in communities different from C is:

$$ND(v|C, \mathcal{L}) = \min_{C' \in C, C' \neq C} \frac{1}{|C'|} \sum_{u \in C'} \sum_{L \in \mathcal{L}} dist_L(u, v) \tag{2}$$

We define the *multiplex silhouette* of $v \in V$ as:

$$MSil(v) = \frac{ND(v|C, \mathcal{L}) - D(v|C, \mathcal{L})}{\max\{ND(v|C, \mathcal{L}), D(v|C, \mathcal{L})\}}. \tag{3}$$

In order to specify the distance function $dist_L(\cdot, \cdot)$, we consider two perspectives: (i) *geodesic-based* and (ii) *homophily-oriented*. As previously discussed, the former has been already considered in the literature, which corresponds to the computation of the shortest path between two nodes in the graph of layer L; the latter perspective, although originally captured in our previous recent work [19], has not been widely studied so far.

We devise a notion of homophily-oriented distance by resorting to any similarity measure that can express the topological affinity of two nodes in a layer graph according to their commonality in terms of their respective neighbors. Formally, given a layer $L \in \mathcal{L}$ and any two nodes $u, v \in \mathcal{V}$, we choose for $dist_L(u, v)$ the generic form as $f(N_L(u), N_L(v))$, where f denotes a function proportional to the similarity of the two node sets $N_L(u), N_L(v)$, with $N_L(x) = \{y \in \mathcal{V}|(y, x) \in E_L\}$ as the set of neighbors of a node x in layer L. To specify the above function, one reasonable choice is a *Jaccard coefficient* based distance, whereby similarity between two nodes is evaluated proportionally to the number of neighbors in common, i.e., $dist_L(u, v) = 1 - \frac{|N_L(u) \cap N_L(v)|}{|N_L(u) \cup N_L(v)|}$. An alternative is *cosine similarity* based distance, whereby the proportionality of shared neighborhood would be smoothed to favor unbalanced neighborhoods to be compared; in this case, we will define $dist_L(u, v) = 1 - \frac{|N_L(u) \cap N_L(v)|}{\sqrt{|N_L(u)||N_L(v)|}}$. Other functions can surely be based on cliques or other substructures. Note that, by exploiting (3) and averaging over all nodes in a community and over all nodes in the multiplex graph, we can compute a multiplex *community silhouette* and a *global silhouette*, respectively.

Approximate multiplex silhouette. Our proposed multiplex silhouette inherits an efficiency issue from the basic silhouette, since it requires a pairwise comparison between nodes in the communities. To overcome this limitation, we identify one representative for every community, and compute the multiplex silhouette more efficiently by only considering distances between one node inside a community and the representatives of the other communities.

Let us denote with $r(C)$ the node in $\mathcal{V}$ that is representative of community $C \in \mathcal{C}$. The simplest definition of $r(C)$ according to graph centrality theory is based on the degree of node, which leads to $r(C) = \mathrm{argmax}_{u \in C} deg(u)$, with $deg(u)$ denoting the total degree of u in the multiplex graph, i.e., $deg(u) = |\{x \in \mathcal{V}|(u, x) \in E_{\mathcal{L}}\}|$. In the case of multiple nodes in C having the maximum degree, $r(C)$ is chosen as the node with the maximum *neighborhood degree*, where a node's neighborhood degree is computed as the sum of the degrees of the nodes's neighbors. It should be noted however that any (reasonably efficient) centrality measure can in principle be used to define $r(C)$ in alternative to degree centrality.

The computation of community representatives enables the following modifications in (1):

$$D(v|C, \mathcal{L}) = \sum_{L \in \mathcal{L}} dist_L(r(C), v) \tag{4}$$

$$ND(v|\mathcal{C}, \mathcal{L}) = \min_{C' \in \mathcal{C}, C' \neq C} \sum_{L \in \mathcal{L}} dist_L(r(C'), v). \tag{5}$$

We hereinafter refer to the multiplex silhouette defined in terms of (4) and (5) as *centrality-based multiplex silhouette*, denoted as $cb\text{-}MSil$.

4 Evaluation Methodology

Datasets. We used six real-world multiplex networks for our evaluation, which are among the most frequently used in recent, relevant studies in multiplex/multilayer community detection: *AUCS* [14], *EU-Airlines* [14], *FAO-Trade* [10], *London* transport [24], FriendFeed-Twitter-YouTube (*FF-TW-YT*) [11], 7thGraders (*VC-Graders*) [24]. Table 1 summarizes main characteristics of the evaluation networks, where the average coverage of the node set is $1/|\mathcal{L}| \sum_{L \in \mathcal{L}} (|V_L|/|\mathcal{V}|)$, and the average coverage of edge set is $1/|\mathcal{L}| \sum_{L \in \mathcal{L}} (|E_L|/ \sum_{L'} |E_{L'}|)$.

Community detection methods. We selected three well-known methods that are representative of the direct approaches and the aggregation-based approaches for community detection in multiplex networks: *Generalized Louvain* (GL) [15], *Multiplex Infomap* (M-Infomap) [9], and *Principal Modularity Maximization* (PMM) [20].

Experimental setting and goals. Our proposed multiplex silhouette $MSil$ and its approximate variant $cb\text{-}MSil$ were evaluated using both the shortest path and the Jaccard coefficient based distances. For each dataset and method, we compared the global values $MSil$ and $cb\text{-}MSil$ in terms of absolute difference as well as in terms of execution time. Moreover, we analyzed the correlation between the node distributions of $MSil$ and $cb\text{-}MSil$. Also at vertex-level, we analyzed the correlation of each of the silhouette measures with the local internal clustering coefficient [22], whereas at community-level, we analyzed the correlation of the two silhouette with the multilayer modularity proposed in [19].

Table 1 Main features of real-world multiplex network datasets used in our evaluation

	#entities ($|\mathcal{V}|$)	#edges	#layers (ℓ)	Node set coverage	Edge set coverage
AUCS	61	620	5	0.73	0.20
EU-Air	417	3588	37	0.13	0.03
FAO-Trade	214	318346	364	0.53	0.003
	#entities ($|\mathcal{V}|$)	#edges	#layers (ℓ)	Node set coverage	Edge set coverage
FF-TW-YT	6407	74836	3	0.58	0.33
London	369	441	3	0.36	0.33
VC-Graders	29	518	3	1.00	0.33

As concerns the community detection methods, we used the default setting for GL and M-Infomap. We varied the number k of communities required as input in PMM, from 5 to 100 (with increments of 5), and finally selected the value corresponding to the highest modularity.

5 Results

Table 2 shows the global values of $MSil$ and $cb\text{-}MSil$ equipped with the Jaccard distance, computed on the community structures obtained by GL, PMM, and M-Infomap on the various networks. One result from the table is the evidence of low values of silhouette obtained by the methods in most cases, especially when the shortest path distance is employed. This is partly expected since silhouette generally behaves as a tough criterion of validity, but also because community detection methods are not designed to optimize their solutions according to criteria that meet silhouette requirements. Also, it can be noted how the employment of the geodesic-based versus homophily-oriented distance notions in the silhouette can actually lead to different behaviors of the methods.

Besides the above remarks, one important finding for our study is that the absolute difference (Δ) between the multiplex silhouette and its centrality-based variant is in general quite small, in particular below 0.1 in 13 out of 18 cases (with Jaccard) and 8 out of 12 cases (with shortest path). Overall, this would indicate that $cb\text{-}MSil$ is a good approximation of $MSil$. In addition, looking at the efficiency results, $cb\text{-}MSil$ always outperforms $MSil$ — as shown in Fig. 1, $cb\text{-}MSil$ obtained a percentage

Table 2 Global values of $MSil$ and $cb\text{-}MSil$ equipped with Jaccard distance (top) and with shortest path distance (bottom). Symbol Δ denotes $|MSil - cb\text{-}MSil|$

	GL			PMM			M-Infomap		
	$MSil$	$cb\text{-}MSil$	Δ	$MSil$	$cb\text{-}MSil$	Δ	$MSil$	$cb\text{-}MSil$	Δ
AUCS	0.20	0.03	0.17	0.09	0.04	0.05	−0.13	−0.11	0.02
EU-Air	0.05	−0.01	0.06	−0.16	−0.01	0.15	−0.19	−0.02	0.17
FAO-Trade	0.05	0.02	0.03	0.01	−0.00	0.01	−0.1	−0.1	0
FF-TW-YT	−0.09	−0.08	0.01	−0.00	−0.01	0.01	−0.08	−0.08	0
London	0.04	−0.00	0.04	0.11	−0.01	0.12	−0.001	−0.001	0
VC-Graders	0.19	0.12	0.07	0.23	0.16	0.07	0.13	−0.18	0.31
	GL			PMM			M-Infomap		
	$MSil$	$cb\text{-}MSil$	Δ	$MSil$	$cb\text{-}MSil$	Δ	$MSil$	$cb\text{-}MSil$	Δ
AUCS	−0.22	−0.27	0.05	0.18	0.16	0.02	−0.31	−0.30	0.01
EU-Air	−0.19	−0.16	0.03	−0.06	−0.05	0.01	0.53	0.50	0.03
London	−0.34	0.06	0.4	−0.09	−0.03	0.06	0.29	0.45	0.16
VC-Graders	0.26	0.18	0.08	0.28	0.17	0.11	0.09	−0.15	0.24

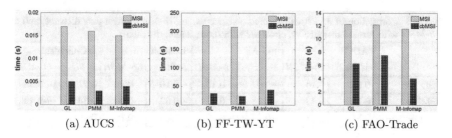

(a) AUCS (b) FF-TW-YT (c) FAO-Trade

Fig. 1 Execution time (in seconds) of $MSil$ (lighter bars) and $cb\text{-}MSil$ (darker bars) with Jaccard distance

Table 3 Correlation of $MSil$ vs. $cb\text{-}MSil$, local clustering coefficient (cc), and multilayer modularity (Q). $MSil$ and $cb\text{-}MSil$ are equipped with Jaccard (upper table) or with shortest path distance (bottom table)

	GL			PMM			M-Infomap		
	$cb\text{-}MSil$	cc	Q	$cb\text{-}MSil$	cc	Q	$cb\text{-}MSil$	cc	Q
AUCS	0.44	0.53	0.48	0.67	0.51	−1	0.62	−0.55	−0.72
EU-Air	0.05	0.02	−0.31	0.09	0.29	−0.71	0.62	0.05	−1
FAO-Trade	0.14	−0.13	−0.41	0.16	0.70	−0.65	1	0.00	1
FF-TW-YT	0.86	−0.01	−0.10	0.50	0.22	−0.48	0.56	−0.12	−0.13
London	0.49	0.00	0.68	0.12	0.37	−0.46	0.92	0.04	−1
VC-Graders	0.26	−0.01	−1	−0.29	0.001	1	0.62	−0.50	−0.48
	GL			PMM			M-Infomap		
	$cb\text{-}MSil$	cc	Q	$cb\text{-}MSil$	cc	Q	$cb\text{-}MSil$	cc	Q
AUCS	0.58	0.16	−0.67	0.10	−0.02	1	0.51	−0.49	−0.86
EU-Air	−0.26	0.06	−0.23	−0.41	−0.22	0.79	−0.005	0.02	1
London	−0.14	0.00	−0.91	−0.02	−0.02	−0.16	0.84	0.01	1
VC-Graders	0.33	−0.27	−1	0.57	−0.29	−1	0.54	−0.36	−0.36

decrease of execution time of 75% on FF-TW-YT, above 67% on AUCS (all methods) and on FAO-Trade.

More insights can be gained by analyzing the correlation between node distributions of the various criteria, as reported in Tables 3–4. Considering $MSil$ versus $cb\text{-}MSil$ with homophily-oriented distance (i.e., Jaccard distance), they exhibit quite a good correlation, which is on average 0.38 for GL, 0.21 for PMM, and 0.72 for M-Infomap (Table 3). Also, $MSil$ tends to be negatively correlated with multilayer modularity in most cases, while it could be positively correlated with the local clustering coefficient (e.g., in PMM solutions).

When the shortest path distance is used, the two multiplex silhouettes have a more varying correlation, which could be positive or negative depending on the network (Table 3). Again, correlation is quite negative between $MSil$ and multilayer modularity and more negative between $MSil$ and the local clustering coefficient for PMM than in the case of Jaccard distance.

Table 4 Correlation of $MSil$ with Jaccard distance vs. $MSil$ with shortest path distance, and of $cb\text{-}MSil$ with Jaccard distance vs. $cb\text{-}MSil$ with shortest path distance

	AUCS		EU-Air		London		VC-Graders	
	$MSil$	$cb\text{-}MSil$	$MSil$	$cb\text{-}MSil$	$MSil$	$cb\text{-}MSil$	$MSil$	$cb\text{-}MSil$
GL	0.62	0.14	−0.07	0.13	−0.10	0.32	0.71	−0.36
PMM	0.19	0.47	−0.28	0.02	−0.02	−0.00	0.34	−0.49
M-Infomap	0.51	0.43	−0.64	−0.29	0.25	0.18	0.90	0.56

Comparing $MSil$ when equipped with Jaccard and with shortest path distance (Table 4), we observe that they can be positively correlated in social multiplex networks, like AUCS and VC-Graders, while the measures tend to be negatively correlated in transport networks. This would suggest that for social environments, the multiplex silhouette is robust to the choice of distance notion, as the shortest path between individuals can provide indicators of homophily that would be consistent with those provided by a neighbor-overlap-based distance. By contrast, in transport networks, the two notions of distance lead to different effects in determining the silhouette of community memberships of nodes.

6 Conclusion

We proposed a definition of silhouette for community structures in multiplex networks, which considers geodesic-based as well as homophily-oriented notions to determine the affinity of nodes according to their community memberships. We also defined an efficient variant of the multiplex silhouette, based on degree centrality, to identify community representatives that are exploited to speed up the computation of the quality measure on large networks. As future work, we plan to investigate alternative layer-aggregation schemes for the definition of multiplex silhouette criteria, and alternative centrality-based approximations of the multiplex silhouette.

References

1. Aldecoa, R., Marín, I.: Surprise maximization reveals the community structure of complex networks. Sci. Rep. **3** (2013)
2. Amelio, A., Tagarelli, A.: Revisiting resolution and inter-layer coupling factors in modularity for multilayer networks. In: Proceedings of the IEEE/ACM International Conference on Advances in Social Networks Analysis and Mining (ASONAM). pp. 266–273 (2017)
3. Berlingerio, M., Coscia, M., Giannotti, F.: Finding and characterizing communities in multidimensional networks. In: Proceedings of the ASONAM. pp. 490–494 (2011)
4. Bródka, P., Kazienko, P.a., Koł oszczyk, B.: Predicting group evolution in the social network. In: Proceedings of the International Conference on Social Informatics (SocInfo). pp. 54–67 (2012)

5. Chakraborty, T., Dalmia, A., Mukherjee, A., Ganguly, N.: Metrics for Community Analysis: A Survey. arXiv:1604.03512 (2016)
6. Chakraborty, T., Srinivasan, S., Ganguly, N., Mukherjee, A., Bhowmick, S.: On the permanence of vertices in network communities. In: Proceedings of the ACM Conference on Knowledge Discovery and Data Mining (KDD). pp. 1396–1405 (2014)
7. Creusefond, J., Largillier, T., Peyronnet, S.: Finding compact communities in large graphs. In: Proceedings of the ASONAM. pp. 1457–1464 (2015)
8. Creusefond, J., Largillier, T., Peyronnet, S.: On the evaluation potential of quality functions in community detection for different contexts. In: Proceedings of the International Conference and School on Advances in Network Science (NetSci-X). pp. 111–125 (2016)
9. De Domenico, M., Lancichinetti, A., Arenas, A., Rosvall, M.: Identifying modular flows on multilayer networks reveals highly overlapping organization in interconnected systems. Phys. Rev. X **5**, 011027 (2015)
10. De Domenico, M., Nicosia, V., Arenas, A., Latora, V.: Structural reducibility of multilayer networks. Nat. Commun. **6**, 6864 (2015)
11. Dickison, M.E., Magnani, M., Rossi, L.: Multilayer Social Networks. Cambridge University Press, UK (2016)
12. Estrada, E.: Community detection based on network communicability. Chaos **21** (2011)
13. Gustafsson, M., Hornquist, M., Lombardi, A.: Comparison and validation of community structures in complex networks. Physica A **367**, 559–576 (2006)
14. Kim, J., Lee, J.: Community detection in multi-layer graphs: a survey. SIGMOD Rec. **44**(3), 37–48 (2015)
15. Mucha, P.J., Richardson, T., Macon, K., Porter, M.A., Onnela, J.P.: Community structure in time-dependent, multiscale, and multiplex networks. Science **328**(5980), 876–878 (2010)
16. Newman, M.E.J., Girvan, M.: Finding and evaluating community structure in networks. Phys. Rev. E **69**, 026113 (2004)
17. Nicosia, V., Mangioni, G., Carchiolo, V., Malgeri, M.: Extending the definition of modularity to directed graphs with overlapping communities. J. Stat. Mech. Theory Exper. (03), P03024 (2009)
18. Rousseeuw, P.J.: Silhouettes: a graphical aid to the interpretation and validation of cluster analysis. J. Comput. Appl. Math. **20**, 53–65 (1987)
19. Tagarelli, A., Amelio, A., Gullo, F.: Ensemble-based community detection in multilayer networks. Data Min. Knowl. Discov. **31**(5), 1506–1543 (2017)
20. Tang, L., Wang, X., Liu, H.: Uncoverning groups via heterogeneous interaction analysis. In: Proceeding of the IEEE International Conference on Data Mining (ICDM). pp. 503–512 (2009)
21. Traag, V.A., Krings, G., Dooren, P.V.: Significant scales in community structure. Sci. Rep. **3** (2013)
22. Watts, D.J., Strogatz, S.H.: Collective dynamics of 'small-world' networks. Nature **393**(6684), 409–10 (1998)
23. Yang, J., Leskovec, J.: Defining and evaluating network communities based on ground-truth. In: Proceedings of the IEEE International Conference on Data Mining (ICDM) (2012)
24. Zhang, H., Wang, C., Lai, J., Yu, P.S.: Modularity in Complex Multilayer Networks with Multiple Aspects: A Static Perspective. CoRR (2016). arXiv:abs/1605.06190

Jaccard Curvature—an Efficient Proxy for Ollivier-Ricci Curvature in Graphs

Siddharth Pal, Feng Yu, Terrence J. Moore, Ram Ramanathan, Amotz Bar-Noy and Ananthram Swami

Abstract The discrete version of the Ollivier-Ricci (OR) curvature, applicable to networks, has recently found utility in diverse fields. OR curvature requires solving an optimal mass transport problem for each edge, which can be computationally expensive for large and/or dense networks. We propose two alternative proxies of curvature to OR that are motivated by the Jaccard index and are demonstrably less computationally intensive. Jaccard curvature (JC) is a simple shift and scaling of the Jaccard index that captures the overlap of edge node neighborhoods. Generalized Jaccard curvature (gJC) captures the shortest path distances in a mass exchange

Research was sponsored by the Army Research Laboratory and was accomplished under Cooperative Agreement Number W911NF-09-2-0053 (the ARL Network Science CTA). The views and conclusions contained in this document are those of the authors and should not be interpreted as representing the official policies, either expressed or implied, of the Army Research Laboratory or the U.S. Government. The U.S. Government is authorized to reproduce and distribute reprints for Government purposes notwithstanding any copyright notation here on. This document does not contain technology or technical data controlled under either the U.S. International Traffic in Arms Regulations or the U.S. Export Administration Regulations.

Ram Ramanathan work done while the author was with Raytheon BBN Technologies.

S. Pal (✉)
Raytheon BBN Technologies, Cambridge, MA, USA
e-mail: siddharth.pal@raytheon.com

F. Yu · A. Bar-Noy
City University of New York, New York City, NY, USA
e-mail: fyu@gc.cuny.edu

A. Bar-Noy
e-mail: amotz@sci.brooklyn.cuny.edu

T. J. Moore · A. Swami
U.S. Army Research Lab, adelphi, MD, USA
e-mail: terrence.j.moore.civ@mail.mil

A. Swami
e-mail: ananthram.swami.civ@mail.mil

R. Ramanathan
Gotenna Inc, New York City, NY, USA
e-mail: Ram@gotenna.com

© Springer International Publishing AG 2018
S. Cornelius et al. (eds.), *Complex Networks IX*, Springer Proceedings in Complexity, https://doi.org/10.1007/978-3-319-73198-8_5

problem. We study the goodness of approximation between the proposed curvatures and an alternative metric, Forman-Ricci curvature, with OR curvature for several network models and real networks. Our results suggest that the gJC exhibits a reasonably good fit to the OR curvature for a wide range of networks, while the JC is shown to be a good proxy only for certain scenarios.

1   Introduction

Various notions of curvature in differential geometry, which measure the curves or bends of tensors on the surface of manifolds [3, 5], have recently been interpreted on graphs and applied to the study of networks [2, 4, 7, 12, 15]. Ollivier-Ricci curvature seems a promising new metric for networks as it has recently been applied to distinguish between cancerous and noncancerous cells [18], to indicate fragility in stock markets [19], and in explaining congestion in wireless networks [22].

The Ollivier-Ricci curvature of a pair of vertices is defined based on the optimal mass transport between their associated mass distributions. When the pairs being considered are restricted to adjacent vertices, Ollivier-Ricci curvature can be viewed as an edge centrality metric, like betweenness or random-walk measures. Informally, positive curvature implies that the neighbors of the two nodes are close (perhaps overlapping or shared); zero (or near-zero) curvature implies that the nodes are locally embeddable in a flat surface (as in a grid or regular lattice); while negative curvature implies that the neighbors of the two nodes are further apart. Unfortunately, computing the Ollivier-Ricci curvature can incur high-computational complexity in high-degree and large networks as solving the transport problem, in the worst case, scales with the quartic of the degree (see Table 2) or, in practice, scales with the product of the two nodes' degrees [14]. This motivates the desire for a less computationally intensive approximation. Jost and Liu [6] demonstrated the significance of overlapping neighborhoods in the Ollivier-Ricci curvature of edges in the formulation of a bound involving the clustering coefficient [23]. Hence, it seems reasonable to build a metric that aims to approximate Ollivier-Ricci from the sets of common and separate neighbors of the nodes in an edge.

Inspired by the Jaccard index, we derive a new curvature metric approximating the Ollivier-Ricci graph curvature metric. The Jaccard index has previously found utility in networks, e.g., as a measure of similarity between nodes [10]. Moreover, the Jaccard index naturally captures the overlapping neighborhood feature found in positively curved edges in a simplistic manner. The notion of set comparison as a curvature metric leads to a more general linear approximation function of Ollivier-Ricci formulated from classes of sets of each node's neighbors that effectively solves a mass exchange problem. The complexity of this new metric is significantly less than that for Ollivier-Ricci. For random graphs, we find that our new metric shares many asymptotic properties of the Ollivier-Ricci curvature [1]. Moreover, comparisons of the Jaccard-inspired curvature with Ollivier-Ricci seem more favorable than the alternative Forman-Ricci curvature metric [2, 20, 21].

2 Curvature Metrics

We first present two adaptations of Ricci curvature to graphs—Ollivier-Ricci curvature [6, 11, 15] and Forman-Ricci curvature [2, 20]. Then, we introduce a new graph curvature metric, which is intuitively similar to Ollivier-Ricci curvature but requires less computational complexity.

Ollivier-Ricci curvature: Consider an undirected graph $G = (V, E)$ on n nodes, i.e., $|V| = n$, with no self-loops. For an edge $(i, j) \in E$, the Ollivier-Ricci (OR) curvature metric is defined as $\kappa(i, j) = 1 - W(\mathbf{m}_i, \mathbf{m}_j)$, where $W(\mathbf{m}_i, \mathbf{m}_j)$ is the Wasserstein distance (see [15] for more details) or optimal mass transport cost between the two probability measures $\mathbf{m}_i$ and $\mathbf{m}_j$. For each vertex i, the probability measure $\mathbf{m}_i$ is set as $m_i(j) = \frac{1}{d_i}$, if $i \sim j$, and 0 otherwise; where $i \sim j$ implies an edge between i and j. The probability measure $\mathbf{m}_i$ shown above distributes a unit weight to all neighbors of i uniformly as in [6, 18]. A more general assignment distributes a mass α to node i and $1 - \alpha$ uniformly among the neighbors of i [11, 14]. But for the distribution considered in this work ($\alpha = 0$), the Ollivier-Ricci curvature for each edge is bounded as $-2 < \kappa < 1$.

Forman-Ricci curvature: Forman discretized the classical Ricci curvature for a broad class of geometric objects, the CW complexes [2]. This was applied to undirected networks by Sreejith et al. [20], and we use that definition. The Forman curvature for an edge $e = (i, j)$ is given by

$$F(e) = w_e \left(\frac{w_i}{w_e} + \frac{w_j}{w_e} - \sum_{e_\ell \in e_i \setminus e} \frac{w_i}{\sqrt{w_e w_{e_\ell}}} - \sum_{e_\ell \in e_j \setminus e} \frac{w_j}{\sqrt{w_e w_{e_\ell}}} \right), \quad (1)$$

where w_e is a weight associated with edge e, w_i is a weight associated with vertex i, and $e_i \setminus e$ denotes the set of edges incident on vertices i excluding the edge e.

For an unweighted graph, two weighting schemes were proposed [20, 21]. The first sets all the node and edge weights to 1, and the second weights the edges by 1 and the nodes by their degree. We implemented both the weighting schemes, and did not find a significant difference in terms of their correlations with the Ollivier-Ricci curvature. The results shown in Sect. 4 follow the second weighting scheme, where the original expression in (1) reduces to

$$F(e) = 4 - d_i - d_j. \quad (2)$$

Other, more involved, weighting schemes have also been proposed [24], but are not considered in this work.

Jaccard Curvatures: As mentioned previously, calculating Ollivier-Ricci curvature can be costly because it involves solving an optimal mass transport problem, or equivalently a linear program [14], for each edge. Especially for large graphs, with high values of maximum degree, calculating OR curvature for all the edges can be prohibitively costly (see Sect. 3). To address this, we introduce an approximation to the OR curvature, which would not require solving the optimal mass transport problem. Toward this end, we revisit the intuition of OR curvature—An edge has positive curvature if the neighborhoods of the two concerned nodes are closer to each other compared to the nodes themselves, zero curvature if the neighborhoods are at the same distance, and negative curvature if the neighborhoods are farther apart. A simple heuristic would be to measure the fraction of common nodes between the neighborhoods of the two concerned nodes. This is related to Jaccard's coefficient, which was introduced in network analysis as a similarity measure between nodes [10].

For an edge (i, j), the set of common neighbors of the nodes i and j is given by $\mathscr{C}(i, j) = \mathscr{N}_i \cap \mathscr{N}_j$, where $\mathscr{N}_k$ is the neighbor set of node k. We let $C(i, j) = |\mathscr{C}(i, j)|$. We also define the set of separate neighbors between i and j as $\mathscr{S}(i, j) = (\mathscr{N}_i \cup \mathscr{N}_j) \setminus \mathscr{C}(i, j)$, with $S(i, j) = |\mathscr{S}(i, j)|$. We define the union of the neighbor sets of i and j as $\mathscr{N}(i, j)$, i.e., $\mathscr{N}(i, j) = \mathscr{C}(i, j) \cup \mathscr{S}(i, j)$, with $N(i, j) = |\mathscr{N}(i, j)|$.

Jaccard's coefficient is defined as the ratio between the intersection of neighborhoods of the two nodes to their union, i.e., $J(i, j) = C(i, j)/N(i, j)$. It is evident that the metric $J(i, j)$ will be closer to 1 if there are more common nodes, and closer to 0 otherwise. However, the range of this metric will be between 0 and 1, as opposed to OR curvature which takes the range $(-2, 1)$. We can scale and shift the Jaccard coefficient so that our desired Jaccard curvature metric for an edge, $JC(i, j)$, approaches a value of 1 when $J(i, j)$ is close to 1 and approaches -2 when $J(i, j) = 0$,

$$JC(i, j) = 1 - \frac{3S(i, j)}{N(i, j)} = -2 + 3J(i, j). \tag{3}$$

See Fig. 1 for an illustrative example. The above expression could be interpreted as subtracting the influence of separate neighbors, with $S(i, j)$ being the total number of separate neighbors and the denominator $N(i, j)$ being the cardinality of the union of the neighbor sets of i and j.

Computing Jaccard curvature is very cheap because it only requires the knowledge of the size of neighborhoods of the two relevant nodes and the common nodes in those neighborhood sets. However, since the Jaccard curvature partitions the set of neighbors into common and separate vertices, the granularities of OR curvature are lost to a great extent. This is best demonstrated by considering edges in canonical graphs. In a complete graph, the OR curvature of each edge will be close to 1, and the Jaccard curvature will be exactly 1. For an edge connecting high-degree nodes in a tree, the OR curvature will be close to -2 and the Jaccard curvature will be exactly -2. However, on a grid or a line, the OR curvature of the edges will be 0, while the Jaccard curvature will still be -2, because there are no common nodes. Clearly, the Jaccard curvature metric should have a more positive value in a grid compared to that on a tree, if we are to obtain a better approximation of the OR curvature. To address this, we now define a generalized version of the Jaccard curvature metric to take into account nodes that are not common, yet closer than 3 hops apart. Note that, nodes

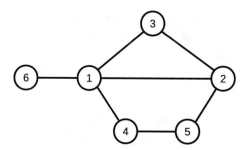

Fig. 1 For the edge $(1, 2)$, $\mathscr{C}(1, 2) = \{3\}$ and $\mathscr{S}(i, j) = \{1, 2, 4, 5, 6\}$. So, from (3), the Jaccard curvature is $JC(1, 2) = 1 - \frac{3 \times 5}{6} = -\frac{3}{2}$. Since $\mathscr{S}_1^{(1)} = \{4\}$, $\mathscr{S}_2^{(1)} = \{5\}$, $\mathscr{S}_1^{(2)} = \{6\}$ (node 6 is 2 hops from a neighbor of node 2, i.e. node 3), and all other expressions are zero, then from (5), the generalized Jaccard curvature is $gJC(1, 2) = 1 - \frac{1+1+2}{6} - 2 \cdot \frac{1}{6} = 0$. In comparison, the Ollivier-Riccci curvature $OR(1, 2) = \frac{1}{4}$, and Forman curvature is -3.

cannot be more than 3 hops apart because there always exists a 3 hop path between them through the edge whose curvature is being computed.

Define $\mathscr{N}_i(i, j)$ as the exclusive neighbors of i with respect to the edge (i, j), i.e., $\mathscr{N}_i(i, j) = \{k \in V \setminus \{j\} \mid (i, k) \in E\}$. Let $d(u, v)$ denote the shortest path length between nodes u and v. Let the set of separate nodes be partitioned into the following sets $\mathscr{S}_i^{(r)} = \{k \in \mathscr{N}_i(i, j) \mid \min_{\ell \in \mathscr{N}_j(i, j)} d(k, \ell) = r\}$, with $S_i^{(r)} = |\mathscr{S}_i^{(r)}|$, for $r = 1, 2, 3$. In other words, $\mathscr{S}_i^{(r)}$ is the set of neighbors of i that have shortest path distance of r from the set of exclusive neighbors of j. If $\mathscr{N}_j(i, j) = \emptyset$, then we set $\mathscr{S}_i^{(1)} = \mathscr{N}_i(i, j)$. Since Ollivier-Ricci curvature includes the nodes of the edge itself, we include i and j in the generalized Jaccard (gJC) metric. Therefore, for an edge $(i, j) \in E$, the generalized Jaccard metric is defined by

$$gJC(i, j) = \alpha + \beta \frac{C(i, j)}{N(i, j)} + \gamma \frac{S_i^{(1)} + S_j^{(1)} + 2}{N(i, j)} + \delta \frac{S_i^{(2)} + S_j^{(2)}}{N(i, j)} + \zeta \frac{S_i^{(3)} + S_j^{(3)}}{N(i, j)},$$

(4)

where the parameters α, β, γ, δ and ζ need to be determined. Since, $i \notin \mathscr{S}_j^{(1)}$ and $j \notin \mathscr{S}_i^{(1)}$, we arrive at the gJC metric by including the two nodes separately in (4). The parameters are determined by considering several cases from canonical graphs.

In a k-complete graph, we would like the generalized Jaccard metric to have a maximum value close to 1, so as to approximate the OR curvature which approaches 1 as k gets large. Therefore we require, as $C(i, j)/N(i, j) \rightarrow 1$, then $gJC(i, j) \rightarrow 1$, which leads to $\alpha + \beta = 1$. Similarly, for edges in a d-dimensional grid, we would like the generalized Jaccard metric to have a value close to 0 as d gets large. This requires that as $d \rightarrow \infty$ and $(S_i^{(1)} + S_j^{(1)} + 2)/N(i, j) \rightarrow 1$, then $gJC(i, j) \rightarrow 0$, which leads to $\alpha + \gamma = 0$. This would best approximate the OR curvature which has a value of 0 for d-dimensional grids. For edges in a tree connecting nodes with degree d, we would like the generalized Jaccard metric to have a value close to a minimum value of -2, approximating the OR curvature which itself approaches -2 as d gets large.

This requires that as $d \to \infty$ and $(S_i^{(3)} + S_j^{(3)})/N(i,j) \to 1$, then $gJC(i,j) \to -2$ which leads to $\alpha + \zeta = -2$. We set, $\alpha = 1, \beta = 0, \gamma = -1, \zeta = -3$, which satisfies the aforementioned requirements. Furthermore, we enforce $\gamma > \delta > \zeta$ so that the effect of $\mathscr{S}^{(2)}$ on the edge curvature falls between that of $\mathscr{S}^{(1)}$ and $\mathscr{S}^{(3)}$. Setting $\delta = (\gamma + \zeta)/2 = -2$, we obtain

$$gJC(i,j) = 1 - \frac{S_i^{(1)} + S_j^{(1)} + 2}{N(i,j)} - 2\frac{S_i^{(2)} + S_j^{(2)}}{N(i,j)} - 3\frac{S_i^{(3)} + S_j^{(3)}}{N(i,j)}; \tag{5}$$

see Fig. 1 for an illustrative example.

It can be shown that the generalized Jaccard expression in (5) is related to the solution of the optimal mass exchange problem between neighborhoods of the two concerned nodes, where the mass distribution at the source is predetermined and fixed and the destination mass distribution is kept flexible, with the constraint that mass from a neighbor of one node needs to be transported to any neighbor of the other node and vice versa (see [16] for further details).

3 Theoretical Results

Study on random graphs: We state results for the behavior of Jaccard and generalized Jaccard curvature of edge (i,j), denoted $JC_n(i,j)$ and $gJC_n(i,j)$ respectively, on a sequence of Erdos–Renyi graphs $\{\mathbb{G}_1, \mathbb{G}_2, \ldots\}$.

Theorem 1 *Let $\{\mathbb{G}_1, \mathbb{G}_2, \ldots\}$ be a sequence of Erdos–Renyi graphs. As $n \to \infty$ and for all $(i,j) \in \mathbb{E}$, we have the following results.*

a. *For $p_n \to p$, $\mathbb{E}\left[JC_n(i,j)\right] \to \frac{5p-4}{2-p}$.*
b. *For $p_n \to 0$, $\mathbb{E}\left[JC_n(i,j)\right] \to -2$.*

Theorem 2 *Let $\{\mathbb{G}_1, \mathbb{G}_2, \ldots\}$ be a sequence of Erdos–Renyi graphs. As $n \to \infty$ and for all $(i,j) \in \mathbb{E}$, we have the following results.*

a. *For $p_n \to p$, $\mathbb{E}\left[gJC_n(i,j)\right] \to \frac{p}{2-p}$.*
b. *For $np_n \to 0$ and $p_n \to 0$, $\mathbb{E}\left[gJC_n(i,j)\right] \to 0$.*
c. *For $np_n^2 \to \infty$ and $p_n \to 0$, $\mathbb{E}\left[gJC_n(i,j)\right] \to 0$.*
d. *For $n^2p_n^3 \to \infty$, $np_n^2 \to 0$ and $p_n \to 0$, $\mathbb{E}\left[gJC_n(i,j)\right] \to -1$.*
e. *For $np_n \to \infty, n^2p_n^3 \to 0$ and $p_n \to 0$, $\mathbb{E}\left[gJC_n(i,j)\right] \to -2$.*

Proofs of Theorems 1–2 can be found in [16]. The two theorems together suggest that gJC is a better asymptotic approximation of OR curvature than the Jaccard curvature. We see that as the scaling changes and the ER graph becomes more dense, gJC increases progressively. Table 1 shows that the asymptotic behavior of gJC matches that of OR curvature closely, while the behavior of JC and Forman curvatures are very different.

Table 1 The asymptotic values for the curvatures under different scalings for the ER graph

	Ollivier-Ricci	Forman	Jaccard	Generalized Jaccard
p constant	p	$-\infty$	$\frac{5p-4}{2-p}$	$\frac{p}{2-p}$
$np_n \to 0$	0	4	-2	0
$np_n \to \infty$ and $n^2 p_n^3 \to 0$	-2	$-\infty$	-2	-2
$n^2 p_n^3 \to \infty$ and $np_n^2 \to 0$	-1	$-\infty$	-2	-1
$np_n^2 \to \infty$	0	$-\infty$	-2	0

Table 2 Computational complexity for OR, Forman, JC, and gJC curvatures for d-regular graphs

	Ollivier-Ricci	Forman	Jaccard	Generalized Jaccard
Complexity	$O(n^4 \log^2 n)$	$O(nd)$	$O(nd^2)$	$O(nd^3)$

Computational Complexity: There is a clear hierarchy in the complexity of the Forman, Jaccard, generalized Jaccard, and Ollivier-Ricci curvatures. See Table 2 for their complexity for d-regular graphs. Intuitively, the Forman curvature is based only on the degrees of i and j so its complexity is $O(m)$ for a graph with m edges which is $O(nd)$ for d-regular graphs. To compute JC, we are looking for the number of common neighbors between i and j while to compute gJC we are looking for the shortest path to get from any exclusive neighbor of i to any exclusive neighbor of j. All of these shortest paths could be to the same neighbor of j. In d-regular graphs, these extra computations require an additional factor of $O(d)$ for JC and an additional factor of $O(d^2)$ for gJC. Finally, in OR, these shortest paths must represent a perfect fractional matching in the sense that one neighbor of y cannot be the target of too many neighbors of x, which is obviously a harder task. In [16], we provide proofs for the above complexity claims.

4 Experimental Results

4.1 Network Models

We consider different network models to investigate the relationship between Jaccard and Forman curvatures in relation to Ollivier-Ricci curvature. This helps provide insight on how the different curvatures are affected by changes in model parameter values that have been shown to characterize certain network properties.

Table 3 Average curvatures shown for different Erdos–Renyi (ER) graphs. Pearson correlation r_p and Kendall's τ between OR curvature and Forman, JC, and gJC curvature are tabulated

Graph	$\overline{OR}$	$\overline{F}$	$\overline{JC}$	$\overline{gJC}$	(OR, F)		(OR, JC)		(OR, gJC)	
					r_p	τ	r_p	τ	r_p	τ
ER(100,0.05)	−0.59	−8	−1.95	−0.86	0.35	0.25	0.40	0.28	**0.77**	**0.55**
ER(100,0.1)	−0.20	−19	−1.83	−0.23	−0.31	−0.21	0.64	0.45	**0.90**	**0.73**
ER(100,0.2)	0.15	−37	−0.77	0.09	−0.46	−0.30	0.89	0.76	**0.94**	**0.80**
ER(100,0.3)	0.26	−57	−1.48	0.17	−0.53	−0.38	**0.97**	**0.86**	**0.97**	**0.86**
ER(100,0.4)	0.35	−74	−1.30	0.23	−0.44	−0.28	**0.97**	**0.86**	**0.97**	**0.86**
ER(100,0.5)	0.47	−97	−1.01	0.35	−0.54	−0.37	**0.96**	**0.83**	**0.96**	**0.83**
ER(100,0.6)	0.55	−113	−0.77	0.41	−0.58	−0.40	**0.93**	**0.77**	**0.93**	**0.77**
ER(100,0.7)	0.67	−137	−0.39	0.54	−0.64	−0.45	**0.92**	**0.75**	**0.92**	**0.75**
ER(100,0.8)	0.77	−154	−0.06	0.65	−0.70	−0.50	**0.91**	**0.74**	**0.91**	0.73
ER(100,0.9)	0.87	−173	0.38	0.80	−0.83	−0.68	**0.91**	**0.77**	**0.91**	**0.77**

Erdös-Rényi (ER) model [13]: $ER(n, p)$ is a network on n nodes that connects every pair of nodes with probability p independently across node pairs. Table 3 shows the average curvatures of different ER graphs as the probability of connection p is varied. When p is small, the graph is more disconnected and tree-like leading to negative OR curvature, and as p increases the density of the graph increases (as does clustering), increasing the average OR curvature as well. Average gJC closely tracks average OR curvature as p increases. On the other hand, the average Forman curvature decreases as p increases due to increasing average degree. Of these lower complexity curvatures, gJC has the best correlation with OR curvature, and this superiority is more pronounced when p is small. The advantage of gJC with respect to JC is small when $p \geq 0.2$. The correlation between the Jaccard curvatures and OR curvature improves as p is increased, deteriorating slightly when $p \geq 0.5$.

Barabási–Albert (BA) model [13]: $BA(n, m)$ is a network growth model, which starts with at least m nodes and new nodes connect preferentially to m existing nodes with probability proportional to their degree. As shown in Table 4, gJC correlates the best with OR curvature for $BA(100,1)$, because all mass transport paths need to pass through any non-leaf edge being considered. Note, when m $= 1$, JC is non-discriminating with a value of -2 for each edge since there are no triangles in the graph. Increasing m from 1 to 2 and keeping number of nodes fixed results in less correlation between gJC and OR curvature because now there potentially exist shorter than 3-hop paths to account for. However, increasing m further leads to improvement, similar to what was observed in ER graphs for moderate p values. Increasing number of nodes n and keeping m fixed decrease the correlation of Jaccard curvatures, probably because the graph becomes more tree-like with larger hubs. The correlations for Forman curvature get worse as m increases.

Watts–Strogatz (WS) model [13]: $WS(n, k, p)$ is a network model on n nodes which first constructs a ring lattice among adjacent nodes such that each node is

Table 4 Average curvatures shown for different Barabasi–Albert graphs. Pearson correlation r_p and Kendall's τ between OR curvature and Forman, JC, and gJC curvature are tabulated

Graph	$\overline{OR}$	$\overline{F}$	$\overline{JC}$	$\overline{gJC}$	(OR, F)		(OR, JC)		(OR, gJC)	
					r_p	τ	r_p	τ	r_p	τ
BA(100,1)	−0.31	−4	−2	−0.54	0.65	0.50	N/A	N/A	**0.92**	**0.94**
BA(100,2)	−0.45	−10	−1.90	−0.85	0.60	**0.50**	0.47	0.24	**0.62**	0.45
BA(100,5)	−0.16	−25	−1.77	−0.19	−0.16	−0.07	0.73	0.54	**0.86**	**0.66**
BA(500,1)	−0.31	−17	−2	−0.55	0.33	0.37	N/A	N/A	**0.93**	**0.72**
BA(500,2)	−0.79	−12	−1.98	−1.32	**0.64**	**0.58**	0.12	−0.05	0.52	0.40
BA(500,5)	−0.58	−32	−1.94	−0.62	−0.08	0.03	0.52	0.34	**0.81**	**0.60**

Table 5 Average curvatures shown for different Watts–Strogatz graph. Pearson correlation r_p and Kendall's τ between OR curvature and Forman, JC, and gJC curvature are tabulated

Graph	$\overline{OR}$	$\overline{F}$	$\overline{JC}$	$\overline{gJC}$	(OR, F)		(OR, JC)		(OR, gJC)	
					r_p	τ	r_p	τ	r_p	τ
WS(100,4,0.1)	−0.04	−4.17	−1.46	−0.22	0.33	0.25	0.89	0.89	**0.96**	**0.90**
WS(200,4,0.1)	−0.08	−4.20	−1.51	−0.28	0.37	0.26	0.88	0.86	**0.96**	**0.90**
WS(500,4,0.02)	0.17	−4.04	−1.33	0.02	0.29	0.21	0.93	**0.95**	**0.98**	**0.95**
WS(500,4,0.05)	0.06	−4.10	−1.41	−0.11	0.29	0.23	0.90	0.91	**0.96**	**0.92**
WS(500,4,0.1)	−0.05	−4.20	−1.47	−0.24	0.42	0.31	0.89	0.87	**0.96**	**0.90**
WS(500,4,0.2)	−0.31	−4.36	−1.65	−0.57	0.39	0.30	0.86	0.79	**0.95**	**0.87**
WS(500,4,0.3)	−0.54	−4.52	−1.78	−0.88	0.42	0.34	0.85	0.73	**0.94**	**0.85**
WS(500,4,0.5)	−0.77	−4.71	−1.91	−1.2	0.49	0.45	0.77	0.55	**0.89**	**0.83**
WS(500,4,0.7)	−0.9	−4.80	−1.98	−1.4	0.66	0.61	0.52	0.30	**0.80**	**0.76**
WS(500,4,0.9)	−0.94	−4.85	−1.99	−1.45	0.76	0.69	0.35	0.19	**0.78**	**0.78**
WS(500,4,0.99)	−0.93	−5.04	−2.00	−1.45	0.76	0.65	0.08	0.06	**0.78**	**0.68**

connected to k closest neighbors and then randomly rewires one end of each edge with probability p. Our simulations, displayed in Table 5, show little variation in the curvatures of edges as the network size grows, when p and k are fixed, since locally the networks appear similar. The curvatures become more negative as the probability of rewiring increases due to the increase in the number of edges that become shortcuts, i.e., edges whose end nodes share no common neighbors. The correlations for JC and gJC decrease for increasing p, whereas the correlations for Forman curvature increase. However, gJC still has better correlation than Forman for the range of values considered. This indicates that gJC is a better approximation overall, but especially so when the graph has more positively curved edges.

Random geometric graph (RGG) model [17]: $RGG(n, r)$ is a network model where all nodes are distributed uniformly on a metric space, e.g., a unit square, and connections between nodes are formed only if the pairwise Euclidean distance is less than a certain radius r, with $0 < r < 1$. From Table 6, note that increasing the

Table 6 Average curvatures shown for different Random Geometric Graphs. Pearson correlation r_p and Kendall's τ between OR curvature and Forman, JC, and gJC curvature are tabulated

Graph	$\overline{OR}$	$\overline{F}$	$\overline{JC}$	$\overline{gJC}$	(OR, F)		(OR, JC)		(OR, gJC)	
					r_p	τ	r_p	τ	r_p	τ
RGG(500,0.05)	0.22	−5	−1.13	0.05	−0.17	−0.15	0.86	**0.83**	**0.88**	0.79
RGG(500,0.1)	0.23	−28	−0.75	0.37	−0.22	−0.12	**0.94**	**0.86**	**0.94**	0.82
RGG(500,0.15)	0.28	−64	−0.64	0.44	0.04	0.04	**0.96**	**0.87**	**0.96**	0.86
RGG(500,0.2)	0.32	−112	−0.61	0.46	0.16	0.09	**0.97**	**0.89**	**0.97**	0.88

radius r increases the clustering and, hence, increases the OR and both Jaccard curvatures slightly. The correlation with Forman curvature becomes more negative as r increases because the average node degree increases. The correlation coefficients of the Jaccard curvatures increase as r increases. This observation agrees with the previously mentioned hypothesis that the Jaccard curvature approximates OR curvature better for positively curved graphs.

4.2 Real-World Networks

We consider several real-world networks: The Gnutella network was obtained from the Stanford large network dataset collection [9], while the rest of the networks were obtained from the Koblenz Network Collection [8]. A brief description of the datasets is provided below. Their network properties are displayed in Table 7.

Table 7 Network properties of real-world networks being considered. The number of nodes n, number of edges m, maximum degree d_{max}, average degree d_{avg}, and other well-known network properties are reported

Dataset	n	m	d_{max}	d_{avg}	Diameter	Mean shortest path length	Clustering coefficient	Assortativity
US power grid	4941	6594	19	2.67	46	20	0.1	0.003
EuroRoad	1174	1417	10	2.41	62	19	0.03	0.13
PGP network	10680	24316	205	4.55	24	7.65	0.38	0.24
p2p-Gnutella	6301	20777	97	6.59	9	4.64	0.01	0.03
Email network	1133	5451	71	9.62	8	3.65	0.17	0.08
Hamsterster	1858	12534	272	13.5	14	3.4	0.09	−0.08
Human protein	3133	6726	129	4.29	13	4.80	0.04	−0.13
Jazz musicians	198	2742	100	27.69	6	2.21	0.52	0.02

4.2.1 Description of Networks

Infrastructure Networks: The US Power Grid network contains information about the power grid of the Western United States. A node in this small-world network is either a generator, a transformer or a power substation, while edges represent high-voltage power supply lines between nodes. The Euroroad network, a road network located mostly in Europe, is undirected with nodes representing cities and an edge between two nodes denotes a physical road between them. This network was observed to be neither scale-free nor small-world.

Online social and communication networks: The PGP web of trust is a social network formed by people that shares confidential information using the Pretty Good Privacy (PGP) encryption algorithm. The PGP network is not a scale-free network but exhibits a bounded degree distribution and a large clustering coefficient. The Gnutella network is a peer-to-peer architecture, where nodes represent Gnutella hosts and edges represent connections between them. It is not a pure power-law network and preserves good fault tolerance characteristics. The email communication network at the University Rovira i Virgili in Tarragona is a network of users where edges indicate at least one mail exchange between the users.

Other miscellaneous networks: Human protein (vidal) is a biological network that was an initial version of a systematic mapping of protein–protein interactions in humans. A collaboration network between Jazz musicians is considered, with each node being a Jazz musician and an edge denoting the two musicians playing together in a band.

4.2.2 Discussion of Results

Table 8 shows mean curvature values and correlations between OR and the other curvature metrics for these networks. Note, gJC tracks OR closest in terms of mean curvature. Since the range of Forman is unbounded, we see large negative average curvatures for many networks. Furthermore, gJC correlates strongly with OR compared to JC and Forman curvatures for almost every network considered. JC correlates with OR curvature better than Forman curvature on PGP Network, p2p-Gnutella, Email network, Hamsterster friendship network, and Human protein network, while Forman correlates stronger only on the EuroRoad network. Table 9 shows that our gJC implementation is at least an order of magnitude faster than the OR implementation, while comparable with our JC implementation. Our Forman implementation is the fastest among all the curvature metrics considered. These results agree with the theoretical analysis in Sect. 3.

Table 8 Average curvatures shown for different real-world networks. Pearson correlation r_p and Kendall's τ between OR curvature and Forman, JC, and gJC curvature are tabulated

Graph	$\overline{OR}$	$\overline{F}$	$\overline{JC}$	$\overline{gJC}$	(OR, F)		(OR, JC)		(OR, gJC)	
					r_p	τ	r_p	τ	r_p	τ
US power grid	−0.34	−3.7	−1.89	−0.78	0.48	0.41	0.40	0.23	**0.80**	**0.69**
EuroRoad	−0.33	−2.0	−1.97	−0.97	**0.76**	**0.69**	0.15	0.09	0.69	0.67
PGP network	−0.10	−33	−1.36	−0.14	0.13	0.08	0.73	0.53	**0.85**	**0.74**
p2p−Gnutella	−1.01	−31	−1.98	−1.16	−0.32	0.08	0.23	0.27	**0.86**	**0.58**
Email network	−0.41	−33	−1.72	−0.38	0.15	0.11	0.73	0.56	**0.81**	**0.69**
Hamsterster friendships	−0.34	−86	−1.87	−0.19	0.13	0.13	0.41	0.23	**0.58**	**0.42**
Human protein	−0.62	−27	−1.93	−0.79	0.35	0.34	0.31	0.10	**0.78**	**0.60**
Jazz musicians	0.27	−73	−0.92	0.32	0.09	0.05	0.91	0.79	**0.92**	**0.80**

Table 9 Running times for computing the OR, Forman, and Jaccard curvatures for different real networks

Graph (s)	OR(LP solver) (s)	Forman (s)	JC (s)	gJC (s)
US power grid	146.011	**0.099**	0.368	0.67
EuroRoad	9.515	**0.023**	0.227	0.432
PGP network	1052.614	**0.419**	2.624	5.258
p2p-Gnutella	219.943	**0.331**	2.146	5.66
Email network	57.279	**0.071**	0.718	1.543
Hamsterster friendships	424.898	**0.267**	5.069	14.354
Human protein	78.64	**0.084**	0.951	1.312
Jazz musicians	72.137	**0.092**	0.957	1.539

5 Conclusion

We introduced two new network curvature metrics, JC and gJC, that are inspired by the Jaccard coefficient. Theoretically, the gJC metric was shown to better approximate OR curvature for Erdos–Renyi graphs compared to the JC metric. We conducted experiments with different classes of network models and real networks, and observed that gJC approximates OR curvature best compared with JC and Forman curvature. Nonetheless, the JC curvature is easier to compute than gJC, and correlates moderately well with OR for positively curved or strongly clustered networks, suggesting that it could be used as a cheap proxy to the OR curvature for such special scenarios. The Forman curvature, while being the cheapest to compute, shows weak correlation with OR curvature for many real networks, demonstrating that at least for the weighting scheme used here it exhibits a different notion of curvature than Ollivier-Ricci.

References

1. Bhattacharya, B.B., Mukherjee, S.: Exact and asymptotic results on coarse Ricci curvature of graphs. Discrete Math. **338**(1), 23–42 (2015)
2. Forman, R.: Bochner's method for cell complexes and combinatorial Ricci curvature. Discrete Comput. Geom. **29**(3), 323–374 (2003)
3. Gallot, S., Lafontaine, J., Hulin, D.: Riemannian Geometry. Springer, New York (1987)
4. Higuchi, Y.: Combinatorial curvature for planar graphs. J. Graph Theory **38**(4), 220–229 (2001)
5. Jost, J.: Riemannian Geometry and Geometric Analysis. Springer, New York (2002)
6. Jost, J., Liu, S.: Ollivier's Ricci curvature, local clustering and curvature-dimension inequalities on graphs. Discrete Comput. Geom. **51**(2), 300–322 (2014)
7. Krioukov, D., Papadopoulos, F., Kitsak, M., Vahdat, A., Boguná, M.: Hyperbolic geometry of complex networks. Phys. Rev. E **82**(3), 036–106 (2010)
8. Kunegis, J.: KONECT: The Koblenz network collection (2013). http://konect.uni-koblenz.de/
9. Leskovec, J., Krevl, A.: SNAP Datasets: Stanford large network dataset collection (2014). http://snap.stanford.edu/data
10. Liben-Nowell, D., Kleinberg, J.: The link-prediction problem for social networks. J. Assoc. Inf. Sci. Technol. **58**(7), 1019–1031 (2007)
11. Lin, Y., Lu, L., Yau, S.T.: Ricci curvature of graphs. Tohoku Math. J. Second Ser. **63**(4), 605–627 (2011)
12. Narayan, O., Saniee, I.: Large-scale curvature of networks. Phys. Rev. E **84**(6), 066–108 (2011)
13. Newman, M.: Networks: An Introduction. Oxford University Press, Oxford (2010)
14. Ni, C.C., Lin, Y.Y., Gao, J., Gu, X.D., Saucan, E.: Ricci curvature of the internet topology. In: IEEE Conference on Computer Communications (INFOCOM), pp. 2758–2766. IEEE (2015)
15. Ollivier, Y.: Ricci curvature of Markov chains on metric spaces. J. Funct. Anal. **256**(3), 810–864 (2009)
16. Pal, S., Yu, F., Moore, T.J., Ramanathan, R., Bar-Noy, A., Swami, A.: An efficient alternative to Ollivier-Ricci curvature based on the Jaccard metric (2017). arXiv:1710.01724
17. Penrose, M.: Random Geometric Graphs, vol. 5. Oxford University Press, Oxford (2003)
18. Sandhu, R., Georgiou, T., Reznik, L.Z., Kolesov, I., Senbabaoglu, Y., Tannenbaum, A.: Graph curvature for differentiating cancer networks. Sci. Rep. **5** (2015)
19. Sandhu, R.S., Georgiou, T.T., Tannenbaum, A.R.: Ricci curvature: an economic indicator for market fragility and systemic risk. Sci. Adv. **2**(5), e1501–495 (2016)
20. Sreejith, R., Mohanraj, K., Jost, J., Saucan, E., Samal, A.: Forman curvature for complex networks. J. Stat. Mech.: Theory Exp. **2016**(6), 063–206 (2016)
21. Sreejith, R., Jost, J., Saucan, E., Samal, A.: Systematic evaluation of a new combinatorial curvature for complex networks. Chaos, Solitons Fractals **101**, 50–67 (2017)
22. Wang, C., Jonckheere, E., Banirazi, R.: Wireless network capacity versus Ollivier-Ricci curvature under heat-diffusion (hd) protocol. In: 2014 American Control Conference (ACC), pp. 3536–3541. IEEE (2014)
23. Watts, D.J., Strogatz, S.H.: Collective dynamics of 'small-world' networks. Nature **393**(6684), 440 (1998)
24. Weber, M., Saucan, E., Jost, J.: Characterizing complex networks with Forman-Ricci curvature and associated geometric flows. J. Complex Netw. p. cnw030 (2017)

Combinatorial Miller–Hagberg Algorithm for Randomization of Dense Networks

Hiroki Sayama

Abstract We propose a slightly revised Miller–Hagberg (MH) algorithm that efficiently generates a random network from a given expected degree sequence. The revision was to replace the approximated edge probability between a pair of nodes with a combinatorically calculated edge probability that better captures the likelihood of edge presence especially, where edges are dense. The computational complexity of this combinatorial MH algorithm is still in the same order as the original one. We evaluated the proposed algorithm through several numerical experiments. The results demonstrated that the proposed algorithm was particularly good at accurately representing high-degree nodes in dense, heterogeneous networks. This algorithm may be a useful alternative to other more established network randomization methods, given that the data are increasingly becoming larger and denser in today's network science research.

1 Introduction

In network science, there are occasions in which one needs to generate random network samples from a given node degree sequence. A typical context for doing this is to conduct a statistical test of whether empirically observed network properties can be explained by a certain degree distribution or not. Several algorithms have already been developed for this purpose, such as the classic Havel–Hakimi algorithm [1], the double edge swap method, the configuration model [2, 3], and the Bayati–Kim–Saberi algorithm [4]. However, they come with respective limitations. The Havel–Hakimi algorithm constructs a network using a heuristic, assortativity-inducing procedure,

H. Sayama (✉)
Center for Collective Dynamics of Complex Systems and Department
of Systems Science and Industrial Engineering, Binghamton University,
Binghamton, NY 13902, USA
e-mail: sayama@binghamton.edu

H. Sayama
Center for Complex Network Research, Northeastern University, Boston,
MA 02115, USA

© Springer International Publishing AG 2018
S. Cornelius et al. (eds.), *Complex Networks IX*, Springer Proceedings
in Complexity, https://doi.org/10.1007/978-3-319-73198-8_6

whose outcomes would not be appropriate to be used as fully randomized controls. The double edge swap method is simple but its randomization process is slow and gradual, with no well-defined termination condition. The configuration model is a systematic, well-defined randomization method, but its outcomes often contain parallel edges and self-loops. The Bayati–Kim–Saberi algorithm can be computationally costly and does not guarantee that it can produce a randomized graph as an output.

The Miller–Hagberg algorithm [5] (called the MH algorithm hereafter) addresses those limitations of the other algorithms mentioned above by relaxing the requirement so that it generates a random network from a given *expected* degree sequence. This relaxation allows for calculation of edge probability *independently* for each pair of nodes. By sorting the nodes according to their expected degrees and implementing an efficient node-skipping mechanism (see [5] for details), the MH algorithm achieves linear computational complexity $O(N + M)$, where N and M are the numbers of nodes and edges, respectively. This property is highly desirable for large-scale network analysis.

While the MH algorithm can be used with any edge probability functions, its original version uses Chung and Lu's random graph model [6] that assumes that an edge probability between two nodes with degrees w_i and w_j can be approximated as $\min(1, w_i w_j / \sum_k w_k)$. It is known that this assumption is invalid if the network is dense (i.e., if w_i is not negligible compared to N). This issue is typically manifested on high-degree nodes whose degrees generated by this algorithm often deviate greatly from their expected degrees specified in the given degree sequence [5].[1] This limitation has not been so critical an issue so far because most real-world networks show significant degree heterogeneity and thus they are fundamentally sparse [9].

With the recent expansion of modeling methodologies and application domains of network science, however, there are now several situations in which one needs to analyze *dense* networks, such as the ego networks in social media data [10], the time/layer aggregations of temporal and multilayer networks [11, 12], and the functional connectivity networks of the brain imaging data [13], to name a few. These networks typically have much higher edge density than other more classical networks, while they still maintain substantial degree heterogeneity. Accurately representing their high-degree nodes in randomized counterparts is, thus, an important methodological challenge.

In this paper, we aim to address the above challenge by implementing a small yet unique revision in the original MH algorithm, by replacing the Chung–Lu edge probability with a combinatorically calculated edge probability that better captures the likelihood of edge presence especially, where edges are dense. In the rest of the paper, we describe technical details of the algorithm revision and then present some results of evaluation of the proposed algorithm through numerical experiments.

---

[1] There have been a couple of modifications of edge probability calculation proposed to address this issue [7, 8], mostly using statistical physics approaches.

2 Revising the MH Algorithm with Combinatorial Edge Probability

We revise the MH algorithm by replacing the approximated edge probability with a combinatorially calculated edge probability. This calculation is done by counting the number of all network configurations in each of the two scenarios: the presence or the absence of an edge between two focal nodes. Let w_i and w_j be the degrees of two nodes, i and j, for which the edge probability between them is to be calculated. Also let N and M be the numbers of nodes and edges in the network, respectively. Assuming that each network configuration occurs randomly with equal probability, the edge probability between the two nodes can be written as

$$p(N, M, w_i, w_j) = \frac{C_c(N, M, w_i, w_j)}{C_c(N, M, w_i, w_j) + C_d(N, M, w_i, w_j)}, \tag{1}$$

where $C_c(N, M, w_i, w_j)$ is the number of network configurations in which the two nodes i and j are connected directly, and $C_d(N, M, w_i, w_j)$ is the number of network configurations in which those nodes are *not* connected directly (Fig. 1). Equation (1) can be rewritten as

$$p(N, M, w_i, w_j) = \left(1 + \frac{C_d(N, M, w_i, w_j)}{C_c(N, M, w_i, w_j)}\right)^{-1}, \tag{2}$$

if $C_c(N, M, w_i, w_j) \neq 0$.

Both C_c and C_d can be calculated as the product of the following three combinatorial quantities (Fig. 1):

- Number of possibilities of placing the edges that emanate from node i to the rest of the network

  – For C_c: $\binom{N-2}{w_i - 1}$ For C_d: $\binom{N-2}{w_i}$

- Number of possibilities of placing the edges that emanate from node j to the rest of the network

  – For C_c: $\binom{N-2}{w_j - 1}$ For C_d: $\binom{N-2}{w_j}$

- Number of possibilities of placing the edges not adjacent to the two nodes among the rest of nodes in the network

  – For C_c: $\binom{\binom{N-2}{2}}{M - w_i - w_j + 1}$ For C_d: $\binom{\binom{N-2}{2}}{M - w_i - w_j}$

By multiplying these three quantities, we obtain

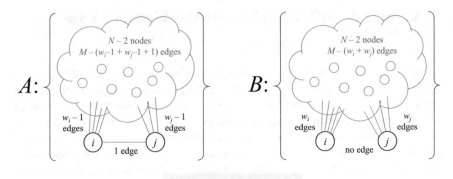

$$p = \frac{|A|}{|A| + |B|}$$

Fig. 1 Schematic illustration of the proposed combinatorial edge probability calculation (1). $|A| = C_c(N, M, w_i, w_j)$: Number of network configurations in which the two focal nodes, i and j, are connected directly. $|B| = C_d(N, M, w_i, w_j)$: Number of network configurations in which the two nodes are not connected directly

$$C_c(N, M, w_i, w_j) = \binom{N-2}{w_i - 1}\binom{N-2}{w_j - 1}\binom{\binom{N-2}{2}}{M - w_i - w_j + 1}, \text{ and} \quad (3)$$

$$C_d(N, M, w_i, w_j) = \binom{N-2}{w_i}\binom{N-2}{w_j}\binom{\binom{N-2}{2}}{M - w_i - w_j}. \quad (4)$$

By applying these combinatorial calculations into (2) and simplifying it, we obtain

$$p(N, M, w_i, w_j) = \left(1 + \frac{N - w_i - 1}{w_i}\frac{N - w_j - 1}{w_j}\frac{M - w_i - w_j + 1}{\binom{N-2}{2} - M + w_i + w_j}\right)^{-1} \quad (5)$$

$$= \left(1 + \frac{2M^*(N - w_i - 1)(N - w_j - 1)}{w_i w_j(N^2 - 5N + 8 - 2M^*)}\right)^{-1}, \quad (6)$$

where $M^* = M - w_i - w_j + 1$. In the actual computation of p, we use the following more straightforward formula that does not involve inversion:

$$p(N, M, w_i, w_j) = \frac{X}{X + Y}, \quad (7)$$

$$X = w_i w_j(N^2 - 5N + 8 - 2M^*), \quad (8)$$

$$Y = 2M^*(N - w_i - 1)(N - w_j - 1). \quad (9)$$

This correctly gives $p = 0$ if w_i or $w_j = 0$, which is convenient for practical purposes.

The formula obtained above is surprisingly simple, involving only a finite, constant number of basic arithmetic operations. Therefore, the revised MH algorithm with this combinatorial edge probability (called the *combinatorial MH algorithm* hereafter) still maintains the original computational complexity $O(N + M)$. Also note that (7)–(9) recovers the original Chung–Lu formula $w_i w_j / (2M) = w_i w_j / \sum_k w_k$, if $N \to \infty$ and $w_i, w_j \ll M \ll N^2$.

Equations (7)–(9) capture the edge probability more accurately where edge density is high. Considering some extreme cases helps illustrate this benefit. For example, in a complete graph made of N nodes, each node has $N - 1$ as its degree, and the total number of edges is $N(N - 1)/2$. Letting $w_i = w_j = N - 1$ and $M = N(N - 1)/2$ (i.e., $M^* = N(N - 1)/2 - (N - 2) - (N - 2) + 1$) in (7)–(9) produces $p = 1$, correctly indicating that any pair of nodes must be connected directly. However, the Chung–Lu model gives $p = (n - 1)/n < 1$ in the same situation. A more extreme case is a star graph made of N nodes and $N - 1$ edges. The edge probability between the central node (with $w_i = N - 1$) and a peripheral node (with $w_j = 1$) is correctly calculated to be $p = 1$ by (7)–(9), while the Chung–Lu model gives $p = 1/2$, which is far off the actual probability 1. Finally, another example that shows the opposite way of deviation is a disconnected graph made of two 6-node star graphs ($N = 12$, $M = 10$). In this graph, the edge probability between the two central nodes ($w_i = w_j = 5$) is calculated to be $p = 125/129$ by (7)–(9), which correctly captures the small possibility that those two central nodes do not have a direct connection to each other. In the meantime, the Chung–Lu model gives $p = \min(1, 5/4) = 1$, which forces the two central nodes to *always* be connected in randomized networks. These examples demonstrate the accuracy of the combinatorial edge probability proposed in this study.

We note that (7)–(9) may not provide accurate edge probabilities for low-degree nodes. For example, they give a nonzero (positive) edge probability between two peripheral nodes in a star graph, since their mandatory connections to the central node are ignored when the edge probability between them is calculated. In general, the proposed algorithm tends to produce slightly higher-than-expected degrees for peripheral nodes in heterogeneous networks (which will be seen in numerical results later). Also, (7)–(9) may malfunction if a graphically impossible input is given, because the formula was derived using combinatorial enumerations under the assumption that the given parameters (N, M, w_i, w_j) are graphically possible. For example, $(N, M, w_i, w_j) = (5, 10, 1, 1)$ (which is graphically impossible) gives a meaningless value $p = -5/76$. However, such a problem will not arise as long as the formula is used for randomizing the topology of an existing network. In what follows, we exclusively consider cases in which the expected degree sequence is always obtained from the degree sequence of another existing network.

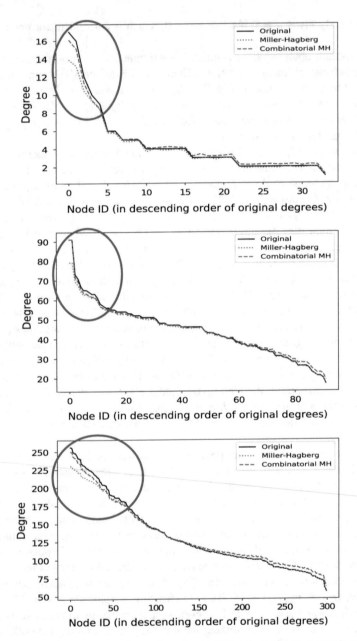

Fig. 2 Comparison of degree sequences among the original network (black, solid lines) and two randomized ones (green, dotted lines: original MH algorithm; red, dashed lines: combinatorial MH algorithm). Top: Zachary's Karate Club network [14]. Middle: Ego network in Leskovec–McAuley Facebook dataset [10]. Bottom: Dense heterogeneous network constructed using the Barabási–Albert model [15]. For each randomization algorithm, the average result of 500 independent randomization trials is shown. Nodes are sorted in descending order of their degrees in the original network. A clear difference between the original and combinatorial MH algorithms is seen on high-degree nodes (highlighted with red circles)

3   Evaluations

We first tested the proposed combinatorial MH algorithm by applying it to several illustrative dense networks. The following three networks were used:

- Zachary's Karate Club network [14] (34 nodes, 78 edges, and density: 0.139)
- Ego network of an arbitrarily chosen user (user '3000' for this example) in Leskovec–McAuley Facebook dataset [10] (92 nodes, 2,044 edges, and density: 0.488)
- Dense heterogeneous network constructed using the Barabási–Albert model [15] (300 nodes, 20,000 edges, and density: 0.446)

Figure 2 shows the results in which the degree sequences among the given original network and two randomized ones (by the original and combinatorial MH algorithms) were compared. For each randomization algorithm, the average result of 500 independent randomization trials is shown. It is clearly seen in these plots that the combinatorial MH algorithm (red, dashed lines) was able to represent high-degree nodes more accurately than the original MH algorithm (green, dotted lines).

We also evaluated the effect of edge density on the performance of randomization algorithms. Figure 3 shows the results of a numerical experiment in which the edge density was systematically varied on Erdős–Rényi and Barabási–Albert networks. The performance of the algorithms was measured by the difference in average node degrees between given and randomized networks. The combinatorial MH algorithm successfully reproduced average node degrees that were closer to the given ones, especially for high edge density cases.

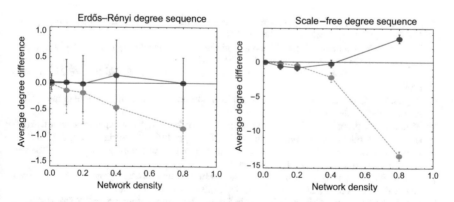

Fig. 3 Performance comparison between the original (gray, dashed lines) and combinatorial (blue, solid lines) MH algorithms. Each algorithm was applied to a randomly generated expected degree sequence (left: sequence generated from Erdős–Rényi networks, right: sequence generated from Barabási–Albert networks) over varying network densities. $N = 1,000$ for all cases. Performance was measured by the difference in average node degrees between given and randomized networks. Each data point was an average of 100 independent simulations. Error bars show standard deviations. Similar trends were observed for $N = 100, 500$ and $2,000$

4   Conclusions

In this paper, we presented the combinatorial MH algorithm in which the edge probability between a pair of nodes was combinatorically calculated. The derived edge probability formula involved only a constant number of basic arithmetic operations, keeping the linear computational complexity of the original MH algorithm. Numerical experiments demonstrated that the proposed algorithm was particularly good at accurately representing high-degree nodes in dense, heterogeneous networks. This algorithm may be a useful alternative of other more established network randomization methods, given that the data are increasingly becoming larger and denser in today's network science research.

What is particularly unique about the proposed algorithm is that it captures, in some sense, certain nonlocal topological dependencies in calculating edge probability (this helps accuracy), even though the probability itself is still calculated independently for each node pair (this helps computational efficiency). In the meantime, such independent calculation of edge probability may also be a limitation of the algorithm because, as noted earlier, it may produce inaccurate results where edges are sparse. This limitation should be taken into account when one decides which network randomization algorithm should be used for a specific network dataset. Proper handling of such interdependency of edge probabilities will require more careful mathematical analysis and algorithm design, which is among our future work.

References

1. Hakimi, S.L.: On realizability of a set of integers as degrees of the vertices of a linear graph. I. J. Soc. Ind. Appl. Math. **10**(3), 496–506 (1962)
2. Bender, E.A., Canfield, E.R.: The asymptotic number of labeled graphs with given degree sequences. J. Comb. Theory Ser. A **24**(3), 296–307 (1978)
3. Newman, M.E.: The structure and function of complex networks. SIAM Rev. **45**(2), 167–256 (2003)
4. Bayati, M., Kim, J.H., Saberi, A.: A sequential algorithm for generating random graphs. Algorithmica **58**(4), 860–910 (2010)
5. Miller, J., Hagberg, A.: Efficient generation of networks with given expected degrees. Algorithms and Models for the Web Graph (WAW 2011), pp. 115–126. Springer, Berlin (2011)
6. Chung, F., Lu, L.: Connected components in random graphs with given expected degree sequences. Ann. Comb. **6**(2), 125–145 (2002)
7. Britton, T., Deijfen, M., Martin-Löf, A.: Generating simple random graphs with prescribed degree distribution. J. Stat. Phys. **124**(6), 1377–1397 (2006)
8. van der Hofstad, R.: Critical behavior in inhomogeneous random graphs. Random Struct. Algorithms **42**(4), 480–508 (2013)
9. Del Genio, C.I., Gross, T., Bassler, K.E.: All scale-free networks are sparse. Phys. Rev. Lett. **107**(17), 178701 (2011)
10. Leskovec, J., McAuley, J.J.: Learning to discover social circles in ego networks. In: Advances in Neural Information Processing Systems, pp. 539–547 (2012)
11. Holme, P., Saramäki, J.: Temporal networks. Phys. Rep. **519**(3), 97–125 (2012)
12. Kivelä, M., Arenas, A., Barthelemy, M., Gleeson, J.P., Moreno, Y., Porter, M.A.: Multilayer networks. J. Complex Netw. **2**(3), 203–271 (2014)

13. Zamani Esfahlani, F., Sayama, H.: A percolation-based thresholding method with applications in functional connectivity analysis (submitted to CompleNet 2018, under review)
14. Zachary, W.W.: An information flow model for conflict and fission in small groups. J. Anthr. Res. **33**(4), 452–473 (1977)
15. Barabási, A.L., Albert, R.: Emergence of scaling in random networks. Science **286**(5439), 509–512 (1999)

Proposal of Strategic Link Addition for Improving the Robustness of Multiplex Networks

Yui Kazawa and Sho Tsugawa

Abstract Recent research trends in network science have shifted from the analysis of single-layer networks to that of multilayer networks. In particular, the robustness of multilayer networks has been actively studied. Two popular multilayer network models exist: interdependent and multiplex. This study proposes link addition strategies to improve the robustness of multiplex networks against layer node-based attack. The proposed strategies extend an existing strategy called random inter degree–degree difference (RIDD), which is proposed for improving the robustness of interdependent networks. While RIDD adds links using inter-layer degree difference, proposed strategies adds links using both inter-layer degree difference and degree of layer nodes. Through several network attack simulations, we show that the proposed link addition strategies can effectively improve the robustness of multiplex networks.

1 Introduction

Recent research trends in network science have shifted from the analysis of single-layer networks to that of multilayer networks [1–17]. A multilayer network is the one having multiple layers of networks that interact with each other [1, 2]. Many real-world complex systems have multilayer structures [1, 3]. For instance, infrastructures such as water supply systems, transportation systems, and power grids are defined as networks that are interdependent on each other, which together can thus be described as a multilayer network [3].

The robustness of multilayer networks has been actively studied [3–7, 9–16]. The robustness of a network is its ability to maintain its connectivity against random fail-

Y. Kazawa (✉) · S. Tsugawa
Graduate School of Systems and Information Engineering, University of Tsukuba,
1-1-1 Tennodai, Tsukuba, Ibaraki 305-8573, Japan
e-mail: kzw-y@mibel.cs.tsukuba.ac.jp; kzw-y@cs.tsukuba.ac.jp

S. Tsugawa
e-mail: s-tugawa@mibel.cs.tsukuba.ac.jp; s-tugawa@cs.tsukuba.ac.jp

© Springer International Publishing AG 2018
S. Cornelius et al. (eds.), *Complex Networks IX*, Springer Proceedings
in Complexity, https://doi.org/10.1007/978-3-319-73198-8_7

ures of nodes and intentional attacks to the network. Because it is desirable that many systems defined as multilayer networks be able to maintain their overall connectivity, even when some nodes fail, the robustness of multilayer networks has attracted extensive research interest.

Two popular multilayer network models exist: interdependent and multiplex networks. An interdependent network consists of multiple layers (i.e., networks) whose nodes are interdependent [2]. In an interdependent network, when a node in a layer fails, nodes in other layers that have a dependent relationship with the node also fail [2, 3]. A multiplex network is a collection of several network layers that contain the same nodes yet different intra-layer connections [2]. In other words, in a multiplex network, each node in each layer has exactly one inter-layer link with a node in each different layer. Each node in a layer is called a layer node, and each set of nodes that are connected via inter-layer links is called a multiplex node [9].

Several methods for improving the robustness of interdependent networks have been proposed [11–16]. To add only a few links to a network is expected to be a promising means of improving the robustness of the network based on its feasibility in actual networks [16]. Ji et al. have proposed two link addition strategies called random inter degree–degree difference (RIDD) and low inter degree–degree difference (LIDD). Because interdependent networks with high level of inter-similarity are known to be robust against random failures of nodes [10], RIDD and LIDD aim to increase the inter-similarity between layers through link addition. Ji et al. show that RIDD and LIDD are more effective at improving the robustness of an interdependent network than are the conventional link addition strategies such as random addition (RA) and low degree (LD).

In our previous work, we have investigated the effectiveness of the link addition strategy of RIDD for improving the robustness of multiplex networks rather than that of interdependent networks [17]. Through simulations, we have shown that RIDD can be applied to multiplex networks. By contrast, we have also shown that the effectiveness of RIDD is comparable to that of the conventional link addition strategy of RA for multiplex networks, which implies a room for improvement in RIDD.

In this study, to improve the robustness of multiplex networks against network attacks, we propose new link addition strategies by extending the existing link addition strategy of RIDD. The main purpose of our proposed strategies is to utilize the degree of layer node in each layer as well as the inter-similarity between layers for link addition. The proposed strategies add links to nodes with low degree and also aim to increase the inter-similarity between layers of multiplex networks. Through network attack simulations, we also investigate the effectiveness of the proposed strategies at improving the robustness of multiplex networks. Consequently, we show that the proposed strategies more effectively maintain the size of mutually connected giant component (MCGC) of multiplex networks than the RIDD which uses only the inter-similarity between layers and the link addition strategies, which uses only degree of layer node.

2 Related Work

Recently, the robustness of interdependent networks against random failures and intentional attacks has been investigated [3, 10]. Buldyrev et al. revealed that an interdependent network can be fragmented by failures of a few nodes in a single layer as a result of cascading failures [3]. Parshani showed that the inter-similarity between layers in an interdependent network affects the network's robustness [10]. The level of the inter-similarity is quantified using two proposed measures: inter degree–degree correlation and interclustering coefficient. Ji et al. also proposed another measure for the inter-similarity called average inter degree–degree difference (AIDD) [16].

In addition, the robustness of multiplex networks have been also studied [5, 7, 9]. Brummit et al. showed that multiplex networks are generally more vulnerable than simple single-layer networks [5]. Min et al. revealed that the inter-similarity between layers in a multiplex network considerably affects their robustness against failures and attacks [7]. Zhao et al. investigated the robustness of multiplex networks against two types of attacks: multiplex node-based and layer node-based attack. Layer node-based attack removes layer nodes in a network, whereas multiplex node-based attack removes multiplex nodes, each of which corresponds to a set of layer nodes. Through theoretical analyses and numerical simulations, these researchers found that multiplex networks consisting of two scale-free networks are vulnerable to layer node-based intentional attacks while simultaneously being robust against random failures of layer nodes. In this study, we propose link addition strategies as countermeasures against layer node-based attack. Note that we particularly focus on layer node-based attack. This is because multiplex node-based attack can be regarded as a special case of layer node-based attack when all the removed nodes or replicas are the same in each network layer [9].

In previous studies, several link addition strategies for single-layer networks were proposed [18–21]. LD, which adds links between low-degree nodes, is one of the popular link addition strategies for single-layer networks [18]. Moreover, RA [19–21], which randomly adds links, is often used as a reference for comparison with other link addition strategies. Zhao et al. showed that LD can improve the robustness of single-layer networks more effectively than RA [18].

As discussed in Sect. 1, link addition strategies for interdependent networks have also been proposed [16]. RIDD and LIDD are shown to be more effective than LD and RA at improving the robustness of interdependent networks [16].

To the best of our knowledge, link addition strategies for multiplex networks have not been proposed in previous studies. However, in our previous work, we applied existing link addition strategies, RA and RIDD, to multiplex networks and examined their effectiveness [17]. Through experiments, we showed that RA and RIDD can be applied to multiplex networks. However, RIDD and RA are also shown to be comparable in most cases. This fact motivates us to extend RIDD to further improve the robustness of multiplex networks. Because RA is a simple baseline strategy, the comparable performance of RIDD with RA suggests that RIDD can be further improved. In the next section, we propose new link addition strategies by extending RIDD.

3   Link Addition Strategy

In this section, we first explain the existing link addition strategies proposed by Ji et al. [16], and then propose link addition strategies for multiplex networks. In this study, following [9], we particularly consider the problem of adding links to multiplex networks with two layers (i.e., duplex networks). Each layer is an undirected unweighted graph, and the two layers are denoted as G_A and G_B. Each link addition strategy has a fixed budget of M' links, and repeat the procedure explained later in this section until the number of added links reaches M'. Note that self-loop and parallel edges are not allowed and the degree of layer node are calculated at each step in every link addition strategy.

3.1   Link Addition Strategies Using IDD

The existing link addition strategies RIDD and LIDD add links based on IDD, which is defined as the degree difference between two interconnecting nodes [16]. Since a network with high inter-similarity (i.e., networks with low AIDD) has been shown to be robust against random failure, both strategies aim to reduce AIDD, which is calculated as the average absolute IDD per node in an interdependent network [16]. Let u be a node in graph G_A, v be a node in graph G_B, and node u have an inter-layer link with node v. Then, the IDD of node u in graph G_A is defined as $IDD(u) = k_u - k_v$, where k_u and k_v are degrees of nodes u and v, respectively. The detailed procedures of RIDD and LIDD are given as follows.

RIDD At each step, IDD of all nodes in G_A and G_B are calculated. For each of graph G_A and G_B, add a link between a pair of nodes selected randomly from the pairs of unconnected nodes with negative IDD. If no pairs of nodes with negative IDD exist, a link is added between randomly selected unconnected nodes.

LIDD At each step, IDD of all nodes in G_A and G_B are calculated. For each of G_A and G_B, add a link between a pair of unconnected nodes with the lowest negative IDD. If no pairs of nodes with negative IDD exist, a link is added between the pair of unconnected nodes with the lowest degree.

3.2   Link Addition Strategies Using IDD and Degree of Layer Node

We propose three link addition strategies that consider both IDD and degree of layer node. These three are called low-degree IDD (LD_IDD), low-degree-product IDD (LDP_IDD), and low-degree-sum IDD (LDS_IDD). Similar to RIDD and LIDD, LD_IDD adds links between nodes with negative IDD, but it prefers to add links to low-degree nodes. Low-degree nodes are vulnerable to attacks and failures [22]. Specifically, low-degree nodes tend to be isolated during attacks and failures because the connectivity of such nodes depends heavily on other relatively high-degree nodes. Therefore, LD_IDD adds links to low-degree nodes while reducing AIDD in the network. Note that adding links to low-degree nodes also occurs in the link addition strategy for single-layer networks, LD [18]. LDP_IDD and LDS_IDD are variants of LD_IDD. LDP_IDD uses layer degree product (i.e., the product of the degrees of two interconnecting nodes) and LDS_IDD uses layer degree sum (i.e., the sum of the degrees of two interconnecting nodes) instead of using degree of layer node only in each layer. Layer degree product is defined as $DP(u) = DP(v) = k_u \times k_v$ and degree sum is defined as $DS(u) = DS(v) = k_u + k_v$, where k_u and k_v are degrees of interconnecting nodes u and v, respectively. These measures are used to measure the importance of nodes in multiplex networks [2, 23].

The detailed procedures of LD_IDD, LDP_IDD, and LDP_IDD are given as follows.

LD_IDD At each step, IDD and degree of all nodes in G_A and G_B are calculated. For each of G_A and G_B, add a link between a pair of unconnected nodes with the lowest degree and negative IDD. If no pairs of nodes with negative IDD exist, a link is added between randomly selected unconnected nodes.

LDP_IDD At each step, IDD and degree product of all nodes in G_A and G_B are calculated. For each of G_A and G_B, add a link between a pair of unconnected nodes with the lowest degree product and negative IDD, except when degree product is 0. If no pairs of nodes with negative IDD exist, a link is added between randomly selected unconnected nodes.

LDS_IDD At each step, IDD and degree sum of all nodes in G_A and G_B are calculated. For each of G_A and G_B, add a link between a pair of unconnected nodes with the lowest degree sum and negative IDD. If no pairs of nodes with negative IDD exist, a link is added between randomly selected unconnected nodes.

To evaluate the effectiveness of combining IDD and degree of layer node, we compare the proposed strategies with those that use only degree of layer node (i.e., LD, LDP, and LDS). LDP adds a link between nodes with the lowest degree product, and LDS adds a link between nodes with the lowest degree sum.

4 Methodology

4.1 Overview

Following [17], we conducted experiments with the following steps: (1) We first generate duplex networks using network generation models. (2) We next add links to both layers of the generated network using the link addition strategies previously described in Sect. 3. (3) We then perform network attack simulation on the network and investigate the connectivity of the remaining network. The details of these three steps are explained in the remainder of this section.

4.2 Generating Duplex Network

We generate a duplex network G by connecting two single-layer networks. We generate two graphs with the same size N, which are denoted as G_A and G_B. For each node in graph G_A, an inter-layer link is created between that node and a randomly selected node in graph G_B to construct an uncorrelated duplex structure. Each node in graphs G_A and G_B is denoted as $v_i^A (i = 1, \ldots, N)$ and $v_i^B (i = 1, \ldots, N)$, respectively. v_i^A and v_i^B are connected via an inter-layer link. In addition, the sets of nodes and links in the graph G_A are denoted by V_A and E_A, respectively, and the sets of nodes and links in the graph G_B are denoted by V_B and E_B, respectively. A duplex network is denoted as $G = (V, E)$, where V is a set of multiplex nodes and E is a set of links connecting multiplex nodes. Layer nodes for multiplex node v_i are v_i^A and v_i^B. Multiplex nodes are regarded as connected if they have links on at least one layer [9]. In other words, $(v_i, v_j) \in E$ when $(v_i^A, v_j^A) \in E_A$ or $(v_i^B, v_j^B) \in E_B$.

We use the following models to generate single-layer networks: Barabási–Albert (BA) model [24], community emergence (CE) model proposed by Kumpula et al. [25], and Erdös–Rényi (ER) model [26]. We generate six types of duplex networks: (1) BA-BA network, (2) CE-CE network, (3) ER-ER network, (4) BA-ER network, (5) CE-BA network, and (6) CE-ER network. BA model that generates scale-free network and ER model that generates random networks are popular models used to generate artificial networks. CE model generates networks that have a skewed degree distribution and tunable community structure. Although the CE model generates weighted graphs, the weight of the generated graph is ignored and the graph is treated as an unweighted undirected graph in this paper.

We generated 10 duplex networks with $N = 1000$. We constructed BA graph with $m = 2.0$, ER graph with $p = 0.004$, and CE graph with $\delta = 2.0$, $p\delta = 0.004$, $p_r = 0.001$, and $p_d = 0.001$. The parameters used to generate the CE and ER graph were determined based on values that make them approximately equal to those of the BA graph. The number of links in the BA graph was 1997, the average number of links in the CE graphs was 2004.8, and the average number of links in the ER graphs was 2006.7.

4.3 Network Attack Simulation

We performed a simulation of layer node-based attack similar to that in [9] to investigate the robustness of multiplex networks with additional links. We first added $M'/2$ links to graph G_A and G_B by link addition strategies previously described in Sect. 3. Now, we define the fraction of links added to duplex network G as $fa = M'/(M_A + M_B)$ where M_A and M_B are the number of links of G_A and G_B, respectively. Next, we removed $N \times \phi_A$ nodes from G_A, and $N \times \phi_B$ nodes from G_B in descending order of their degree. We then removed those nodes that are not part of the largest component of G_A and G_B. Finally, we calculated the size of MCGC used as a common measure to evaluate the robustness of multiplex networks [2, 3, 9]. Note that MCGC is a set of connected multiplex nodes [9]. Multiplex nodes are regarded as connected if they have links on at least one layer [9]. We used the normalized size of MCGC, R, which is defined as the number of nodes belonging to the MCGC normalized by the number of nodes N in the network.

For each network and link addition strategy, we performed 10 independent simulations of link addition and node removal. We then obtained the average of R from the 100 independent simulation runs for each link addition strategy.

5 Results and Discussion

To evaluate the effectiveness of the proposed strategies, we compared the robustness of multiplex networks when using the proposed link addition strategies and when using RIDD and LIDD. Figure 1 shows the relative size of MCGC R against the fraction of removed nodes ϕ_A. We used $fa = 0.10$ and $\phi_B = 0.40$. For comparison purposes, these figures include the results when using RA to randomly add $M'/2$ links to graphs G_A and G_B and those without link addition (denoted as *NONE* in the figures). These results show that the proposed strategies LD_IDD, LDP_IDD, and LDS_IDD outperform existing link addition strategies (Fig. 1). The size of MCGC R when using the proposed strategies are larger than those when using LIDD, RIDD, and RA. This confirms the effectiveness of the proposed strategies to improve the robustness of multiplex networks. The results also show that the differences among the three proposed strategies are marginal.

We next compared the proposed strategies with those that employ only degree of layer node (i.e., LD, LDP, and LDS) in order to investigate the effectiveness of using both IDD and degree of layer node. Figure 2 shows the relative size of MCGC R

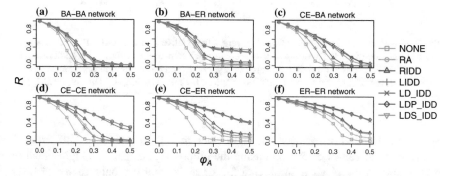

Fig. 1 Fraction of removed nodes from G_A versus the relative size of MCGC R under a degree-based attack when $fa = 0.10$ and $\phi_B = 0.40$. Performance comparison between link addition strategies that add links based on only the IDD of each node and those that add links based on both IDD and degree of each node

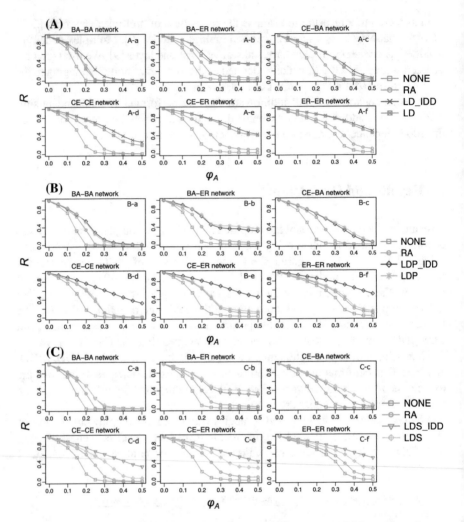

Fig. 2 Fraction of removed nodes from G_A versus the relative size of MCGC R under a degree-based attack when $fa = 0.10$ and $\phi_B = 0.40$. Performance comparison between link addition strategies that add links based on only the degree of layer node and those that add links based on both IDD and degree of layer node. **a** LD_IDD versus LD, **b** LDP_IDD versus LDP, and **c** LDS_IDD versus LDS

against the fraction of removed nodes ϕ_A when $fa = 0.10$ and $\phi_B = 0.40$. Figure 2 shows that the values of R when using the proposed strategies are higher than or comparable to those when employing the strategies that use only degree of layer node. This confirms the effectiveness of using both IDD and degree of layer node.

6 Conclusion and Future Works

We proposed three link addition strategies that add links based on both IDD and degree of each node: LD_IDD, LDP_IDD, and LDS_IDD. Through extensive simulations, we showed that the proposed link addition strategies are more effective than the existing strategies (i.e., those that use only IDD or only degree of layer node) at improving the robustness of multiplex networks against layer node-based attack.

In future works, we will propose a more effective link addition strategy as well as analyze the effectiveness of link addition strategies at improving the robustness of multiplex networks consist of three or more layers and real-world multiplex networks.

References

1. Kivelä, M., Arenas, A., Barthelemy, M., Gleeson, J.P., Moreno, Y., Porter, M.A.: Multilayer networks. J. Complex Netw. **2**(3), 203–271 (2014)
2. Boccaletti, S., Bianconi, G., Criado, R., Del Genio, C.I., Gómez-Gardenes, J., Romance, M., Sendina-Nadal, I., Wang, Z., Zanin, M.: The structure and dynamics of multilayer networks. Phys. Rep. **544**(1), 1–122 (2014)
3. Buldyrev, S.V., Parshani, R., Paul, G., Stanley, H.E., Havlin, S.: Catastrophic cascade of failures in interdependent networks. Nature **464**(7291), 1025–1028 (2010)
4. Huang, X., Gao, J., Buldyrev, S.V., Havlin, S., Stanley, H.E.: Robustness of interdependent networks under targeted attack. Phys. Rev. E **83**(6), 065101 (2011)
5. Brummitt, C.D., Lee, K.M., Goh, K.I.: Multiplexity-facilitated cascades in networks. Phys. Rev. E **85**(4), 045102 (2012)
6. Lee, K.M., Kim, J.Y.: Cho, Wk, Goh, K.I., Kim, I.: Correlated multiplexity and connectivity of multiplex random networks. New J. Phys. **14**(3), 033027 (2012)
7. Min, B., Do Yi, S., Lee, K.M., Goh, K.I.: Network robustness of multiplex networks with interlayer degree correlations. Phys. Rev. E **89**(4), 042811 (2014)
8. Ouyang, M.: Review on modeling and simulation of interdependent critical infrastructure systems. Reliab. Eng. Syst. Saf. **121**, 43–60 (2014)
9. Zhao, Dw, Wang, Lh, Zhi, Yf, Zhang, J., Wang, Z.: The robustness of multiplex networks under layer node-based attack. Sci. Rep. **6** (2016)
10. Parshani, R., Rozenblat, C., Ietri, D., Ducruet, C., Havlin, S.: Inter-similarity between coupled networks. Europhys. Lett. **92**(6), 68002 (2011)
11. Shao, J., Buldyrev, S.V., Havlin, S., Stanley, H.E.: Cascade of failures in coupled network systems with multiple support-dependence relations. Phys. Rev. E **83**(3), 036116 (2011)
12. Zhou, D., Stanley, H.E.: DAgostino, G., Scala, A.: Assortativity decreases the robustness of interdependent networks. Phys. Rev. E **86**(6), 066103 (2012)
13. Nguyen, D.T., Shen, Y., Thai, M.T.: Detecting critical nodes in interdependent power networks for vulnerability assessment. IEEE Trans. Smart Grid **4**(1), 151–159 (2013)
14. Ruj, S., Pal, A.: Analyzing cascading failures in smart grids under random and targeted attacks. In: Proceedings of AINA'14, pp. 226–233. IEEE (2014)
15. Reis, S.D., Hu, Y., Babino, A., Andrade Jr., J.S., Canals, S., Sigman, M., Makse, H.A.: Avoiding catastrophic failure in correlated networks of networks. Nat. Phys. **10**(10), 762–767 (2014)
16. Ji, X., Wang, B., Liu, D., Chen, G., Tang, F., Wei, D., Tu, L.: Improving interdependent networks robustness by adding connectivity links. Phys. A Stat. Mech. Appl. **444**, 9–19 (2016)
17. Kazawa, Y., Tsugawa, S.: On the effectiveness of link addition for improving robustness of multiplex networks against layer node-based attack. In: Proceedings of the 41st Annual IEEE

International Computers, Software, and Applications Conference (Student Research Symposium), pp. 697–700 (2017)

18. Zhao, J., Xu, K.: Enhancing the robustness of scale-free networks. J. Phys. A Math. Theor. **42**(19), 195003 (2009)

19. Beygelzimer, A., Grinstein, G., Linsker, R., Rish, I.: Improving network robustness by edge modification. Phys. A Stat. Mech. Appl. **357**(3), 593–612 (2005)

20. Jiang, Z., Liang, M., Guo, D.: Enhancing network performance by edge addition. Int. J. Mod. Phys. C **22**(11), 1211–1226 (2011)

21. Cao, X.B., Hong, C., Du, W.B., Zhang, J.: Improving the network robustness against cascading failures by adding links. Chaos Solitons Fractals **57**, 35–40 (2013)

22. Barabási, A.L., Jennifer, F.: Linked: The New Science of Networks Science of Networks. Basic Books, Cambridge (2002)

23. Pu, C., Li, S., Yang, X., Yang, J., Wang, K.: Information transport in multiplex networks. Phys. A Stat. Mech. Appl. **447**, 261–269 (2016)

24. Barabási, A.L., Albert, R.: Emergence of scaling in random networks. Science **286**(5439), 509–512 (1999)

25. Kumpula, J.M., Onnela, J.P., Saramäki, J., Kertesz, J., Kaski, K.: Model of community emergence in weighted social networks. Comput. Phys. Commun. **180**(4), 517–522 (2009)

26. Erdös, P., Rényi, A.: On random graphs I. Publ. Math. Debr. **6**, 290–297 (1959)

Part II
Graph Embeddings

Embedding-Centrality: Generic Centrality Computation Using Neural Networks

Rami Puzis, Zion Sofer, Dvir Cohen and Matan Hugi

Abstract Deriving vector representations of vertices in graphs, a.k.a. vertex embedding, is an active field of research. Vertex embedding enables the application of relational data mining techniques to network data. Unintended use of vertex embedding unveils a novel generic method for centrality computation using neural networks. The new centrality measure, termed Embedding Centrality, proposed in this paper is defined as the dot product of a vertex and the center of mass of the graph. Simulation results confirm the validity of Embedding Centrality which correlates well with other commonly used centrality measures. Embedding Centrality can be tailored to specific applications by devising the appropriate context for vertex embedding and can facilitate further understanding of supervised and unsupervised learning methods on graph data.

1 Introduction

Centrality indices are an important tool for studying the role and function of entities in interconnected complex systems represented by networks, i.e., vertices and edges. Centrality indices are used to rank vertices according to their importance for the proper functioning of the complex system represented by the network. For example, the most central vertices can be those that keep the network together, vertices that monitor traffic, or spread information, and alike [1]. Together with other structural properties such as the local clustering coefficient, centrality indices can be used to create the vertex profiles for behavioral studies or classification tasks of sorts.

Majority of structural properties[1] of vertices were designed by researchers to meet some specific goal. Similar to feature extraction in classical machine learning, it takes a human domain expert to define a proper centrality measure.

---

[1]Properties stemming from the network topology

R. Puzis (✉) · Z. Sofer · D. Cohen · M. Hugi
Software and Information Systems Engineering, Ben-Gurion University of the Negev,
Beersheba, Israel
e-mail: puzis@bgu.ac.il

© Springer International Publishing AG 2018
S. Cornelius et al. (eds.), *Complex Networks IX*, Springer Proceedings
in Complexity, https://doi.org/10.1007/978-3-319-73198-8_8

Recently many efforts have been made to develop vertex embedding techniques that automatically generate a numeric profile to represent vertices in a vector space [2]. The first major advance in vertex embedding was made by Perozzi et al. [3], who used Word2Vec style embedding treating vertices as words and random walks as sentences. Intuitively speaking, the skip-grams approach trains a neural network that highlights the vicinity of a vertex at the output layer when its identity is provided as an input. CBOW trains a neural network that can infer the identity of a vertex from its vicinity (see Sect. 2 for additional details). In both approaches, there is a strong affinity between each vertex and its vicinity and the hidden layer of the neural network is used for the embedding. As the result, neighbor vertices are usually embedded near each other and the community structure of the network is maintained.

State-of-the-art vertex embedding techniques result in a coarse-grained representation of the network in a condensed matrix form. Yet similar to other applications of neural networks, the values of the hidden neurons are uninterpretable. The embedding may sketch the position of a vertex relative to other vertices in the network, but it does not encode vertex centrality.

In this paper, we propose a new point of view on vertex embedding.[2] We note that affinity between the input layer and the output layer is preserved also for sets of vertices–an observation that makes Doc2Vec possible in textual applications. Thus, we can investigate the affinity between a vertex and any region of the network, let it be a path, a community, or even the network as a whole. Intuitively, a vertex having the strongest affinity to all other vertices in the network is a central vertex. In this paper, we hypothesize that the values of the output layer of a neural network trained for vertex embedding encode a centrality measure when the set of all vertices is provided as an input, i.e., all input neurons are set on. We elaborate on this approach in Sect. 3.

Centrality measures proposed in the past were manually designed by researchers or tailor-made for a specific purpose. To the best of our knowledge, this is the first paper that proposed a method for *learning* a general-purpose centrality measure. As such, we face the challenge of providing a sound evaluation of arbitrary centrality measures. A set of small intuitive tests on which all existing centrality measures agree is a starting point for a proof of concept, but unfortunately it is not sufficient for convincing evaluation. In the past, researchers that proposed variants of existing measures have shown the similarity of their measure to the prototype and demonstrated that the differences are more intuitive according to their approach [5, 6]. Since this paper focuses on an unsupervised approach for learning a general centrality measure, we evaluate its agreement with a set of widely known standard centrality measures and show that it falls within the bounds of agreement between the standard measures and themselves. The evaluation method is elaborated in Sect. 4 and results of the evaluation of centrality measure learnt by an artificial neural network are summarized in Sect. 4.

---

[2]An online tool for visualization of Word2Vec, named WEVI [4], played a major role in the ideation of this point of view.

The contribution of this paper is twofold: (1) We propose an evaluation method for centrality measures developed using unsupervised machine learning techniques. (2) We propose an unsupervised centrality measure learnt by a neural network and demonstrate its validity.

2 Background

Since word2vec was introduced by Mikolov et al. in [7], the use of neural networks for embedding of entities (words, vertices, genes, etc.) has gained a lot of traction. We refer to the taxonomy recently published by Hamilton et al. [2] for a variety of vertex embedding techniques. The basic assumption behind word2vec and its successors is that similar entities are used in similar contexts. Mikolov et al. define the context of a word as 5–10 preceding and 5–10 following words. Perozzi et al. [3] define the context of a vertex v_i as 10 preceding and 10 following vertices along a random walk passing through v_i. See Fig. 1a for an example of a graph and a random walk. Other approaches build vertex's context from its neighborhood with closer vertices being more relevant than the farther [8–10].

We will describe the basic embedding concepts in terms of vertices and their contexts. Let $G = (V, E)$ be a graph where V is the set of n vertices and $E \subseteq V^2$

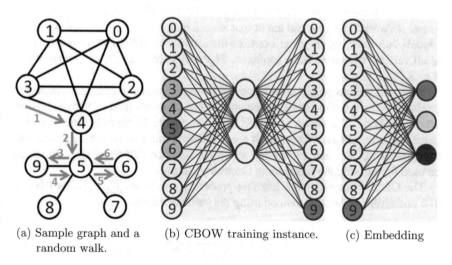

(a) Sample graph and a random walk. (b) CBOW training instance. (c) Embedding

Fig. 1 Sample graph with 10 vertices (**a**), where the vertices 4 and 5 are the most central according to all centrality measures. The random walk 3, 4, 5, 9, 5, 6, 5 is used as one of the training instances for the CBOW neural network (**b**), where three preceding and three following vertices are the context of the vertex 9. The vertex 5 is more emphasized because it appears three times in the context. The rightmost figure **c** represents the embedding of the vertex 9, where the neurons of the hidden layer correspond to the dimensions and the weights (synapses) between 9 and the neurons of the hidden layer correspond to the coordinates of 9 in $\mathbb{R}^3$

is the set of edges. Consider an n-dimensional vector $\overline{v} \in \mathbb{R}^n$, where each index i corresponds to a particular vertex. In the rest of this paper, we will use over-line ($\overline{\circ}$) to denote vectors. Vector dimensions will be specified as the vector domain ($\overline{\circ} \in \mathbb{R}^x$) when needed.

The n-dimensional representation of a single vertex has the entry at the corresponding index set to 1 and the rest of the entries set to 0. The representation of a context $C_v \subseteq V$ is the weighted sum of the respective vertex-vectors $\overline{C_v} = \sum_{u \in C_v} \overline{u} \cdot \omega(v, u)$ where $\omega(v, u)$ encodes the relevance of a context vertex u to v. Relevance is application dependent and can be defined, for example, as the hop distance between u and v, the number of times u appears in a particular context of v, etc. For example, in Fig. 1b the output vector corresponds to the vertex 9 and the input vector corresponds to its context 3, 4, 5, $\bullet$, 5, 6, 5. In this example, 5 is three times more relevant to 9 than 3, 4, and 6.

The objective of vertex embedding is finding a low- dimensional representation $\overline{z}_v \in \mathbb{R}^d$ for every vertex $v \in V$ such that $d << n$ and similar vertices have similar representations. Similarity of vertices is loosely defined by similarity of their contexts. For example, structurally equivalent vertices[3] should have the same coordinates in the low dimensional space $\mathbb{R}^d$.

Let $D = \{(v, C)\}$ be a dataset containing vertices and their respective contexts in a particular graph. There could be multiple contexts for each vertex as in the case of random walks. Following word2vec, many researchers use one of the two log-linear models for learning the d-dimensional vector representations of vertices using neural networks: Continuous Bag-of-Words Model (CBOW) where the input layer of the neural network is set to c and the output should match v, or Continuous skip-gram Model (skip-gram) where given a vertex v the network should output the probabilities of all vertices appearing in its context c. Figure 1b is an example of CBOW where $d = 3$.

Based on various empirical evidence, skip-gram based embedding is claimed to produce better, more meaningful, representations of words (and of vertices). However, there is no clear theoretical basis for these claims. Further discussions in this paper will refer to CBOW because it provides better intuition on the suggested centrality measure. Yet, any embedding approach based on CBOW or skip-grams can be used for deriving the Embedding Centrality measure.

The CBOW representation learning processes maximize $\sum_{(v,C) \in D} \log P(v|C)$. The probability $P(v|C)$ is defined using the softmax model

$$P(v|C) = \frac{e^{\overline{z}_C \cdot \overline{z}_v}}{\sum_{v' \in V} e^{\overline{z}_C \cdot \overline{z}_{v'}}}, \tag{1}$$

where $\overline{z}_v, \overline{z}_C \in \mathbb{R}^d$ are low-dimensional vector representations of a vertex and a context respectively. $\overline{z}_C$ is obtained by averaging the vector representations of the vertices in C:

---

[3] Vertices having exactly the same set of neighbors [11].

$$\bar{z}_C = \frac{1}{|C|} \sum_{u \in C} \bar{z}_u. \tag{2}$$

3 Embedding Centrality

Trained neural network, either CBOW or skip-gram, can be used to derive the vector representation of vertices (or sets of vertices according to 2). However, the output layer is always ignored for any practical application of the trained network. Only the weights which encode the vertex (or word) coordinates in $\mathbb{R}^d$ are maintained. Nevertheless, the classical neural network discussed here can be used for a variety of supervised machine learning tasks. As such, given previously unseen context, a trained CBOW model should predict the respective vertex with high accuracy. Equation 1 can be regarded as the confidence of the model. $P(v|C)$ ranks the vertices according to their affinity with the given context.

What would happen if the context is set to all vertices in the network?
Intuitively, the output layer would encode the affinity of each vertex with the network as a whole. Next, we derive the Embedding Centrality (EmbC) based on this intuition. In Sect. 4 we will provide an empirical evaluation of EmbC versus common centrality measures.

Let $\bar{z}_V$ be the vector representation of the whole set of vertices in the network. According to (2), $\bar{z}_V = \frac{1}{n} \sum_{v \in V} \bar{z}_v$, which is the center of mass of the vertices in $\mathbb{R}^d$ assuming that the mass of each vertex is unity. We will use softmax (Eq. 1) to derive the log probability of a vertex v to be associated with the center of the mass of the network. The softmax normalization factor, is constant for a network and thus, can be ignored for the purpose of vertex ranking.

$$\log P(v|V) = \bar{z}_V \cdot \bar{z}_v - \log \left(\sum_{v' \in V} e^{\bar{z}_V \cdot \bar{z}_{v'}} \right) = \bar{z}_V \cdot \bar{z}_v + const \tag{3}$$

Following Eq. 3, we will define Embedding Centrality (EmbC) of the vertex v as its dot product with the center of the mass of the network.

Definition 1 (*Embedding Centrality*) The Embedding Centrality (EmbC) of a vertex $v \in V$ is defined as:

$$EmbC(v) = \bar{z}_v \cdot \bar{z}_V,$$

where $\bar{z}_v \in \mathbb{R}^d$ is an embedding of v in the d-dimensional Euclidean space $\mathbb{R}^d$ and $\bar{z}_V$ is the center of mass of the graph.

Although we derived the definition of EmbC from the probability $P(v|V)$, it is valid for both CBOW and skip-gram style embeddings due to the symmetry of dot product $\bar{z}_v \cdot \bar{z}_C$.

Figure 2 depicts the result of setting the input context to all vertices in the sample network from Fig. 1a. The three hidden neurons in Fig. 2a represent the center of the

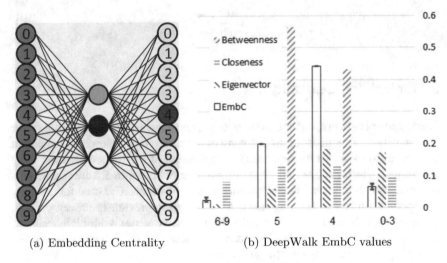

(a) Embedding Centrality   (b) DeepWalk EmbC values

Fig. 2 EmbC computed in the sample graph from Fig. 1a

mass coordinates and the output neurons represent EmbC (darker color corresponds to higher values). The EmbC values are compared to other centrality measures in Fig. 2b. Vertex 4 is the most central according to all measures except betweenness. 5 has higher betweenness because it controls the paths between the leafs 0–3. Since EmbC computation is stochastic, the centrality of similar vertices (such as 0–3 or 6–9) may vary. Error bars indicate the minimal and maximal values of EmbC for these sets of vertices. Overall, in this example, the values of EmbC fit well between the values of other centrality measures.

Algorithm 1: Embedding Centrality

Input: $G = (V, E), d$
Output: $\forall_{v \in V} EmbC(v)$
1 $Z^{[n \times d]} \leftarrow$ VertexEmbedding(G, d);
2 $\bar{r} \leftarrow \sum_{\bar{z}_i \in Z} \bar{z}_i$;
3 **for** $\bar{z}_v \in Z$ **do**
4 $\quad | \quad EmbC(v) \leftarrow \bar{z}_v \cdot \bar{r}$;
5 **end**

3.1 Complexity

Algorithm 1 briefly outlines the steps for computing EmbC according to Definition 1. Computational complexity of EmbC depends primarily on *VertexEmbedding* (steps 2–5 can be executed in $O(nd)$ time). Although, the computational complexity of some embedding techniques is linear in the number of vertices [12], vertex embedding can be quite expensive in a general case. The neural network optimization can be performed in $O(d \log n)$ time per training sample when using hierarchical softmax. However, the number of training samples $|D|$ plays the major role.

3.2 Discussion

Definition 1 suggests that a vertex with high EmbC would have high average affinity for all vertices in the graph.

$$EmbC(v) = \bar{z}_v \cdot \bar{z}_V = \frac{1}{n} \sum_{u \in V} \bar{z}_v \cdot \bar{z}_u$$

In this sense EmbC is similar to other centrality measures such as Valued Centrality (VC) [13]—a variant of closeness centrality defined as the average of reciprocal distances:

$$VC(v) = \frac{1}{n} \sum_{u \in V \setminus \{v\}} \frac{1}{dist(v, u)},$$

where $dist(v, u)$ is the distance between vertices v and u in a valued network. Here the reciprocal distance represents the affinity between v and each other vertex in the graph.

EmbC is also in line with centrality taxonomies such as the one proposed by Borgatti and Everett [14]. The authors classify centrality measures according to type of walk (walk, trail, or path), position of the vertex within the walk (radial or medial), and property of the walk being aggregated (volume/count or length). EmbC provides a generic method to construct a centrality measure according to this classification. For example, EmbC with DeepWalk [3] considers medial random walks while EmbC with node2vec [15] considers radial walks. Various embedding techniques [2] consider various contexts including vertex neighborhoods, graphlets, or subgraphs of sorts. Understanding that centrality can be defined based on vast variety of contexts helps extending the Borgatti and Everett's taxonomy of centrality measures way beyond the concept of walks.

Similar classification is relevant also for walk-based vertex embedding techniques. For example, DeepWalk [3] used medial random walks, volume is captured by the number of train instances involving a particular vertex, and the length property can be captured by the relevance weight ω. Another example is node2vec [15], which employs radial walks while the type of walk and length are controlled using tunable parameters. In general, vertex embedding, and as a result also EmbC, relies on a more general concept of *contexts* which could be any type of subgraphs including walks, paths, or graphlets.

4   Evaluation

In this paper we claim that EmbC, which encodes the affinity of a vertex with the network as a whole, is a centrality measure. EmbC can be derived from arbitrary embedding schemes in fully unsupervised, task-independent manner. Therefore, we refrain from task-specific evaluation. Instead, we present a proof-of-concept evaluation of the correlation between EmbC and common centrality measures.

The proof-of-concept evaluation of centrality agreement is performed on artificial networks generated using the Barabasi–Albert [16] model with 100 or 500 vertices and average degree ranging from 4 to 10. We used DeepWalk as the vertex embedding scheme for EmbC with the length of the walk ranging from 15 to 100, 5–20 walks starting at each vertex, and the context size ranging from 2 to 10. Contexts were padded with zeroes in cases of insufficient vertices—a common practice with Word2Vec and DeepWalk. The size of the hidden layer (d) was ranging from 5 to 20 and the neural network was trained for eight epochs.

Correlation analysis was employed in the past to evaluate the tolerance of centrality measures to noise in the data [17]. Here we will analyze the correlations between three common centrality measures and themselves vs. their correlation to EmbC. We choose closeness (CC), betweenness (BC), and eigenvector centrality (EC) as the representative centrality measures. We also include Walk Count (WC)—the number of random walks which pass through the vertex and Pass Count (PC)—the total number of times the sampled walks passed through a vertex.

Since centrality measures are used mostly for ranking vertices and pinpointing the most central ones, we will use Spearman correlation and average precision (AP) correlation [18]. We choose AP because it values correct ranking of high centrality vertices more than ranking of low centrality vertices in agreement with the common practice in centrality analysis. Let r_1^k r_2^k be the top k vertices according to two different centrality measures. Precision@k of the either one of the measures with respect to the other one is $\frac{|r_1^k \cap r_2^k|}{k}$. AP is defined as the average precision@k for $1 \leq k \leq n$. Vertices with high centrality are affected the AP.

For every network we compute AP, and Spearman correlation between every pair of centrality measures. It should be noted that all centrality measures differ significantly from each other with Spearman correlation ranging from 0.1 to 0.9. Figure 3 presents the distribution of correlations between EmbC and CC, BC, EC denoted as *others* and between the *other* measures and themselves. We observe that more than 60–65% of cases have correlation between EmbC and the common centrality measures of 0.5 or higher and AP of 0.7 or higher.

The performance of EmbC depends on the quality of embedding. Table 1 displays the average performance of EmbC for several sets of DeepWalk parameters. Specifically the size of the context and the number of hidden neurons. The results clearly show that larger contexts and higher dimensionality of the embedding (i.e., the number of hidden neurons) results in EmbC which better correlates with all other measures. Highest average AP of 0.816 and Spearman correlation of 0.836 are obtained for $d = 20$ and context size 10.

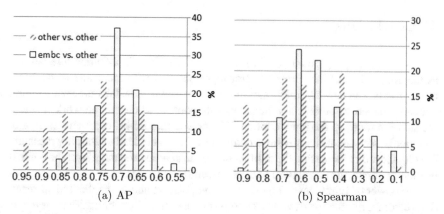

(a) AP   (b) Spearman

Fig. 3 The distributions of correlations between EmbC and other measures and between all other measures

Table 1 Average AP and Spearman correlation of EmbC with all other measures (BC,CC,EC,WC, and PC) for different context sizes and different number of hidden neurons

(a) AP

context	d 5	10	20
2	0.592	0.692	0.764
5	0.65	0.758	0.791
10	0.749	0.796	0.816

(b) Spearman

context	d 5	10	20
2	0.28	0.547	0.704
5	0.424	0.709	0.777
10	0.647	0.794	0.836

Table 2 AP and Spearman correlation between all vertex centrality measures in one of the networks. Vertex embedding was obtained with 20 hidden neurons and context size of 10

(a) AP

	BC	CC	EC	WC	PC	EmbC	AVG
BC	1.00	0.74	0.70	0.83	0.83	0.78	0.78
CC	0.75	1.00	0.90	0.78	0.73	0.82	0.80
EC	0.73	0.90	1.00	0.78	0.74	0.82	0.79
WC	0.84	0.78	0.79	1.00	0.91	0.85	0.83
PC	0.84	0.74	0.75	0.91	1.00	0.81	0.81
EmbC	0.78	0.82	0.82	0.85	0.81	1.00	0.82

(b) Spearman

	BC	CC	EC	WC	PC	EmbC	AVG
BC	1.00	0.60	0.53	0.82	0.82	0.72	0.70
CC	0.60	1.00	0.96	0.79	0.65	0.88	0.78
EC	0.53	0.96	1.00	0.79	0.66	0.86	0.76
WC	0.82	0.79	0.79	1.00	0.95	0.91	0.85
PC	0.82	0.65	0.66	0.95	1.00	0.81	0.78
EmbC	0.72	0.88	0.86	0.91	0.81	1.00	0.84

Are these correlation values sufficient? Table 2 presents the AP and Spearman correction between all evaluated centrality measures. Here the DeepWalk parameters were set to their optimal values. We observe average AP of 0.78–0.83 for all tested centrality measures and average Spearman correlation of 0.70–0.85. EmbC performance fits well into these ranges allowing us to conclude that EmbC is a centrality measure.

5   Conclusions

In this paper, we proposed a novel Embedding Centrality (EmbC) measure defined as the dot product of the vertex's vector representation with the center of the mass of the graph. The new centrality measure can be explained as the output of a neural network trained to predict a vertex given its context when this network receives the set of vertices as the input context. We showed that EmbC correlates well with commonly used centrality measures confirming that it is indeed a centrality measure. Results show that increasing the number of hidden neurons and the size of the context positively affects the correlation with other centrality measures.

The power of EmbC lays in its generic computation and in the theoretical result that vector representation of vertices preserves the notion of centrality. Further investigation of EmbC is required to understand the relation between the configuration parameters used during embedding and interpretation of EmbC. Similar to neural network based word embedding whose parameter optimization facilitated the development of simpler and more accurate embedding schemes, we believe that EmbC will facilitate further understanding of supervised and unsupervised learning methods on graph data. If successful, this could also project the concept of centrality measures to various domains without the necessity for explicit network construction.

References

1. Newman, M.: Networks: An Introduction. Oxford university press, Oxford (2010)
2. Hamilton, W.L., Ying, R., Leskovec, J.: Representation learning on graphs: methods and applications (2017). arXiv:1709.05584
3. Perozzi, B., Al-Rfou, R., Skiena, S.: Deepwalk: Online learning of social representations. In: Proceedings of the 20th ACM SIGKDD International Conference on Knowledge Discovery and Data Mining, pp. 701–710. ACM (2014)
4. Rong, X.: word2vec parameter learning explained (2014). arXiv:1411.2738
5. Newman, M.E.J.: A measure of betweenness centrality based on random walks. Soc. Netw. 27(1), 39–54 (2005)
6. Freeman, L.C., Borgatti, S.P., White, D.R.: Centrality in valued graphs: a measure of betweenness based on network flow. Soc. Netw. 13(2), 141–154 (1991)
7. Mikolov, T., Chen, K., Corrado, G., Dean, J.: Efficient estimation of word representations in vector space (2013). arXiv:1301.3781
8. Cao, S., Lu, W., Xu, Q.: Deep neural networks for learning graph representations. In: AAAI, pp. 1145–1152 (2016)
9. Wang, D., Cui, P., Zhu,W.: Structural deep network embedding. In: Proceedings of the 22nd ACM SIGKDD International Conference on Knowledge Discovery and Data Mining, pp. 1225–1234. ACM (2016)
10. Niepert, M., Ahmed, M., Kutzkov, K.: Learning convolutional neural networks for graphs. In: International Conference on Machine Learning, pp. 2014–2023 (2016)
11. Sailer, L.D.: Structural equivalence: meaning and definition, computation and application. Soc. Netw. 1(1), 73–90 (1978)
12. Tian, F., Gao, B., Cui, Q., Chen, E., Liu, T.-Y.: Learning deep representations for graph clustering. In: AAAI, pp. 1293–1299 (2014)
13. Dekker, A.: Conceptual distance in social network analysis. J. Soc. Struct. (JOSS) 6 (2005)

14. Borgatti, S.P., Everett, M.G.: A graph-theoretic perspective on centrality. Soc. Netw. **28**(4), 466–484 (2006)
15. Grover, A., Leskovec, J.: node2vec: scalable feature learning for networks. In: Proceedings of the 22nd ACM SIGKDD International Conference on Knowledge Discovery and Data Mining, pp. 855–864. ACM (2016)
16. Barabási, Albert-László, Albert, Réka: Emergence of scaling in random networks. Science **286**(5439), 509–512 (1999)
17. Borgatti, S.P., Carley, K.M., Krackhardt, D.: On the robustness of centrality measures under conditions of imperfect data. Soc. Netw. **28**(2), 124–136 (2006)
18. Yilmaz, E., Aslam, J.A., Robertson, S.: A new rank correlation coefficient for information retrieval. In: Proceedings of the 31st Annual International ACM SIGIR Conference on Research and Development in Information Retrieval, pp. 587–594. ACM (2008)

Fast Sequence-Based Embedding with Diffusion Graphs

Benedek Rozemberczki and Rik Sarkar

Abstract A graph embedding is a representation of graph vertices in a low- dimensional space, which approximately preserves properties such as distances between nodes. Vertex sequence-based embedding procedures use features extracted from linear sequences of nodes to create embeddings using a neural network. In this paper, we propose diffusion graphs as a method to rapidly generate vertex sequences for network embedding. Its computational efficiency is superior to previous methods due to simpler sequence generation, and it produces more accurate results. In experiments, we found that the performance relative to other methods improves with increasing edge density in the graph. In a community detection task, clustering nodes in the embedding space produces better results compared to other sequence-based embedding methods.

1 Introduction

Embedding graphs into a low dimensional Euclidean spaces is a way of simplifying the graph information by associating each node with a point in the space. Thus, various methods of graph embedding have been developed and applied to different domains, such as visualization [8], community and cluster identification [18], localisation of wireless devices [16], network routing [15], etc. Graph embeddings usually aim to preserve proximity—nearby nodes on the graph should have similar coordinates—in addition to properties specific to the application.

In recent years, sequence-based graph embedding methods have been developed as a way of generating Euclidean representations using sequence of vertices obtained

B. Rozemberczki (✉) · R. Sarkar
School of Informatics, University of Edinburgh, Edinburgh, U.K.
e-mail: benedek.rozemberczki@ed.ac.uk

R. Sarkar
e-mail: rsarkar@inf.ed.ac.uk

© Springer International Publishing AG 2018
S. Cornelius et al. (eds.), *Complex Networks IX*, Springer Proceedings
in Complexity, https://doi.org/10.1007/978-3-319-73198-8_9

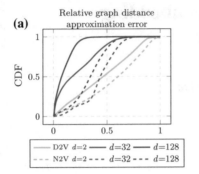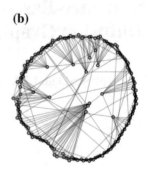

Fig. 1 a Cumulative distribution of the shortest path distance approximation error on the PPI network [3] ($|V| = 3{,}890$) for embedding dimension d. The distortion for our method (D2V) is much smaller than the state of the art (N2V). The distortion error for nodes u and v is defined as $e_{u,v} = |d(u, v) - \gamma \cdot \|\mathbf{X}_v - \mathbf{X}_u\|| /d(u, v)$. Embeddings were created with parameter settings such that $n = 10$, $\widehat{w} = 10$, $\alpha = 0.025$, $k = 1$, $l = 40$ (D2V) and $l = 80$ (N2V). The best inout and return parameters of N2V were chosen with grid search over $\{0.25, 0.5, 1, 2, 4\}$. **b** Visualization of a Watts–Strogatz graph with our embedding procedure

from random walks. These methods are inspired by Word2Vec—a method to embed words into Euclidean space based on sequences in which they occur. Word2vec takes short sequences of words from a document and uses them to train a neural network; in the process it obtains an embedding for the words. The embedding space acts as an abstract latent space of *features*, and places two words close if they frequently occur nearby in the sequences [11]. Sequence-based graph embedding methods on the other hand obtain their vertex sequences by random walk on graphs and then apply analogous neural network methods for the embedding. The random walk has the advantage that it obtains a view of the neighborhood, without having to compute and store complete neighborhoods, which can be expensive in a large graph with many high-degree vertices.

However, random walks are inefficient for generating proximity statistics. They are known to spread slowly, and revisit a vertex many times producing redundant information [2]. As a result, they require many steps or many restarts to cover the neighborhood of a node. Methods like Node2vec [7] try to bias the walks away from recently visited nodes, but in the process they incur a cost due to the complexity of modifying transition probabilities with each step. We instead use a diffusion process that samples a subgraph of the neighborhood, from which several walks can be generated more efficiently.

Our contributions. In our method, we extract a subgraph of the neighborhood of a node using a diffusion-like process, and call it a *diffusion graph*. On this subgraph, we compute an Euler tour to use as a sequence. By covering all adjacencies in the graph, the Euler tour contains a more complete view of the local neighborhood than random walks. We refer to this sequence-generating method as *Diff2Vec* (D2V). The sequences generated by Diff2Vec are then used to train a neural network with one hidden layer containing d neurons. The input weights of the neurons determine the embedding of the nodes.

Due to its better coverage of neighborhoods, Diff2Vec can operate with smaller neighborhood samples. As a result, it is more efficient than existing methods. In our experiments with a basic implementation, it turned out to be several times faster. In particular, it scales better with increasing density (vertex degrees) of graphs. Our experiments also show that the embedding preserves graph distances to a high accuracy. On experiments of community detection, we found that clustering applied to the embedding produces communities of high quality—verified by the high modularity of the clusters.

2   Related Works

Well-known embedding techniques use a matrix that describes the graph and factorize it in order to create the embedding of the network. One can factorize the adjacency, neighbourhood overlap, or Laplacian matrices. Based on the properties of the matrix either eigenvalue decomposition or some variant of stochastic gradient descent is used to obtain the graph embedding. These embedding methods all have a weakness, namely that they are computationally expensive. We refer the reader to the recent survey in [6] for a broader overview of graph embedding, and focus here on relevant neural network-based embeddings.

Sequence-based embedding. The generation of node sequence based graph embeddings consists of three phases. First, the algorithm creates vertex sequences—usually by a random process. Second, features that are extracted from the synthetic sequences describe the approximated proximities of nodes. Finally, the embedding itself is learned using the extracted node-specific features with a neural network which has a single hidden layer. Sequence-based embedding originates from the *DeepWalk* model [13], which uses random walks to generate node sequences. This approach was improved upon by *Node2Vec* (henceforth N2V) [7], which uses second-order random walks to generate the vertex sequences. Second-order random walks alternate between depth-first and breadth-first search on the graph in a random, but somewhat in a controlled way. In this attempt to have greater control on random walks, N2V introduces parameters that affect the embedding quality and are hard to optimize.

3   Feature Extraction and Neural Network Embedding

Feature extraction. We start with extracting features called hitting frequency vectors—denoting frequencies with which vertices occur near each other. The graph is denoted by $\mathcal{G}(V, E)$. The set of vertices is V and the edge set is E. We assume that the graph is undirected and unweighted. Let us consider an example to see how an embedding is generated.

Consider the example in Fig. 2a. The vertex set contains nodes a, b, c, d, e and nodes are indexed, respectively, from 1 to 5, and suppose we are given the 3 node

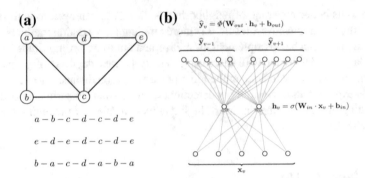

Fig. 2 **a** Graph with linear vertex sequences. The three vertex sequences listed are used for feature extraction in our example. **b** Architecture of the example neural network

sequences in the figure. To generate features from the sequences we choose a sliding window size denoted by $\widehat{w}$ which limits the maximal graph proximity among nodes that we are going to approximate. In this case we choose $\widehat{w} = 1$. We calculate the co-occurrence frequencies for node c as follows—we count how many times other nodes appeared at given positions before and after c limited by the window's size. In this toy example, it means positions at maximal 1 step before or after c in the sequence. Counts at different positions are stored in separate vectors for each node. The resulting frequency vectors are as follows: $\mathbf{y}_{c,-1} = \begin{bmatrix} 1 & 1 & 0 & 2 & 0 \end{bmatrix}$ and $\mathbf{y}_{c,+1} = \begin{bmatrix} 0 & 0 & 0 & 4 & 0 \end{bmatrix}$. Components of the vectors can be interpreted as noisy proximity statistics in the graph. The idea is that nearby nodes will have higher values in each other's vectors. We concatenate these vectors to form a vector of $2 \cdot \widehat{w} \cdot |V|$ components and call it the hitting frequency vector $\mathbf{y}_v$ of a node v. We construct such a hitting frequency vector for each node from the given sequences.

Learning an embedding from the features. For each vertex $v \in V$, we wish to compute a coordinate in $\mathbb{R}^d$. The set of hitting frequency vectors is a representation of the graph in $\mathbb{R}^{|V| \times 2 \cdot \widehat{w} \cdot |V|}$, which we have to reduce to a $\mathbb{R}^{|V| \times d}$ space. We write as $\mathbf{x}_v$ the indicator (sometimes called hot-one) vector for v, which has $|V|$ elements, all of which are zero, except the element at index of v, which is set to 1. A schematic of the neural network architecture is in Fig. 2b. The neural network has d hidden neurons, each with $|V|$ inputs and $2 \cdot \widehat{w} \cdot |V|$ outputs. The incoming and outgoing weight matrices of the hidden neurons are written as $\mathbf{W}_{in}$ and $\mathbf{W}_{out}$. To train the neural network, the training algorithm uses input–output pairs of the form $(\mathbf{x}_v, \mathbf{y}_v)$ corresponding to each vertex v. Thus, the neural network learns to associate with each vertex, an output that is its hitting frequency vector. After the training, the incoming weight matrix $\mathbf{W}_{in}$ (of dimension $d \times |V|$) gives the d dimensional embedding of the vertices.

The weight matrix is used to approximately reconstruct the hitting frequencies of a node. If two nodes have similar hitting frequency vectors, meaning that their proximity is high, they will also have a similar latent space representation. Our goal is the efficient and scalable learning of the embedding so we use asynchronous gradient descent (ASGD). Analogous to previous works [7, 13], we used hierarchical softmax activation with multinomial logloss, with which the computational complexity of a training epoch (while we decrease the learning rate from starting value to zero) is $\mathcal{O}(|V|\log(|V|))$. We refer to the embedding as $\mathbf{X}$, and the embedding of node v is noted by $\mathbf{X}_v$.

4 Sequence Generation Algorithm and Design

To generate sequences in the neighborhood of a node, we first compute a *diffusion graph*, and then use it to compute vertex sequences.

Diffusion graph generation: We emulate a simple diffusion-like random process starting from a vertex v to sample a subgraph of l vertices near v. The diffusion graph $\widetilde{\mathcal{G}}$ is initialized with $\{v\}$. Next, at each step, we sample a random node u from $\widetilde{\mathcal{G}}$ and from the neighbors of u in the original graph $\mathcal{G}$, we select w. We add w to the set of vertices in $\widetilde{\mathcal{G}}$, and add the edge (u, w) to $\widetilde{\mathcal{G}}$. This process is repeated until $\widetilde{\mathcal{G}}$ has l nodes.

Data: $\mathcal{G}$ – Graph object.
       l – Number of nodes sampled.
       v – Starting node .
Result: P – Eulerian sequence from v.

1 $V_{\widetilde{\mathcal{G}}} \leftarrow \{v\}$
2 **while** $|V_{\widetilde{\mathcal{G}}}| < l$ **do**
3      $w \leftarrow$ Random Sample$(V_{\widetilde{\mathcal{G}}})$
4      $u \leftarrow$ Random Sample$(N_{\mathcal{G}}(w))$
5   **if** $u \notin V_{\widetilde{\mathcal{G}}}$ **then**
6         $V_{\widetilde{\mathcal{G}}} \leftarrow V_{\widetilde{\mathcal{G}}} \cup \{u\}$
7         $E_{\widetilde{\mathcal{G}}} \leftarrow E_{\widetilde{\mathcal{G}}} \cup \{(u, w)\}$
8   **end**
9 **end**
10 $\widetilde{\mathcal{G}} \leftarrow$ Duplicate Edges$(\widetilde{\mathcal{G}})$
11 $P \leftarrow$ Random Eulerian Circuit$(\widetilde{\mathcal{G}}, v)$

Algorithm 1: Graph sampling

Node sequence sampling: To generate sequences from the subgraph $\widetilde{\mathcal{G}}$, we take the following approach. We convert $\widetilde{\mathcal{G}}$ into a multigraph by doubling each edge into two edges. A connected graph where every node has an even degree is Eulerian, and the Euler walk is easy to find [17]. We use this method to find the Euler walk and use that as a vertex sequence. Observe that this diffusion graph sampling and sequence generation can be performed in parallel across many machines, since each diffusion graph can be generated independent of others. The generated sequences are then used to produce graph embedding using neural networks as seen in the previous section. Note that an Euler walk has the nice property that it captures every adjacency relation in the subgraph into a linear sequence using asymptotically optimal space.

This property then helps our method perform better both in the sense of efficiency and quality of results.

Data: $\mathcal{G}$ – Graph embedded.
       p – Sequence samples per node.
       l – Number of nodes per sample.
       d – Dimension of embedding.
       k – Number of epochs.
       $\widehat{w}$ – Size of sliding window.
       α – Learning rate.
Result: $\mathbf{X}$ – Embedding of graph $\mathcal{G}$.

1 $\mathcal{G}_1, \ldots \mathcal{G}_S \leftarrow$ Component Extraction$(\mathcal{G})$
2 Samples $\leftarrow$ []
3 **for** i in $1 : p$ **do**
4   Walks $\leftarrow \{\}$
5      $l' \leftarrow l$
6   **for** j in $1:|\{\mathcal{G}_1, \mathcal{G}_2, \ldots \mathcal{G}_S\}|$ **do**
7   **if** $\left|V_{\mathcal{G}_j}\right| < l'$ **then**
8             $l' \leftarrow \left|V_{\mathcal{G}_j}\right|$
9   **end**
10   **for** v in V **do**
11   Walks$(v) \leftarrow$ Traceback$(\mathcal{G}_j, v, l')$
12   **end**
13   **end**
14   Samples$(i) \leftarrow$ Walks
15 **end**
16 $\mathbf{X} \leftarrow$ Learn Emb.(Samples, $d, \widehat{w}, \alpha, k$)

Algorithm 2: Learning from sequences

5 Experiments

In our experiments we compare our method D2V with the state-of-the- art N2V [7] method. We look at quality of embeddings and the computational performance. The main observations from the experiments are the following:

- With increasing size of graphs, efficiency of D2V scales better than that of N2V.
- The D2V embedding preserves distances well between most pairs of nodes: in 128-dimensional embedding, over 90% pairs suffer a distortion smaller than 20%. In any dimensions, it performs better than N2V.
- Clustering of the D2V embedding works well for community detection, and performs better than N2V measured by the modularity of clusters.

Computational efficiency. In the first series of experiments we measured the average graph preprocessing and sequence generation times on a number of real-world networks. Preprocessing in this case involves reading the graph and creating suitable

Table 1 Computation time on real-life graphs. **BlogCatalog**: is a social network of bloggers, nodes are bloggers and links are social relationships [1]. **PPI**: is a protein–protein interaction network of humans [3]. **Wikipedia**: is a word co-occurrence network based on a chunk of the Wikipedia corpus [9]. Columns report running time in seconds extracted from 100 experiments on the datasets. Bold numbers mark the fastest mean preprocessing— sequence generation times on a given dataset

	BLOGCATALOG		PPI		WIKIPEDIA	
	$|V| = 10,312$		$|V| = 3,890$		$|V| = 4,777$	
	$|E| = 333,982$		$|E| = 38,739$		$|E| = 92,517$	
	N2V	D2V	N2V	D2V	N2V	D2V
Sequence generation	59.089	**19.983**	**4.253**	4.684	12.135	**6.879**
Preprocessing	784.899	**3.231**	12.797	**0.362**	185.287	**0.667**

data structures. N2V in particular requires data structures to regularly update the random walk probabilities. Note that it is the preprocessing and sequence generation, where these two methods differ, as they use similar methods for training neural networks. Our results in Table 1 show that on larger networks D2V has a consistent advantage performance wise.

Node distance approximation. Using the PPI network, we measure how well the shortest path distance of nodes $d(u, v)$ can be approximated by the Euclidean distance of nodes in the embedding space. The relative approximation error $e_{u,v}$ for a given pair of nodes u, v is defined by $e_{u,v} = |d(u, v) - \gamma \cdot \|\mathbf{X}_v - \mathbf{X}_u\| | /d(u, v)$. Essentially, the absolute difference between $d(u, v)$ and the scaled Euclidean distance to $d(u, v)$. The factor γ adjusts for the uniform scaling over the graph. We take the γ that minimizes the sum of errors.

We plotted cumulative distribution of the relative approximation error for different embedding dimensions on Fig. 1a. With a 32-dimensional D2V embedding one can approximate half of the shortest path distances with a relative error below 20%. Increasing the embedding dimension to 128 allows to approximate 90% of shortest paths with an approximation error below 20%. Finally, we also plotted the approximation error obtained with N2V embeddings. A 32-dimensional N2V embedding can only approximate roughly 10% of the shortest path distances with a relative error below 20%. Moreover, increasing the N2V embedding dimension does not decrease the distortion considerably. We conclude that on this graph D2V approximates graph distances better than N2V (Fig. 2).

Community detection. We evaluated the utility of the embedding in community detection. We clustered the embedded nodes in the embedding space using k-means clustering, and then computed modularity [12] as a quality measure. The experiments involved six different datasets with number of vertices ranging from few thousands to millions and we compared our results to clusterings obtained with standard community detection methods. Results are seen in Table 2. Our results show that k-means clustering of the embeddings outperforms all other methods on most of the datasets. Moreover, D2V (our method) results in clusterings that are higher quality than clusters created with N2V.

Table 2 Clustering quality measured by modularity. The baseline community detection algorithms can be found in [4, 12, 14]. Bold numbers note the highest modularity value obtained on the dataset. Dashes denote missing modularity values when obtaining a clustering was not feasible due to computational complexity of the algorithm. Embeddings were created with baseline parameter settings such that $d = 128$, $n = 10$, $\hat{w} = 10$, $\alpha = 0.025$, $k = 1$, $l = 40$ (D2V) and $l = 80$ (N2V). The best input and return parameters of N2V were chosen with grid search over $\{0.25, 0.5, 1, 2, 4\}$ while the cluster number varied between 2 and 50. The distance measure was the Euclidean distance in the latent space. Besides the earlier used datasets we chose 3 additional social networks to asses the representation quality. **Flickr**: a network of Flickr users [10]. **YouTube**: is a friendship network of YouTube users [19]. **Markercafe**: is data from an Israeli social network [5]

Algorithm	Blogcatalog	PPI	Wikipedia	Flickr	YouTube	Markercafe
Fast Greedy	0.2069	0.3029	**0.1456**	0.4517	–	0.2597
Walktrap	0.1766	0.2571	0.0553	0.4873	–	0.2026
Eigenvector	0.2035	0.2262	0.0915	0.4810	–	0.2455
K-means D2V	**0.2225**	**0.3365**	0.1420	**0.5078**	**0.6265**	**0.2818**
K-means N2V	0.2184	0.3270	0.1376	0.3647	0.4862	0.2630

6 Conclusions

In this work we proposed *Diff2Vec* a node sequence-based graph embedding model that uses diffusion processes on graphs to create vertex sequences. We demonstrated that the design of the algorithm results in fast sequence creation in realistic settings. It also allows parallel vertex sequence generation which leads to additional speed up. We confirmed that node features created with *Diff2Vec* are useful features for downstream machine learning tasks. We gave a detailed evaluation of the representation quality of embeddings on shortest path distance approximation and the machine learning task of community detection. Our findings show that besides the favorable computational performance the representation quality itself is competitive with other methods.

Acknowledgements Benedek Rozemberczki was supported by the Centre for Doctoral Training in Data Science, funded by EPSRC (grant EP/L016427/1).

References

1. Agarwal, N., Liu, H., Murthy, S., Sen, A., Wang, X.: A social identity approach to identify familiar strangers in a social network. In: ICWSM (2009)
2. Alon, N., Avin, C., Koucký, M., Kozma, G., Lotker, Z., Tuttle, M.R.: Many random walks are faster than one. Comb. Probab. Comput. **20**(4), 481–502 (2011)
3. Chatr-Aryamontri, A., Breitkreutz, B.J., Oughtred, R., Boucher, L., et al.: The biogrid interaction database: 2015 update. Nucleic Acids Res. **43**(D1), D470–D478 (2014)
4. Clauset, A., Newman, M.E.J., Moore, C.: Finding community structure in very large networks. Phys. Rev. E **70**(6), 066111 (2004)

5. Fire, M., Tenenboim, L., Lesser, O., Puzis, R., Rokach, L., Elovici, Y.: Link prediction in social networks using computationally efficient topological features. In: IEEE Third Inernational Conference on Social Computing (SocialCom), pp. 73–80. IEEE (2011)
6. Goyal, P., Ferrara, E.: Graph embedding techniques, applications, and performance: A survey. arXiv:1705.02801 (2017)
7. Grover, A., Leskovec, J.: node2vec: scalable feature learning for networks. In: Proceedings of the 22nd ACM SIGKDD International Conference on Knowledge Discovery and Data Mining, pp. 855–864. ACM (2016)
8. Herman, I., Melançon, G., Marshall, M.S.: Graph visualization and navigation in information visualization: a survey. IEEE Trans. Vis. Comput. Graph. **6**(1), 24–43 (2000)
9. Mahoney, M.: Large text compression benchmark (2011)
10. McAuley, J., Leskovec, J.: Image labeling on a network: using social-network metadata for image classification. In: Computer Vision-ECCV, pp. 828–841 (2012)
11. Mikolov, T., Chen, K., Corrado, G., Dean, J.: Efficient estimation of word representations in vector space. arXiv:1301.3781 (2013)
12. Newman, M.E.J.: Modularity and community structure in networks. Proc. Natl. Acad. Sci. **103**(23), 8577–8582 (2006)
13. Perozzi, B., Al-Rfou, R., Skiena, S.: Deepwalk: online learning of social representations. In Proceedings of the 20th ACM SIGKDD International Conference on Knowledge Discovery and Data Mining, pp. 701–710. ACM (2014)
14. Pons, P., Latapy, M.: Computing communities in large networks using random walks. J. Graph Algorithms Appl. **10**(2), 191–218 (2006)
15. R Sarkar, Yin, X., Gao, J., Luo, F., Gu, X.D.: Greedy routing with guaranteed delivery using ricci flows. In: International Conference on Information Processing in Sensor Networks (IPSN), pp. 121–132. ACM (2009)
16. Shang, Y., Ruml, W., Zhang, Y., Fromherz, M.P.J.: Localization from mere connectivity. In: Proceedings of the 4th ACM International Symposium on Mobile Ad Hoc Networking and Computing, pp. 201–212. ACM (2003)
17. West, D.B., et al.: Introduction to Graph Theory. Prentice hall, Upper Saddle River (2001)
18. White, S., Smyth, P.: A spectral clustering approach to finding communities in graphs. In: Proceedings of the 2005 SIAM International Conference on Data Mining, pp. 274–285. SIAM (2005)
19. Yang, J., Leskovec, J.: Defining and evaluating network communities based on ground-truth. Knowl. Inf. Syst. **42**(1), 181–213 (2015)

Semi-supervised Graph Embedding Approach to Dynamic Link Prediction

Ryohei Hisano

Abstract We propose a simple discrete-time semi-supervised graph embedding approach to link prediction in dynamic networks. The learned embedding reflects information from both the temporal and cross-sectional network structures, which is performed by defining the loss function as a weighted sum of the supervised loss from the past dynamics and the unsupervised loss of predicting the neighborhood context in the current network. Our model is also capable of learning different embeddings for both formation and dissolution dynamics. These key aspects contribute to the predictive performance of our model and we provide experiments with four real-world dynamic networks showing that our method is comparable to state of the art methods in link formation prediction and outperforms state-of-the-art baseline methods in link dissolution prediction.

1 Introduction

One of the central tasks concerning network data is the problem of link prediction. Link prediction can be roughly divided into two types: static link prediction and temporal link prediction. Static link prediction is concerned with the problem of predicting the overall structure of a network. The goal is to predict missing links in partially observed network data that are absent from the dataset but that should in fact exist. Example applications include knowledge graph completion, predicting relationships among participants in social networking services and protein–protein interactions. We refer to [3, 5, 9] for excellent reviews of the field. In a temporal link prediction problem, the goal is to predict the future network state given previous linkage patterns. Example applications include recommender systems, where users and products are modeled as a bipartite graph and user purchases are modeled as linkages over time. The goal here is to predict future purchase patterns of users from past purchase patterns [4, 6, 12].

R. Hisano (✉)
Social ICT center, University of Tokyo, Tokyo, Japan
e-mail: hisano.ryohei@sict.i.u-tokyo.ac.jp

© Springer International Publishing AG 2018
S. Cornelius et al. (eds.), *Complex Networks IX*, Springer Proceedings
in Complexity, https://doi.org/10.1007/978-3-319-73198-8_10

In this paper, we focus on a slight variation of the temporal link prediction problem. Given a sequence of network snapshots from time 1 to time t, our problem is to predict the *transition* of a network from time t to time $t + 1$. A *transition* of a network can be summarized using two networks, a link formation network and a link dissolution network. We choose to predict the *transition* of a network instead of a network at the next time step for three main reasons. First, by predicting a network only at the next time step, one cannot distinguish whether the prediction of link formation is successful, whether the prediction of link dissolution is successful or whether the network itself did not change much between different time steps, and whether simply using the network information from the last time step might suffice for prediction. We want to avoid this redundancy by focusing on predicting the *transition*. Secondly, different forces might govern link formation and link dissolution. This is true in many domains, where the behavior governing link formation and link dissolution are different. In social network, attitudes towards friending and unfriending are different [8] and in economic networks, it is shown that the network adopts gradually to exogenous productivity shocks, where the timescales to form and dissolve a link are significantly different [7].[1] Our hope is that by separately modeling these forces we might obtain better predictive accuracy. Thirdly, predicting link dissolution is important in its own right. For instance, in the financial crisis of 2008, many banks were reported to dissolve their relationships with poorly performing firms while forming new links with better performing firms. Being able to predict the formation and dissolution dynamics of a network separately in this setting is an important issue in risk management. This is true even in social networks, where important dissolutions in links might prevent the spread of good or bad influences in a community [2].

Our modeling approach is a variant of semi-supervised graph embedding [17]. The supervised part consists of a complex-valued latent feature bilinear model [14], where past link formation and link dissolution information play the role of target values in the training data. The unsupervised part consists of a graph embedding predicting the neighborhood context in the current network [10]. The same complex-valued vectors are used in both tasks, and the weighted sum of these two losses is the total loss in our model. Semi-supervised graph embedding [17] was originally intended for use in node classification, but we extend the idea to learning complex-valued vectors capable of predicting the *transition* of a network.

To gain a better understanding of our model, we suggest the following intuitive interpretation. While the temporal information concerning past link formation and link dissolution networks provides a direct target signal for which nodes were more likely to form or dissolve links with each other, these networks are usually much sparser than the current network. Thus, by only using the past network information we may not have enough information to learn the complex-valued vector bilinear model sufficiently. On the other hand, the current network can be seen as providing a different dimension, such as a spatial dimension in spatiotemporal modeling, which is independent of the temporal information. Our strategy is to leverage this extra

---

[1] In short, it takes more time to form a link than dissolve a link.

dimension to enhance the model learned from our supervised task. Thus the power of graph embedding to effectively learn a distributional context capable of predicting nearby nodes is used in our model to force nearby nodes in the network to have similar complex-valued vectors [10]. We show that our semi-supervised approach gives better predictive performance than using a supervised or an unsupervised approach alone.

The rest of the paper is organized as follows. We present our proposed model in Sect. 2. Our training methodology is presented in Sect. 3. We give empirical results in Sect. 4, followed by conclusions in Sect. 5.

2 Proposed Method

We refer to our link prediction method as *SemiGraph*, which has the objective functions in (9) and (10) for link formation and link dissolution, respectively. Predictions are made using (13) and (14).

2.1 Notations

Consider a sequence of directed networks defined as a set of adjacency matrices $G = \{G_1, G_2, \ldots, G_t\}$, where G_{jkt} equals 1 if the link $j \to k$ exists at time t and equals 0 otherwise. Let V denote the set of nodes in the union of each snapshot of the network $G_1 \cup G_2 \cup \cdots \cup G_t$, and let $|V|$ denote the number of nodes in the union of all the networks. The goal of this paper is to predict the transition of the network from G_t to G_{t+1} using the information up to G_t.

We define three kinds of network. The *current network* is the network state just before prediction. With the above definitions, this is simply G_t. The past *formation networks* are defined by concatenating all the link formation adjacency matrices until time t. The adjacency matrix describing the link formation network at time t is defined as

$$\begin{cases} F_{jkt} = 1 \ \ if \ \ G_{jkt} - G_{jkt-1} = 1 \\ F_{jkt} = 0 \ \ otherwise. \end{cases}$$

The past *dissolution networks* are defined similarly, where the adjacency matrix describing the link dissolution network at time t is defined as

$$\begin{cases} D_{jkt} = 1 \ \ if \ \ G_{jkt} - G_{jkt-1} = -1 \\ D_{jkt} = 0 \ \ otherwise. \end{cases}$$

2.2 Learning from Past Formation and Dissolution Networks

We start with the supervised part, which consists of learning a complex-valued vector bilinear model with past link formation and link dissolution information playing the role of target values in the training data. The complex-valued matrix of the node representations (i.e., $C^{|V| \times d}$, where $|V|$ denotes the number of nodes in the network and d the dimension of the learned representations) are learned separately for link formation and link dissolution. These are learned in an identical manner, and we focus on the link formation case.

Formally, let (j, k) be a set of links in the past formation networks. The set of past formation networks is restricted to the information from link formation networks for a time window F_t, F_{t-1}, F_{t-p}. The loss function can be written as

$$\Sigma_{j,k \in (j,k)} log p(k|j) = \Sigma_{j,k \in (j,k)} (Re(\overline{v}_{fj}^T W_f v_{fk}) - log \Sigma_{k' \in Ne(j)} exp(Re(\overline{v}_{fj}^T W_f v_{fk'}))), \quad (1)$$

where $Ne(j)$ is the set of all edges that did not form links with j in the past formation networks, W_f is a diagonal complex-valued matrix defining the scaling of the basis, v_{fj} is the complex vector representation for node j with dimension d, $\overline{v}$ denotes the conjugate of v (i.e., $\overline{v} = Re(v) - iIm(v)$) and Re() is a function keeping only the real part of a complex values. The use of Re() is a simple trick to make the resulting value interpretable as a probability. The use of a complex-valued vector instead of a real-valued vector is to take into account symmetric as well as antisymmetric relations in both linear space and time complexity [14]. This could be confirmed by the fact that $\overline{v}_{fj}^T W_f v_{fk} \neq \overline{v}_{fk}^T W_f v_{fj}$ holds without forcing W_f to be non-diagonal. We take the complex conjugate in (1) to interpret it as a Hermitian inner product. We also restrict each diagonal element of W_f and W_d to have an absolute value of 1 to make the model identifiable.

It is often intractable to directly optimize (1) due to the normalization constant, and we use negative sampling to address this issue. Formally, given a triple (j, k, γ_f), where j and k are nodes (we assume that $j \neq k$) and γ_f is a binary label indicating whether a node pair exists in the past link formation networks (this is positive when links exists), we minimize the cross entropy loss of classifying the pair j, k with a binary label γ_f:

$$I(\gamma_f = 1) log \sigma(Re(\overline{v}_{fj}^T W_f v_{fk})) + I(\gamma_f = -1) log \sigma(-Re(\overline{v}_{fj}^T W_f v_{fk})), \quad (2)$$

where $I(.)$ is an indicator function that outputs 1 when the argument is true and 0 otherwise and σ is a sigmoid function defined as $\sigma(x) = 1/(1 + e^{-x})$. Therefore, the supervised loss with negative sampling can be written more succinctly as

$$L_{fs} = E_{j,k,\gamma_f} log \sigma(\gamma_f Re(\overline{v}_{fj}^T W_f v_{fk})). \quad (3)$$

The supervised loss for past dissolution networks is defined in an identical manner, resulting in

$$L_{ds} = E_{j,k,\gamma_d} log\sigma(\gamma_d Re(\bar{v}_{dj}^T W_d v_{dj})). \tag{4}$$

2.3 Graph Embedding from the Current Network

The unsupervised part of our model consists of a graph embedding defined by the current network. In previous works, a skip-gram model is used to learn the embedding and we adhere to this approach. Given a pair of an instance and its context (i.e., (j, c)), the loss function can be written as

$$\Sigma_{j,c\in(j,c)} log p(c|j) = \Sigma_{j,c\in(j,c)} (Re(\bar{v}_{fj}^T u_{fc}) - log \Sigma_{c'\in Ne(j)} exp(Re(\bar{v}_{fj}^T u_{fc'}))), \tag{5}$$

where v_{fj} is the complex vector representation for node j as used in (1) and u_{fc} is a parameter for the skip-gram model. A context for each node is generated by performing a truncated random walk (i.e., deep walk) starting from the instance node [10]. Although other types of walk beside the simple random walk (such as a breadth-first walk) are possible, preliminary experiments showed that the difference is marginal and we use the simple deep walk in this paper. As in (1), (5) is intractable due to the normalization constants and we again resort to negative sampling, resulting in

$$L_{fu} = E_{j,c,\gamma_c} log\sigma(\gamma_c Re(\bar{v}_{fj}^T u_{fc})). \tag{6}$$

The unsupervised loss for link dissolution is developed in an identical manner, resulting in

$$L_{du} = E_{j,c,\gamma_c} log\sigma(\gamma_c Re(\bar{v}_{dj}^T u_{dc})). \tag{7}$$

2.4 Semi-supervised Graph Embedding Approach

Given the loss functions defined in the previous sections, the loss functions for our framework can be expressed as

$$L_f = L_{fs} + \lambda_f L_{fu} \tag{8}$$

for learning link formation and

$$L_d = L_{ds} + \lambda_d L_{du} \tag{9}$$

for learning link dissolution. The L_{fs} and L_{ds} terms are the supervised losses for predicting past formation or dissolution networks, respectively, and L_{fu} and L_{du} are the unsupervised losses for predicting the graph context from the current network. The loss function is similar in spirit to graph-based semi-supervised learning, where graph embedding was used instead of the graph Laplacian [17].

2.5 Prediction

Prediction is made by using the learned complex-valued vectors and matrices v_f, v_d, W_f and W_d. A straightforward approach is to predict

$$p(G_{jkt+1} = 1|G_{jkt} = 0) = \sigma(Re(\overline{v}_{fj}^T W_f v_{fk})) \tag{10}$$

for link formation and

$$p(G_{jkt+1} = 0|G_{jkt} = 1) = \sigma(Re(\overline{v}_{dj}^T W_d v_{dk})) \tag{11}$$

for link dissolution. Although this simple prediction works quite well in practice, the predictive performance can be further improved by combining the predictions as

$$p(G_{jkt+1} = 1|G_{jkt} = 0) = \sigma(Re(\overline{v}_{fj}^T W_f v_{fk})$$
$$+ Re(\overline{v}_{dj}^T W_d v_{dk})) \tag{12}$$

for link formation and

$$p(G_{jkt+1} = 0|G_{jkt} = 1) = \sigma(Re(\overline{v}_{dj}^T W_d v_{dk})$$
$$+ Re(\overline{v}_{fj}^T W_f v_{fk})) \tag{13}$$

for link dissolution. The underlying understanding of this prediction is that link formation and link dissolution are more likely to be driven by a rewiring process: Thus the more likely a node is to form new links, the more likely the node is to dissolve an existing link at the same time. Although subtracting the two effects, as in

$$p(G_{jkt+1} = 1|G_{jkt} = 0) = \sigma(Re(\overline{v}_{fj}^T W_f v_{fk})$$
$$- Re(\overline{v}_{dj}^T W_d v_{dk})) \tag{14}$$

for link formation and

$$p(G_{jkt+1} = 0 | G_{jkt} = 1) = \sigma(Re(\bar{v}_{dj}^T W_d v_{dk}))$$
$$- Re(\bar{v}_{fj}^T W_f v_{fk})) \tag{15}$$

for link dissolution, is also reasonable (i.e., a growing network where the more likely a node is to form links the less likely the node is to lose a link), in our experiments (12) and (13) outperform the other prediction method, so we use this prediction in our experiments.

3 Training

We use stochastic gradient descent to train our model. We first sample a node and perform a deep walk [10] to sample the context nodes from a network. We then sample negative samples from the current network, past formation networks, and past dissolution networks. Equipped with these positive and negative samples, we take a gradient step with learning rate η_1 for v_f, v_d, u_f, and u_d.

Each diagonal element of W_f and W_d is learned in a different manner. As noted before, to make the model identifiable we restrict each diagonal element of W_f and W_d to take an absolute value of 1. Thus, each diagonal element of W_f can be rewritten as

$$W_f(j, j) = cos(\theta_j) + i sin(\theta_j), \tag{16}$$

for $j = 1, \ldots, d$. We take a gradient step with learning rate η_2 in θ instead. All the off-diagonal elements are set to 0.

4 Experiments

Our empirical investigations are based on four real-world networks: a world trade network, a bipartite customs data between Japan and the US (Japan to US exports only), a small size interfirm buyer–seller network and a larger size interfirm buyer–seller network. Code and a subset of the data to reproduce our results will be available on the authors website.

4.1 Data

We next give a brief outline of the data used.

Table 1 Statistics for datasets. Num Edges denotes the total number of interactions, Num Unique Edges denotes the number of distinct interactions, Ave Form denotes the average number of formed edges, Ave Diss denotes the average number of dissolved edges, and Snapshots denotes the number of discrete time points observed in our datasets

Dataset	Num nodes	Num edges	Num unique edges	Ave form	Ave diss	Snapshots
WorldTrade	50	6,620	477	16.7	16.7	20
Customs	1,043	7,825	1,488	113.9	126	12
FirmSmall	690	13,108	1995	118.9	126.3	10
Firm	32,475	1,282,562	251,061	42,487.1	41,400.4	10

- WorldTrade is a network of world trade relationships among 50 countries from 1981 to 2000 [15]. We define two countries to be linked if the trading volume was above the 90th percentile for all trade in a given year.
- Customs is a bipartite network dataset that records the names of exporters and consignees of trade from Japan to the US. The data was obtained from the US customs office and covers the period from January 2003 to December 2014. We focus on firms that had more than 500 transactions during the time period, which results in 431 Japanese firms and 603 US firms. To adjust for seasonal effects, we aggregate the network data on a yearly basis resulting in snapshots of 12 networks. Two firms are linked if there was a trade relation more than once a year.
- FirmSmall is an interfirm buyer–seller network for Japan from 2003 to 2012. Two firms are connected if firms have a buyer–seller relationship. These data are based on questionnaires and is obtained from the Teikoku Data Bank.[2] We use a subset of this dataset, restricting our attention to firms in Hokkaido, which is the northern part of Japan.
- Firm is an interfirm buyer–seller network for Japan from 2003 to 2012. This is a larger version of FirmSmall including firms from all parts of Japan.

The basic statistics for each dataset are reported in Table 1.

4.2 Evaluation Criteria

Given a training set $G_{1:t}$, we predict the transition from time t to time $t + 1$ which consists of a link formation network (i.e., F_{t+1}) and a link dissolution network (i.e., D_{t+1}). We predict the last network in each datasets using all the past transitions as training. For link prediction accuracy, we use the area under the receiver operating characteristic curve (AUCROC), where the value is calculated for both link dissolution networks and link formation networks. We also use the area under the precision

---

[2]http://www.tdb.co.jp/index.html.

and recall curve (AUCPR) which provides an alternative view to AUCROC when the class distribution is highly imbalanced [16].

4.3   Baseline Methods

We compare our prediction algorithm with the following baselines.

- Adamic-Adar (AA): scores are calculated as the weighted variation of common neighbors [1] using the current network only. Since AA only generates scores for node pairs within 2 hops geodesic distance, we only report results for non-bipartite networks.
- Preferential Attachment (PA): scores are calculated as the product of the degree of each node from the current network.
- Last time of Linkage (LL): scores are calculated by ranking pairs in ascending order according to the last time of linkage.
- Longitudinal Mixed Effects Models (LME): is a mixed effects model for modeling dynamic networks [15]. Although it is computationally expensive, it is one of the state-of-the-art methods in modeling dynamic networks. We used the same hyperparameter settings as in [15].

We also compute AA-all and PA-all, which are computed over the union of all networks until the current network. The graph heuristic approaches presented here are simple, but have been shown to be surprisingly hard to be at in practice, making them good baselines for comparison [13]. In particular, LL has been shown to often be among the best heuristic measures for link prediction [13]. When predicting link dissolution, we use the complementary score method (i.e., which is basically taking the negative value of the original score) as in [11]. We also compare our model with unsupervised graph embedding and supervised approach (i.e., our model without the graph embedding term) to clarify the improvement in semi-supervised learning. Throughout all of the experiments, we set the dimension of the learned complex vectors as $d = 15$, the number of walks as five, $\lambda_f = \lambda_d = 0.05$, initial learning rates $\eta_1 = 0.1$, $\eta_2 = 10^{-5}$, and $p = t - 1$ (i.e., using all past information). The learning rate is decreased linearly with the number of nodes that have been used for training.

4.4   Experimental Results

Results for link dissolution prediction are presented in Table 2. We make the following observations. Basically for all the experiments our method performs better than the state-of-the-art methods in both AUCROC and AUCPR except for the AUCPR in the Customs data where the added value of our semi-supervised approach is small due to the bipartite nature of the network. It is worth noting that our method outperforms better than the other methods quite significantly for the FirmSmall and Firm

Table 2 AUC for link dissolution prediction. Since AA only generates scores for node pairs within 2 hops geodesic distance, we only report results for non-bipartite networks. Omissions are indicated with '-'. Failure runs are indicated with '*'

Dataset method	WorldTrade ROC	PR	Customs ROC	PR	FirmSmall ROC	PR	Firm ROC	PR
AA	0.638	0.091	-	-	0.478	0.100	0.473	0.129
PA	0.711	0.099	0.679	**0.397**	0.496	0.110	0.464	0.127
AA-all	0.676	0.100	-	-	0.521	0.111	0.483	0.132
PA-all	0.703	0.101	0.649	0.307	0.538	0.121	0.489	0.134
LL	0.535	0.075	0.573	0.278	0.650	0.229	0.620	0.232
LME	0.672	0.120	0.587	0.266	0.538	0.126	*	*
Supervised	0.679	0.074	0.635	0.255	0.653	0.171	0.612	0.197
GraphEmb	0.480	0.058	0.609	0.234	0.514	0.108	0.497	0.131
SemiGraph	**0.715**	**0.126**	**0.686**	0.291	**0.736**	**0.254**	**0.674**	**0.251**

dataset, whereas graph embedding shows almost no signs of predictability. Also in this experiment, supervised learning is outperformed by our method by around 5–13% in AUCROC, suggesting again the added value of our semi-supervised approach. We define an experimental run as failure (indicated by a '*' in the tables) if it does not complete within 7 days on a standard workstation with a 2.33 GHz processor and 64 GB of memory. Since the longitudinal mixed effects model involves inversion of a matrix, it could not handle the large scale Firm data. Our proposed method takes more time to run compared with the baseline state of the art methods, but nevertheless returns results in a fair amount of time.

Results for the link formation prediction task are presented in Table 3. We make the following observations. Although our proposed method often shows superior performance when evaluating with AUCROC, the performance becomes subtle when evaluating with AUCPR. The reason behind this behavior is that for all the data studied here the same links appear and disappear multiple times. While methods like LL naturally restricts its link prediction to links that have been formed before, our method does not involve any type of restriction to past formed links, and hence making it inferior in performance. On the other hand, LL could not predict links which was not formed before having its own shortcomings and for all the networks studied here, our proposed method is among the top performing methods. Our method also shows significant improvements over graph embedding and supervised learning. In this experiment, supervised learning is outperformed by our method by around 10–23% in AUCROC, while graph embedding is outperformed by more than 13–39%, suggesting the added value of our semi-supervised approach.

To see how an increase in past information affects the performance of our proposed model, we report results on predicting the transition of a network for the years 2005–2012 for the FirmSmall dataset.[3] Because we only have ten snapshots of the network,

---

[3] We omit AA, PA and LME since it did not show superior performance and for visual clarity.

Table 3 AUC for link formation prediction. Due to heavy computational costs, AUCPR for the Firm data is calculated over 10^6 negative samples making the values higher than that of FirmSmall

Dataset method	WorldTrade ROC	PR	Customs ROC	PR	FirmSmall ROC	PR	Firm ROC	PR
AA	0.666	0.013	-	-	0.617	6.64×10^{-4}	0.614	0.066
PA	0.774	0.038	0.521	1.30×10^{-4}	0.707	1.43×10^{-3}	0.815	0.178
AA-all	0.659	0.013	-	-	0.692	8.40×10^{-4}	0.716	0.078
PA-all	0.864	0.094	0.756	2.39×10^{-4}	0.790	1.94×10^{-3}	**0.847**	0.204
LL	**0.873**	**0.209**	0.836	**0.125**	0.781	**0.145**	0.744	**0.511**
LME	0.670	0.012	0.718	1.76×10^{-4}	0.720	2.11×10^{-3}	*	*
Supervised	0.683	0.003	0.683	2.62×10^{-4}	0.703	6.18×10^{-4}	0.727	0.108
GraphEmb	0.706	0.016	0.581	7.95×10^{-5}	0.672	6.00×10^{-4}	0.624	0.069
SemiGraph	0.800	0.027	**0.842**	4.90×10^{-4}	**0.860**	0.011	0.802	0.222

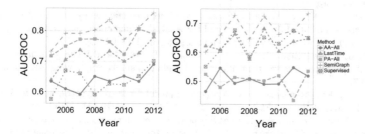

Fig. 1 AUC for link formation and link dissolution prediction for the Firm dataset

the prediction in 2005 is based on only one past transition and the last network before prediction. We observe that for link formation prediction, almost all the methods including our proposed method show improved accuracy with an increase in past information. In this dataset, our method is the best performing methods. Comparing our performance with supervised learning, we clearly see the benefit of our semi-supervised approach. For link dissolution, although less clear than link formation prediction we also observe that our method show improved accuracy with an increase in past information (Fig. 1).

5 Conclusions

We have proposed *SemiGraph*, a simple discrete-time semi-supervised graph embedding approach to link prediction in dynamic networks. Our model is capable of learning different embeddings for both formation and dissolution dynamics. To show the effectiveness of our approach, we focused on predicting the *transition* of a network, including both link formation prediction and link dissolution prediction. We have shown that our method outperforms previous state-of-the-art baseline methods in predicting link dissolution and is comparable to state of the art methods in predicting link formation through experiments using a variety of real-world networks.

References

1. Adamic, L.A., Adar, E.: Friends and neighbors on the web. Soc. Netw. **25**(3), 211–230 (2003)
2. Christakis, N.A.A., Fowler, J.H.H.: The spread of obesity in a large social network over 32 years. New Engl. J. Med. **357**(4), 370–379 (2007). https://doi.org/10.1056/nejmsa066082
3. Clauset, A., Moore, C., Newman, M.E.J.: Hierarchical structure and the prediction of missing links in networks. Nature **453**, 98–101 (2008). https://doi.org/10.1038/nature06830
4. Dunlavy, D.M., Kolda, T.G., Acar, E.: Temporal link prediction using matrix and tensor factorizations. ACM Trans. Knowl. Discov. Data **5**(2), 10:1–10:27 (2011). https://doi.org/10.1145/1921632.1921636
5. Getoor, L., Diehl, C.P.: Link mining: a survey. SIGKDD Explor. Newsl. **7**(2), 3–12 (2005). https://doi.org/10.1145/1117454.1117456
6. Hasan, M.A., Chaoji, V., Salem, S., Zaki, M.: Link prediction using supervised learning. In: Proceedings of SDM 06 Workshop on Link Analysis, Counterterrorism and Security (2006). http://citeseerx.ist.psu.edu/viewdoc/summary?doi=10.1.1.61.1225
7. Hisano, R., Watanabe, T., Mizuno, T., Ohnishi, T., Sornette, D.: The gradual evolution of buyer-seller networks and their role in aggregate fluctuations. Appl. Netw. Sci. Forthcoming (2017)
8. Krivitsky, P.N., Handcock, M.S.: A separable model for dynamic networks. J. R. Stat. Soc.: Ser. B (Stat. Methodol.) **76**(1), 29–46 (2014). https://doi.org/10.1111/rssb.12014
9. Liben-Nowell, D., Kleinberg, J.: The link prediction problem for social networks. In: Proceedings of the Twelfth International Conference on Information and Knowledge Management, CIKM '03, pp. 556–559. ACM, New York, NY, USA (2003). https://doi.org/10.1145/956863.956972
10. Perozzi, B., Al-Rfou, R., Skiena, S.: Deepwalk: online learning of social representations. In: Proceedings of the 20th ACM SIGKDD International Conference on Knowledge Discovery and Data Mining, KDD '14, pp. 701–710. ACM, New York, NY, USA (2014). https://doi.org/10.1145/2623330.2623732
11. Preusse, J., Kunegis, J., Thimm, M., Gottron, T., Staab, S.: Structural dynamics of knowledge networks. In: ICWSM'13: Proceedings of the 7th International AAAI Conference on Weblogs and Social Media (2013). http://dl.dropboxusercontent.com/u/20411070/Publications/2013-ICWSM-Preusse-KTG.pdf
12. Sarkar, P., Siddiqi, S.M., Gordon, G.J.: A latent space approach to dynamic embedding of co-occurrence data. In: M. Meila, X. Shen (eds.) Proceedings of the Eleventh International Conference on Artificial Intelligence and Statistics (AISTATS 2007) (2007). http://www.stat.umn.edu/\~ {}aistat/proceedings/start.htm
13. Sarkar, P., Chakrabarti, D., Jordan, M.: Nonparametric link prediction in large scale dynamic networks. Electron. J. Stat. **8**(2), 2022–2065 (2014). https://doi.org/10.1214/14-EJS943

14. Trouillon, T., Welbl, J., Riedel, S., Gaussier, É., Bouchard, G.: Complex Embeddings for Simple Link Prediction, pp. 1–2 (2016). http://jmlr.org/proceedings/papers/v48/trouillon16.html
15. Westveld, A.H., Hoff, P.D.: A mixed effects model for longitudinal relational and network data, with applications to international trade and conflict. Ann. Appl. Stat. 5(2A), 843–872 (2011). https://doi.org/10.1214/10-AOAS403
16. Yang, Y., Lichtenwalter, R.N., Chawla, N.V.: Evaluating link prediction methods. Knowl. Inf. Syst. 45(3), 751–782 (2015). https://doi.org/10.1007/s10115-014-0789-0
17. Yang, Z., Cohen, W.W., Salakhutdinov, R.: Revisiting semi-supervised learning with graph embeddings. In: Proceedings of the 33nd International Conference on Machine Learning, ICML 2016, New York City, NY, USA, June 19–24, 2016, pp. 40–48 (2016). http://jmlr. org/proceedings/papers/v48/yanga16.html

Modularity Optimization as a Training Criterion for Graph Neural Networks

Tsuyoshi Murata and Naveed Afzal

Abstract Graph convolution is a recent scalable method for performing deep feature learning on attributed graphs by aggregating local node information over multiple layers. Such layers only consider attribute information of node neighbors in the forward model and do not incorporate knowledge of global network structure in the learning task. In particular, the modularity function provides a convenient source of information about the community structure of networks. In this work, we investigate the effect on the quality of learned representations by the incorporation of community structure preservation objectives of networks in the graph convolutional model. We incorporate the objectives in two ways, through an explicit regularization term in the cost function in the output layer and as an additional loss term computed via an auxiliary layer. We report the effect of community-structure-preserving terms in the graph convolutional architectures. Experimental evaluation on two attributed bibliographic networks showed that the incorporation of the community-preserving objective improves semi-supervised node classification accuracy in the sparse label regime.

1 Introduction

In recent years, convolutional neural networks (CNNs) [8] have successfully exploited the statistical regularities on several domains and have shown state-of-the-art results in image classification [7], speech recognition [4], and related tasks. CNNs achieve this by making two important assumptions about the statistical properties of the data, locality, and translation invariance. These assumptions are exploited by learning local feature extractors that are shared across the domain. Defining such feature extractors requires clear notions of translation and locality. These are clear on domains where

T. Murata (✉) · N. Afzal
Department of Computer Science, School of Computing Tokyo Institute
of Technology, W8-59 2-12-1 Ookayama, Meguro, Tokyo 152-8552, Japan
e-mail: murata@c.titech.ac.jp
URL: http://www.net.c.titech.ac.jp/

© Springer International Publishing AG 2018
S. Cornelius et al. (eds.), *Complex Networks IX*, Springer Proceedings
in Complexity, https://doi.org/10.1007/978-3-319-73198-8_11

the sample dependencies have a regular Euclidean structure (i.e., images, videos, text, and speech). However, in many data domains the relationships between the samples are arbitrary and cannot be expressed as a regular Euclidean structure. The fields of social networks analysis, bioinformatics, and information science all involve data with various complex relationships between the entities. These relationships can be represented mathematically with a graph $G(V, E)$ where V is a set of vertices and E is a set of pair-wise relationships between them. Each vertex in V represents a single training sample in $\mathbf{X}$, and an edge between two vertices can represent some domain-specific relationship.

Starting with [1], recent efforts have attempted to generalize the main assumptions of CNNs to arbitrary graph domains. With [2], we now have scalable formulations of convolutional layers applicable to general graphs and are denoted in the literature as graph convolutional layers. Given a definition of proximity of graph vertices as defined by simple k-hop distances, the graph convolutional layer extracts local features for each node. These feature extractors are replicated over every node of the graph, effectively limiting the number of learnable parameter to be independent of graph size. This model has been shown to be very effective in various graph-based learning tasks as demonstrated in [5, 6].

The assumption of locality on networks can be used to learn very useful models but real networks also tend to exhibit various global properties. The field of Network Science [11] is generally concerned with studying such high-level properties of networks. An important feature of real networks is the existence of dense clusters or communities where the vertices of a network have a higher density of edges among them that between clusters. [12] defined graph clustering as an optimization problem with an objective function called modularity. Modularity is a global measure of the networks' structure and can thus provide a higher-level view of the network's properties than considering local neighborhood information alone.

This work is based on the assumption that injecting additional information about higher-level network structure into a neural architecture trained on an appropriate graph-based learning task can improve performance on that task. Recent attempts [14, 15, 17] have successfully integrated community structure information into various graph representation learning methods. Similar work has not yet been attempted for graph convolutional models. In situations where the number of labeled examples for training is very sparse, the model should be able to benefit from leveraging more higher-level network structure information as compared to learning from labels alone. In this work, we incorporate community structure information by integrating modularity score optimization into the framework of graph convolutional neural networks. For evaluation of our approach, we explore the case of semi-supervised node classification on graph datasets in the sparse label regime.

2 Background

Community or Mesoscopic Structure A graph G can be quantified as an $n \times n$ adjacency matrix A where n is the number of nodes. If nodes i and j have an edge between them, then entry A_{ij} of the matrix is 1 else it is 0.

Networks tend to be organized in a higher structural level into clusters. A cluster or community in a network is a group of vertices that has a higher density of edges among nodes within the group but sparser connections to nodes outside.

Reference [12] proposed modularity as a score to measure the goodness of partition of a given network. Statistically, it is the fraction of edges within a group minus the expected fraction for a random graph with same degree distribution, summed over every group. It is the most commonly used objective function for community detection. Modularity optimization provides additional information not normally available from optimizing neighborhood structure alone. Given a group assignment Matrix H, the score of the partition can be computed as:

$$Q = tr(H^T B H), \tag{1}$$

where $B_{ij} = A_{ij} - \frac{k_i k_j}{2e}$ is the modularity matrix.

2.1 Graph Neural Network

The standard convolutional filters are not directly applicable to arbitrary graph structures due to lack of clear notion of translation and ordering of the node neighborhoods.

Locality and Translational Invariance A CNN layer is simply a filter of learnable parameters that is convolved on the input pixels of an image. Naturally, these layers possess the properties of locality and translational invariance. The locality property is based on the assumption that the statistical dependence of pixels in an image is inversely dependent on the distance between them. In practice, this is exploited by making the filtering step consider only pixel values in a fixed local neighborhood which is defined by the size of the filter. The second assumption is that of spatial model invariance, which is based on the observation that the identity of an object in an image does not change regardless of its translation in the image. The CNN filter exploits this property by sharing the same set of weights for every image patch. Through these assumptions, the number of parameters of the convolutional layer can be made independent of the input size. A suitable definition of convolutional layers on graphs should possess these two properties.

Polynomial Graph Convolutional Filters [2] defined localized graph convolutional filters directly computed in the spatial domain. The filter is defined as a polynomial function of the graph Laplacian matrix. Besides a simple polynomial formulation, a model computed via Chebyshev polynomials [3] has also been defined in the same work. These formulations have computational complexity of order equal to number of graph edges due to sparse multiplications. An important property of polynomial convolutions is that the locality of the filter is equal to the order of the polynomial. [6] defined a simplified version of this by restricting the model to first-order filters. In this document, we will refer to Kipf's model as GCN and Defferrard's Chebyshev formulation as ChebNet. These classes of models posses both locality and translation invariance properties of graph convolutions and have scalable computation time. The model described in this paper is also based on the polynomial graph convolutions of Defferard and Kipf and will be described in greater detail (Fig. 1).

Regularizing Deep Models via Graph Structure [16] described an approach for incorporating graph structural information directly into any arbitrary deep architecture during training. The structural information is incorporated as an unsupervised loss term directly computed over the representations in any layer of the deep architecture. The final loss function is then composed of two parts, the supervised loss over labeled examples and the unsupervised loss over all example regardless of labels. The domain structure can be incorporated in three ways as illustrated in Fig. 2, either by (a) directly adding the unsupervised term to the output layer loss of the architecture, by (b) computing the loss directly on the representations on any non-output layer, or by (c) attaching a new auxiliary feed-forward layer to the architecture and computing the loss on the output of that layer. In this work, the modularity optimization term is incorporated into a graph convolutional architecture based on these techniques. The resulting models will be described in detail in the next section.

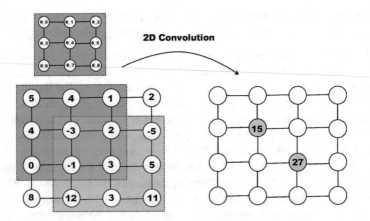

Fig. 1 Standard convolutional layer

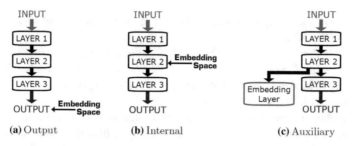

Fig. 2 Semi-supervised embedding [16]

3 Model Description

3.1 Encoding Model

This section will describe the encoding layers used in the experiments. The inputs are the attribute and adjacency matrices of the graph to be encoded, and the output is a latent representation Z of any arbitrary dimensionality. The latent representations depend on the weights and the biases of the encoder model which can be trained using gradient descent. The encoders can also be made arbitrarily deeper by stacking additional layers. Given a graph convolutional layer with C input channels (features) and F output features:

GCN: W^t should be a $C \times F \times 1$-dimensional weight matrix. Even though GCN incorporates first-order neighborhood, it also enforces parameter sharing between first- and zeroth-order neighborhood weights, thus fixing its weight matrix size to $C \times F$.

ChebNet: For a Kth-order neighborhood model, W^t should have $C \times F \times (K + 1)$ parameters. This is because the model learns a separate set of $C \times F$ weights for each order degree ranging from $k = 0$ to K. For example, for a second-order model W^t would be a $C \times F \times 3$-dimensional tensor with a forward matrix corresponding to the zeroth-, first-, and second-order degrees.

Graph Convolutional Network (GCN)

This is the first-order layer used in the forward models of [5, 6]. This layer also enforces weight sharing between the first-order and self-weights, effectively acting as a regularizer and restricting the model to only one set of parameter weights. Given the adjacency matrix A, the filter support is computed in the following steps:

1. Addition of self-connections: $\tilde{A} = A + I_N$
2. Computation of degree matrix: $\tilde{D} = \sum \tilde{A}$
3. Normalization Step: $\hat{A} = \tilde{D}^{-\frac{1}{2}} \tilde{A} \tilde{D}^{-\frac{1}{2}}$

The normalization step is done to prevent numerical instabilities and is computed as a pre-processing step. Using the normalized adjacency matrix effectively makes averaging the aggregation strategy. The layer-wise model then becomes: $H^l = \sigma(\hat{A} H^{l-1} W^l)$.

Now considering input $X \in \mathbb{R}^{N \times C}$ with C input channels and convolved output $Z \in \mathbb{R}^{N \times F}$ with F features. The final model with a two-layer GCN with ReLU nonlinearity in hidden layer and softmax in output layer is given as: $Z = f(X, A) = softmax(\hat{A}\ ReLU(\hat{A} X W^{(0)})\ W^{(1)})$.

The number of input channels C is equal to the dimensionality of the features of the data (such as the attributes of the node). For example, in case of Cora and CiteSeer, it is equal to the size of the bag-of-words feature vector. In case of a featureless approach, C will be equal to the number of nodes in the graph, as each node is then represented with a one-hot representation.

Even though the layer-wise model is first-order, stacking multiple layers increases the locality of the filter. For example, a two-layer GCN would incorporate information in the second-order neighborhood of the node for computing the filter.

ChebNet This is the graph convolutional layer originally described in [2]. Unlike the GCN model of Kipf, the Chebnet can incorporate multiple-hop neighborhood information in a single layer. There is also no parameter sharing, as a separate set of parameters are learned for the original node, and each hop neighborhood. For example, a second-order Chebyshev layer would have three sets of weights and biases in the forward model. The support matrices are computed as Chebyshev polynomials of the scaled graph Laplacian matrix. The Laplacian matrix is scaled by division with its largest Eigenvalue λ_{max} to prevent numerical instabilities when stacking multiple layers. The Chebyshev polynomials themselves are computed recursively as a preprocessing step, and the largest Eigenvalue of the Laplacian is computed via efficient power iterations.

Given the symmetric normalized graph Laplacian matrix:

$$L = I_n - D^{-\frac{1}{2}} A D^{-\frac{1}{2}}, \tag{2}$$

its rescaled version is given as:

$$\tilde{L} = \frac{2L}{\lambda_{max}} - I_n \tag{3}$$

With $T_0 = I$ and $T_1 = \tilde{L}$, the Chebyshev polynomials are defined recursively as:

$$T_k(x) = 2x T_{k-1}(x) - T_{k-2}(x). \tag{4}$$

Given the rescaled Laplacian $\tilde{L}$ and arbitrary model degree K, the filter supports can be pre-computed from T_0 to T_K as follows:

$$T_0(\tilde{L}) = I$$
$$T_1(\tilde{L}) = \tilde{L}$$
$$T_2(\tilde{L}) = 2\tilde{L}^2 - 1 \tag{5}$$
$$T_3(\tilde{L}) = 4\tilde{L}^3 - 3\tilde{L}$$
$$T_4(\tilde{L}) = 8\tilde{L}^4 - 8\tilde{L}^2 + 1$$

The layer-wise model can then be defined as a weighted sum of these components.

$$H^t = \sigma \left(\sum_{k=0}^{K} T_k(L) H^{t-1} W_k^t \right) \tag{6}$$

Since $\tilde{L}$ is a sparse matrix with $O(|E|)$ elements, the filtering step involves multiplications with $\tilde{L}$ only. The Kth-order ChebNet layer will then have K+1 sets of parameter weights and computational complexity $O(|E|)$ where $|E|$ is the number of edges in the graph.

3.2 Task-Specific Cost Functions

Here we briefly describe the optimization objective of graph-based learning task used in the experimental evaluation. For the task of semi-supervised node classification, we use the cross entropy over all labeled examples. The cross entropy loss is used as a supervised training signal for semi-supervised node classification. Given k prediction classes and $Z = GCN(X, A)$ as the matrix of normalized prediction probabilities for each class as computed by the model. The cross entropy loss over all labeled examples $y_L \in Y$ is then given as: $\mathcal{L} = -\sum_{\ell \in y_L} Y_\ell \ln Z_\ell$

For graph convolution models, even though the loss is only computed over labeled examples, the predictions depend on the unlabeled examples as well due to the nature of encoding model. The model is tested via prediction accuracy over a hold-out test set.

3.3 Modularity Optimization Term

The main contribution of this paper is the incorporation of a modularity-preserving constraint on the embeddings of the graph convolutional network. This constraint is imposed by the addition of a new loss term into the task-specific cost functions of the previous section. We can describe the modularity optimization as a secondary objective that is jointly optimized with the primary objectives. As described in the previous section, modularity is a quality function that scores the partition of a given network into k clusters or partitions. Given a cluster assignment matrix H, we can simply compute this score as:

$$Q = tr(H^T B H), \tag{7}$$

where $B_{ij} = A_{ij} - \frac{k_i k_j}{2e}$ is the modularity matrix.

Setting the model embeddings as the cluster assignments, we can compute a modularity score for the embeddings using the term given above. Network modularity can then be optimized using gradient descent in any neural network pipeline. In this work, we follow the approaches in [16] to jointly optimize for both modularity score and the task-specific in the semi-supervised embedding frame with the graph convolutional model as the deep architecture.

Regularization Term We simply subtract the modularity optimization term from the cost function of the output layer of the architecture. The term simply acts as a regularizer that encourages the optimizer to favor weight values that maximize this term. We use a trade-off parameter α to balance between the unsupervised loss and task-specific objectives. Essentially, the two objectives share the same layers and model parameters, and the model is trained to jointly optimize them. The modified cost function is given as: $\mathcal{L}_{total} = (1 - \alpha)\mathcal{L}_{supervised} - \alpha tr(H^T B H) * (\frac{1}{2e})$ where e is the number of edges in the graph.

The first term of this loss function depends on all labeled examples in the case of node classification. The modularity optimization term, however, is computed over all embeddings regardless of the number of training samples. α then becomes a hyper-parameter to be optimized.

Auxiliary Layer: An issue with regularizing the output layer of the architecture is that the number of community partitions to be optimized over is fixed to the output layer size. We introduce an auxiliary layer to compute a separate set of representations which dimensionality equal to the number of network partitions to be optimized over.

Considering a two-layer encoder model given by two graph convolutional layers denoted by GCN_1 and GCN_2, the forward model is given as:

$$\begin{aligned} H_{hid} &= GCN_1(X, A) \\ H_{out} &= GCN_2(H_{hid}, A) \end{aligned} \tag{8}$$

We regularize the intermediate hidden layer representations via an auxiliary feed-forward layer denoted by MLP_{aux}, whose input is the hidden layer embeddings and the output is an $n \times k$ cluster assignment matrix H_{aux} where k is the number of partitions.

$$H_{aux} = MLP_{aux}(H_{hid}) \tag{9}$$

The task-specific loss is then evaluated over H_{out} and the modularity optimization term over H_{aux}. The total loss is then taken to be a weighted sum of these two.

$$\mathcal{L}_{total} = (1 - \alpha)\mathcal{L}_{supervised}(H_{out}) - \alpha\mathcal{L}_{modularity}(H_{aux}) \tag{10}$$

In this type of architecture, the layers before the branching are shared between the two objectives and receive training signals from both gradients. After branching, the two terms have their own sets of weight parameters to optimize.

At each iteration, the layers unique to the supervised layer (after branching) are only updated via gradients from the supervised loss, and the auxiliary layer only receives gradient updates from the modularity loss. The layers before branching should receive a linear combination of the gradients from both output layers (based on α).

4 Experiments

4.1 Experimental Setup

Evaluation Metrics For node classification, we use accuracy score as evaluation metrics. The accuracy score is simply the percentage of correctly classified nodes.

Datasets We also use two bibliographic network datasets, Cora and CiteSeer, initially introduced in [13]. These possess node features and classification labels for analysis on the semi-supervised classification task. These have been extensively used in the semi-supervised learning literature, including [6, 18]. The node features consist of bag-of-word representations of the document text, and the edges represent the citation links. The links are assumed to be undirected in this case.

4.2 Semi-supervised Node Classification

Given the network structure, node features, and label values of a subset of nodes, the task of semi-supervised node classification is to predict labels for the remaining nodes. The task is different from standard supervised classification because the pairwise affinity structure of the samples is available as an input. The feature values of both labeled and unlabeled examples can be used for prediction. The basic distinction from pure community detection is that the latter involves inferring topological clusters from structure alone. On the other hand, the labels for node classification have a semantic meaning and might not necessarily correspond to structural communities. However, many information network domains do tend to exhibit correspondence between semantic and topological groups according to the cluster hypothesis, and this can be attributed to the success of many semi-supervised learning methods in recent literature.

Experiment Design Since the modularity optimization term provides additional information about network community structure not available from the labels themselves, we expect it to be most effective when the number of labeled examples for training is very small. To verify this, we train the baseline models and our community-enhanced variants on different numbers of training examples sampled from the label set. To avoid issues with label imbalance, we sample an equal number of labels from each class. We use uniform random sampling in our experiments but other sampling strategies like PageRank and degree have been explored in [9]. We avoid centrality and degree-based sampling because they tend to return a clustered set of nodes that is not representative of the full label set. We also sample 1000 addition nodes as a test set in each run. We do not use a validation set for early stopping for a fair comparison in the sparse label regime. Instead, we let each model run for 100 epochs. Since the accuracy is highly dependent on the initial training sample, we average the results for 20 times with different train–test splits to get a good estimate of the average performance.

We test with two graph convolutional encoders, GCN and ChebNet. For the community-enhancement term, we use two approaches, regularization via auxiliary layer and direct regularization in the output layer. All models not counting the auxiliary layers are stacked two-layer graph convolutional architectures. This brings us to a total of six different architecture choices. We test all models for 5, 8, 11, 14, 17, and 20 training labels per class and measure performance for 20 runs. We also report results for the Iterative Classification Algorithm [10], a semi-supervised classification baseline, averaged over 20 runs. For Cora, the accuracy scores for node classification for all six neural architectures and the ICA baseline are detailed in Table 1. The results for CiteSeer are omitted because of space limitations, but they are similar to the results for Cora. We also report the standard error for each instance. Results in the top row of the table represent the baseline methods, and the second row shows the proposed architectures in this paper.

Results We note that the community-enhanced ChebNet without the auxiliary layer is the best-performing model on both datasets. ChebNet benefits more from the modularity optimization term because GCN's first-order filter with weight sharing already acts as a regularizer and therefore addition of another unsupervised regularization term has little additional benefit. ChebNet is a more general model with higher number of parameters and can incorporate higher-order neighborhood information, allowing it to more easily optimize for secondary tasks. Figures 3 and 4 show the relative effect of the community-enhanced loss term on accuracy for both GCN and ChebNet.

Embedding Visualization We also visualize the effect of the community enhancement on the quality of learned model representations in the sparse label regime. We create three instances of a two-layer ChebNet with values of α set to 0, 0.5, and 1.0 respectively. We train these models on Cora with five training labels per class until convergence. We visualize the hidden layer representations of each using T-SNE. These are shown in Fig. 5. We note that joint optimization with $\alpha = 0.5$ leads to better separation of ground-truth labels as compared to optimizing either term alone. Setting $\alpha = 1.0$ leads to some visible clustering patterns but this does not correspond to a good accuracy score on the given task. We attribute this due to the model converging to local minima of the modularity score loss.

Table 1 Accuracy score and standard error on Cora with different number of labeled examples per class, averaged over 20 runs

Model	5	8	11	14	17	20
ICA	0.571 ± 0.013	0.641 ± 0.011	0.700 ± 0.010	0.713 ± 0.007	0.716 ± 0.007	0.744 ± 0.007
GCN	0.664 ± 0.012	0.715 ± 0.010	0.747 ± 0.007	0.768 ± 0.005	0.772 ± 0.006	0.785 ± 0.004
ChebNet	0.612 ± 0.013	0.685 ± 0.010	0.732 ± 0.008	0.735 ± 0.006	0.762 ± 0.007	0.783 ± 0.005
GCN-mod	**0.699 ± 0.013**	0.738 ± 0.007	0.755 ± 0.010	0.775 ± 0.005	0.791 ± 0.005	0.786 ± 0.007
Chebnet-mod	0.652 ± 0.024	**0.745 ± 0.012**	**0.768 ± 0.007**	**0.791 ± 0.004**	**0.813 ± 0.003**	**0.811 ± 0.004**
GCN-aux	0.670 ± 0.011	0.703 ± 0.012	0.748 ± 0.006	0.754 ± 0.006	0.780 ± 0.005	0.775 ± 0.006
Chebnet-aux	0.633 ± 0.010	0.682 ± 0.011	0.738 ± 0.006	0.747 ± 0.005	0.779 ± 0.005	0.784 ± 0.005

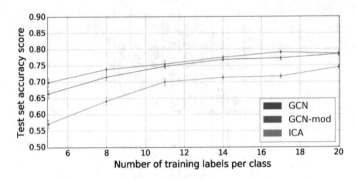

Fig. 3 Node classification accuracy score for GCN on Cora

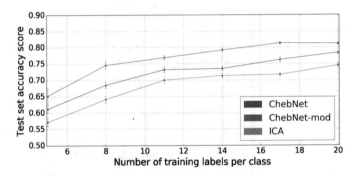

Fig. 4 Node classification accuracy score for ChebNet on Cora

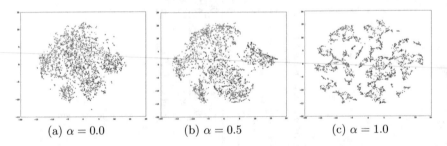

(a) $\alpha = 0.0$ (b) $\alpha = 0.5$ (c) $\alpha = 1.0$

Fig. 5 T-SNE visualization of hidden layer embeddings of second-order two-layer ChebNet trained on Cora with varying values of α

5 Conclusion

We successfully incorporate a community-structure-preserving objective in the graph convolutional semi-supervised learning framework. To the best of our knowledge, this is the first such attempt in this area. We showed that the incorporation of higher-level structural information can improve the quality of learned representations for node classification in sparse label regime. We also identified that the specific choice of filter support has a significant impact on the result. Higher-order filters tend to benefit more from the additional network-structure-preserving terms in the loss function.

Acknowledgements This work was supported by Tokyo Tech - Fuji Xerox Cooperative Research (Project Code KY260195), JSPS Grant-in-Aid for Scientific Research(B) (Grant Number 17H01785), and JST CREST (Grant Number JPMJCR1687).

References

1. Bruna, J., Zaremba, W., Szlam, A., LeCun, Y.: Spectral networks and locally connected networks on graphs (2013). arXiv:1312.6203
2. Defferrard, M., Bresson, X., Vandergheynst, P.: Convolutional neural networks on graphs with fast localized spectral filtering. In: Advances in Neural Information Processing Systems, pp. 3844–3852 (2016)
3. Hammond, D.K., Vandergheynst, P., Gribonval, R.: Wavelets on graphs via spectral graph theory. Appl. Comput. Harmon. Anal. **30**(2), 129–150 (2011)
4. Hinton, G., Deng, L., Dong, Y., Dahl, G.E., Mohamed, A., Jaitly, N., Senior, A., Vanhoucke, V., Nguyen, P., Sainath, T.N., et al.: Deep neural networks for acoustic modeling in speech recognition: the shared views of four research groups. IEEE Signal Process. Maga. **29**(6), 82–97 (2012)
5. Kipf, T.N., Welling, M.: Variational graph auto-encoders (2016). arXiv:1611.07308
6. Kipf, T.N., Welling, M.: Semi-supervised classification with graph convolutional networks (2017)
7. Krizhevsky, A., Sutskever, I., Hinton, G.E.: Imagenet classification with deep convolutional neural networks. In *Advances in Neural Information Processing Systems*, pp. 1097–1105 (2012)
8. LeCun, Y., Bottou, L., Bengio, Y., Haffner, P.: Gradient-based learning applied to document recognition. Proc. IEEE **86**(11), 2278–2324 (1998)
9. Lin, F., Cohen, W.W.: Semi-supervised classification of network data using very few labels. In: 2010 International Conference on Advances in Social Networks Analysis and Mining (ASONAM), pp. 192–199. IEEE (2010)
10. Neville, J., Jensen, D.: Iterative classification in relational data. In: Proceedings of AAAI-2000 Workshop on Learning Statistical Models from Relational Data, pp. 13–20 (2000)
11. Newman, M.E.J.: The structure and function of complex networks. SIAM Rev. **45**(2), 167–256 (2003)
12. Newman, M.E.J.: Modularity and community structure in networks. Proc. Natl. Acad. Sci. **103**(23), 8577–8582 (2006)
13. Sen, P., Namata, G., Bilgic, M., Getoor, L., Galligher, B., Eliassi-Rad, T.: Collective classification in network data. AI Mag. **29**(3), 93 (2008)
14. Tu, C., Wang, H., Zeng, X., Liu, Z., Sun, M.: Community-enhanced network representation learning for network analysis (2016). arXiv:1611.06645
15. Wang, X., Cui, P., Wang, J., Pei, J., Zhu, W., Yang, S.: Community preserving network embedding. In: AAAI, pp. 203–209 (2017)
16. Weston, J., Ratle, F., Mobahi, H., Collobert, R.: Deep learning via semi-supervised embedding. In: Neural Networks: Tricks of the Trade, pp. 639–655. Springer, Berlin (2012)
17. Yang, L., Cao, X., He, D., Wang, C., Wang, X., Zhang, W.: Modularity based community detection with deep learning. In: IJCAI, pp. 2252–2258 (2016)
18. Yang, Z., Cohen, W.W., Salakhutdinov, R.: Revisiting semi-supervised learning with graph embeddings (2016). arXiv:1603.08861

Part III
Network Dynamics

Outer Synchronization for General Weighted Complex Dynamical Networks Considering Incomplete Measurements of Transmitted Information

Xinwei Wang, Guo-Ping Jiang and Xu Wu

Abstract Outer synchronization for general weighted complex dynamical networks with randomly incomplete measurements of transmitted state variables is studied in this paper. The incomplete measurements of control information, always occurring during the transmission, should be considered seriously since it would cause the failure of outer synchronization process. Different from previous methods, we develop a new method to handle the incomplete measurements, which cannot only balance well the overly deviated controllers affected by the incomplete measurements, but also has no particular restriction on the node dynamics. Using the Lyapunov stability theory along with the stochastic analysis method, sufficient criteria are deduced rigorously to obtain the adaptive control law. Illustrative simulations are given to verify that our proposed controllers are effective and efficient dealing with the incomplete measurements.

1 Introduction

In recent years, studies on the complex dynamical network have attracted growing attention due to its ubiquity in real world [1]. Since the small-world [2] and scale-free [3] network models were proposed, more complex networks in nature can be described as simple models with their own features across many scientific and engineering fields, such as social networks [4], the Internet [5], electrical power grids [6]. Along with in-depth researches, understanding better the complexity of a network has been further divided into many detailed aspects, which include the evolution of network nodes [7], the diversity of topological structures [8], the phenomenon of

X. Wang · G.-P. Jiang (✉) · X. Wu
Nanjing University of Posts and Telecommunications,
No. 9, Wenyuan Road, Nanjing 210023, China
e-mail: jianggp@njupt.edu.cn

X. Wang
e-mail: wxw1415@163.com

X. Wu
e-mail: wxnupt@126.com

© Springer International Publishing AG 2018
S. Cornelius et al. (eds.), *Complex Networks IX*, Springer Proceedings
in Complexity, https://doi.org/10.1007/978-3-319-73198-8_12

network synchronization [9] and so on. It has been noticed that the synchronization of complex dynamical networks is worth studying emphatically. Coherent behaviors in complex dynamical networks can explain many observed phenomena and induce some underlying characteristics.

Abundant works on the synchronization of complex dynamical networks have been made so far. Employing the master stability function method, Pecora et al. first investigated the synchronization stability for a number of coupled systems [10]. Afterward, based on previous studies, it was feasible to realize the inner synchronization of a complex dynamical network only by regulating the network coupling strength [11, 12]. Additionally, applying suitable coupling strength to the network, one single controller could pin the whole complex dynamical network to a synchronous state [13]. Besides the above research results under the ideal conditions, many studies are devoted to the synchronization problem under the influence of unreliable factors, such as coupling delays [14], stochastic noise [15], incomplete measurements [16–19]. Owing to physical limitations or measurement cost, the control information, like state variables, transmitted through the channel cannot always be perfectly measured. In fact, unlike the influence brought by time delays or noise disturbance, the damage to the synchronization process caused by incomplete measurements has been seriously underestimated. The unexpected absence of control information will intensify the unbalance of controllers from a normal state and lead to a total failure of the synchronization process. Therefore, it is necessary to consider the synchronization problem with the incompletely measured control information.

Existing studies regarding the incomplete measurements have been mainly concentrated on the stability analysis [16], synchronization [17], and state estimation [18, 19] of complex dynamical networks. The Bernoulli probability distribution [16–18] was usually used to depict the incomplete measurements of transmitted information. If the sent information is incompletely measured at the receiver, the common method is just ignoring the absent information without any replacement [17–19] or replacing the vacancy with the most recently received information [16]. It is worth noting that the mentioned methods [16–19] seem to apply only to the stable networked systems or complex dynamical networks with stationary nodes, but not to the general complex dynamical networks. In their stability analysis, the Lyapunov function is designed by error states along with system states. In this way, if the error dynamical network and every single system are stabilized asymptotically at the same time, it means that the values of all the state variables must reach constant (even zero) themselves without any external control. Otherwise, the influence brought by incomplete measurements will be directly reflected on the excess deviation of controllers from the normal states where they should be, which will lead to the failure of synchronization process. Furthermore, the condition is too ideal to be satisfied in most real complex networks.

Motivated by the above discussions, we investigate the outer synchronization problem for two general weighted complex dynamical networks with the incompletely measured information in this paper. If the sent state variables are measured incompletely by the response network for some time periods, the corresponding states of the response network will be replaced during those time periods. It is effective to fix the excess deviation of controller inputs caused by the incomplete measurements. The proposed method for the outer synchronization has no special restriction on the node dynamics. Resorting to the Lyapunov stability theory together with the stochastic analysis method, sufficient criteria are deduced rigorously to design the novel adaptive controllers.

The rest parts of this paper are organized as follows. Preliminaries and problem description are given in Sect. 2. The outer synchronization for two general weighted complex dynamical networks with incomplete measurements of transmitted information is further discussed in Sect. 3. In Sect. 4, illustrative simulations are provided to verify the effectiveness of the novel adaptive controllers. Some conclusions are drawn in Sect. 5.

2 Preliminaries and Problem Description

Some useful notations are introduced first. Suppose that p, q are constant matrices of proper dimensions. $\|p\|$ denotes the Euclidean norm of p. $p \otimes q$ denotes the Kronecker product of p and q. $\lambda_{\max}(p + p^T)$ denotes the maximum eigenvalue of $(p + p^T)$. I denotes the identity matrix of proper dimensions.

Consider a general weighted complex dynamical network consisting of N non-identical nodes, which is described as

$$\dot{x}_i(t) = A_i x_i(t) + f_i(x_i(t)) + \varepsilon \sum_{j=1}^{N} c_{ij} \Gamma x_j(t), \tag{1}$$

where $i = 1, 2, \ldots, N$, $x_i(t) = [x_{i1}(t), x_{i2}(t), \cdots, x_{in}(t)]^T \in R^n$ is the state vector of the ith node. $A_i \in R^{n \times n}$ is the system matrix of the ith node, and $f_i : R^n \to R^n$ is a continuously vector-valued function of that. The full dynamics of the ith isolated node is governed by A_i and f_i. ε is the coupling strength of the complex dynamical network, and $\Gamma \in R^{n \times n}$ is the inner coupling matrix of that. $C = (c_{ij})_{N \times N} \in R^{N \times N}$ denotes the configuration matrix which represents the topological structure of the complex dynamical network. If there exists a directed connection from node j to node i ($i \neq j$), then $c_{ij} \neq 0$; otherwise, $c_{ij} = 0$. The diagonal elements $\{c_{ii} \mid i = 1, 2, \ldots, N\}$ of C are assumed to satisfy $c_{ii} = -\sum_{j=1, j \neq i}^{N} c_{ij}$.

Consider another weighted complex dynamical network (2) consisting of N non-identical nodes, in which the node dynamics is assumed to be the same as that of the complex dynamical network (1).

$$\dot{y}_i(t) = A_i y_i(t) + f_i(y_i(t)) + \varepsilon \sum_{j=1}^{N} c_{ij} \Gamma y_j(t), \tag{2}$$

where $y_i(t) = [y_{i1}(t), y_{i2}(t), \cdots, y_{in}(t)]^T \in R^n$ is the state vector of the ith node in the network (2). The other network parameters are the same as the complex dynamical network (1). In order to achieve the outer synchronization between the complex dynamical networks (1) and (2), we take the network (1) as the drive network and accordingly the network (2) as the response one. The adaptive feedback controllers $u_i(t)$ are designed using the state variables of the drive and response networks, which could be expressed by [16–19]

$$\begin{cases} \dot{y}_i(t) = A_i y_i(t) + f_i(y_i(t)) + \varepsilon \sum_{j=1}^{N} c_{ij} \Gamma y_j(t) + u_i(t), \\ u_i(t) = -k_i \left(y_i(t) - x_i^\alpha(t) \right), \end{cases} \tag{3}$$

where $\{k_i\}$ are the adaptive feedback gains of synchronization controllers. $x_i^\alpha(t)$ is the state variables which is received by the response network (3). It is different from the original states $x_i(t)$ sent from the drive network (1) since the incomplete measurements have taken place in the transmission process. For simulating the common perturbation of transmitted information occurred in real circumstances, a random variable $\alpha_i(t)$ is introduced as follows.

$$\begin{cases} \text{Prob}\,\{\alpha_i(t) = 1\} = E\,\{\alpha_i(t)\} = \bar{\alpha}_i, \\ \text{Prob}\,\{\alpha_i(t) = 0\} = 1 - E\,\{\alpha_i(t)\} = \bar{\beta}_i, \end{cases} \tag{4}$$

where $\alpha_i(t) \in R$ is the random Bernoulli-distributed variable. For instance, as shown in Fig. 1, $\alpha_i(t) = 1$ represents that the sent state variable of the ith node is measured completely by the response network during the time period $t \in (t_1, t_2) \bigcup (t_3, t_4)$. Otherwise, if the sent information is measured incompletely, then $\alpha_i(t) = 0$ during the time period $t \in (t_0, t_1) \bigcup (t_2, t_3) \bigcup (t_4, \infty)$. $\{\alpha_i(t) \,|\, i = 1, 2, \ldots, N\}$ are mutually independent and identically distributed to each other. $\bar{\alpha}_i \in R$ denotes the mathematical expectation of the random variable $\alpha_i(t)$. In real networks, $\alpha_i(t)$ can be detected at any time since there always exists a mechanism for detecting whether the transmitted information is received or not. In the existing studies [16–19], $x_i^\alpha(t)$ was described as

$$x_i^\alpha(t) = \alpha_i(t) x_i(t). \tag{5}$$

Fig. 1 An illustrative example of the random variable $\alpha_i(t)$

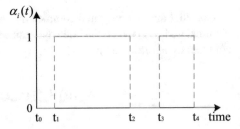

For the purpose of eliminating the influence brought by the incomplete measurements, most previous methods [16–19] employed the controllers shown in (3) for the outer synchronization. However, the random incomplete measurements for some time periods would make these controllers deviate overly from the normal states unless the node dynamics reaches a stationary state. The excess quantity of input variables to the controllers would prevent the outer synchronization between different complex dynamical networks.

Motivated by the above discussions, we present a novel adaptive controller in the following (6), which can handle well with the incomplete measurements of transmitted information for outer synchronization between different complex dynamical networks:

$$
\begin{cases}
u_i(t) = -k_i \left(y_i(t) - \bar{x}_i(t) \right), \\
\bar{x}_i(t) = \alpha_i(t) x_i(t) + (1 - \alpha_i(t)) y_i(t).
\end{cases}
\tag{6}
$$

For instance, as shown in Fig. 1, if the sent state variable $x_i(t)$ from the drive network (1) is incompletely measured during the time period (t_2, t_3), the state variable $y_i(t)$ in the response network (3) will replace the absent $x_i(t)$ right for the time period (t_2, t_3). It will fix in time the excess deviation of synchronization controllers from the normal state owing to the incomplete measurements.

Let $e_i(t) = y_i(t) - x_i(t)$, and then the error dynamical network (7) could be obtained from the drive network (1) and response network (3) with the novel controller (6).

$$
\begin{aligned}
\dot{e}_i(t) &= \dot{y}_i(t) - \dot{x}_i(t) \\
&= A_i e_i(t) + f_i \left(y_i(t) \right) - f_i \left(x_i(t) \right) + \sum_{j=1}^{N} c_{ij} \Gamma e_j(t) - k_i \alpha_i(t) e_i(t).
\end{aligned}
\tag{7}
$$

In order to stabilize the error dynamical network (7) as well as obtain the outer synchronization, one suitable assumption is introduced here. The nonlinear functions $\{ f_i(\cdot) \,|\, i = 1, 2, \ldots, N \}$ are assumed to be continuous and satisfy the following condition: there exist positive constants $\{ \mu_i \,|\, i = 1, 2, \ldots, N \}$ such that

$$
\| f_i \left(z_1(t) \right) - f_i \left(z_2(t) \right) \| \leq \mu_i \| z_1(t) - z_2(t) \|,
\tag{8}
$$

which hold for any vectors $z_1(t), z_2(t) \in R^n$.

3 Main Results

In this section, by employing the Lyapunov stability theory and stochastic analysis method, the main results of the outer synchronization between different weighted complex dynamical networks with incompletely measured information are given in the following.

Theorem 1 *Suppose that the assumption (8) holds. Use the following adaptive law*

$$\dot{k}_i = \delta_i \bar{\alpha}_i e_i^T(t) e_i(t).$$ (9)

If there exists a positive constant k^* that is sufficiently large, then the response network (3) will synchronize with the drive network (1) under the proposed controllers (6), i.e.,

$$\lim_{t \to \infty} \| e_i(t) \| = \lim_{t \to \infty} \| y_i(t) - x_i(t) \| = 0,$$

where δ_i is a positive constant which is used to adjust the amplitude of the error variables and the speed of the outer synchronization process.

Proof Choose the scalar Lyapunov candidate function V as follows.

$$V = \sum_{i=1}^{N} e_i^T(t) e_i(t) + \sum_{i=1}^{N} \frac{1}{\delta_i} (k_i - k^*)^2.$$ (10)

Taking the form of mathematical expectation, the derivative of V is deduced in (11) with the novel adaptive controller (6), and one gets

$$
\begin{aligned}
E(\dot{V}) &= \sum_{i=1}^{N} \left(e_i^T(t) \dot{e}_i(t) + \dot{e}_i^T(t) e_i(t) \right) + 2 \sum_{i=1}^{N} \frac{1}{\delta_i} (k_i - k^*) \dot{k}_i \\
&= \sum_{i=1}^{N} e_i^T(t) \left(A_i e_i(t) + f_i(y_i(t)) - f_i(x_i(t)) + \sum_{j=1}^{N} c_{ij} \Gamma e_j(t) \right) \\
&\quad + \sum_{i=1}^{N} \left(A_i e_i(t) + f_i(y_i(t)) - f_i(x_i(t)) + \sum_{j=1}^{N} c_{ij} \Gamma e_j(t) \right)^T e_i(t) \\
&\quad - 2 \sum_{i=1}^{N} k_i \bar{\alpha}_i e_i^T(t) e_i(t) + 2 \sum_{i=1}^{N} \frac{1}{\delta_i} k_i \dot{k}_i - 2 \sum_{i=1}^{N} \frac{1}{\delta_i} k^* \dot{k}_i,
\end{aligned}
$$ (11)

Together with the assumption (8) and the adaptive law (9), one has

$$
\begin{aligned}
E(\dot{V}) &= \sum_{i=1}^{N} \left(e_i^T(t) \dot{e}_i(t) + \dot{e}_i^T(t) e_i(t) \right) + 2 \sum_{i=1}^{N} \frac{1}{\delta_i} (k_i - k^*) \dot{k}_i \\
&= \sum_{i=1}^{N} e_i^T(t) \left(A_i + A_i^T \right) e_i(t) - 2 \sum_{i=1}^{N} e_i^T(t) k_i \bar{\alpha}_i e_i(t)
\end{aligned}
$$

$$+ 2 \sum_{i=1}^{N} \frac{1}{\delta_i} k_i \dot{k}_i - 2 \sum_{i=1}^{N} \frac{1}{\delta_i} k^* \dot{k}_i + \sum_{i=1}^{N} e_i^T(t) \left(\sum_{j=1}^{N} c_{ij} \Gamma e_j(t) \right)$$

$$+ \sum_{i=1}^{N} \left(\sum_{j=1}^{N} c_{ij} \Gamma e_j(t) \right)^T e_i(t) + \sum_{i=1}^{N} e_i^T(t) \left(f_i(y_i(t)) - f_i(x_i(t)) \right) \quad (12)$$

$$+ \sum_{i=1}^{N} \left(f_i(y_i(t)) - f_i(x_i(t)) \right)^T e_i(t)$$

$$\leq \sum_{i=1}^{N} e_i^T(t) \left(A_i + A_i^T + 2\mu_i I - 2k^* \bar{\alpha}_i I \right) e_i(t)$$

$$+ \sum_{i=1}^{N} \left(e_i^T(t) \left(\sum_{j=1}^{N} c_{ij} \Gamma e_j(t) \right) + \left(\sum_{j=1}^{N} c_{ij} \Gamma e_j(t) \right)^T e_i(t) \right),$$

Let $e(t) = \left[e_1^T(t), e_2^T(t), \ldots, e_N^T(t) \right]^T \in R^N$, $P = C \otimes \Gamma$, and one obtains

$$E(\dot{V}) \leq e^T(t) \left(A + A^T + 2\mu I - 2k^* \alpha I + C \otimes \Gamma + C^T \otimes \Gamma^T \right) e(t)$$
$$\leq \left(\lambda_{\max}(A + A^T) + 2\mu + \lambda_{\max}(P + P^T) - 2k^* \alpha \right) e^T(t) e(t), \quad (13)$$

where

$$A = diag(A_1, A_2, \ldots, A_N),$$
$$\mu = diag(\mu_1, \mu_2, \ldots, \mu_N) \otimes I,$$
$$\alpha = diag(\bar{\alpha}_1, \bar{\alpha}_2, \ldots, \bar{\alpha}_N) \otimes I.$$

Taking $k^* = \frac{1}{2\alpha} \left(\lambda_{\max}(A + A^T) + 2\mu + \lambda_{\max}(P + P^T) + 1 \right)$, it is proved that the sufficiently large k^* does exist. Then, one has $E(\dot{V}) \leq -e^T(t) e(t) < 0$ holding for any $e(t) \neq 0$. Only if $e(t) = 0$, then $E(\dot{V}) = 0$. Based on the Lyapunov stability theory, the error dynamical network (7) is asymptotically stabilized at the origin, which means the drive network (1) and response network (3) achieve the outer synchronization with the incomplete measurements of transmitted state variables under the proposed controllers (6). The proof is completed.

4 Numerical Simulations

In this section, numerical simulations are given to demonstrate the effectiveness of the novel adaptive controllers (6) that handle well the random incomplete measurements of transmitted information. The Lorenz system (14) is taken as the node dynamics since its irregular behavior could increase the difficulty of the outer synchronization.

Additionally, as a chaotic system, the Lorenz system is extremely sensitive to initial values, which is convenient to construct different networks for the simulation:

$$\begin{cases} \dot{x}_{i1}(t) = a\,(x_{i2}(t) - x_{i1}(t)), \\ \dot{x}_{i2}(t) = cx_{i1}(t) - x_{i2}(t) - x_{i1}(t)x_{i3}(t), \\ \dot{x}_{i3}(t) = x_{i1}(t)x_{i2}(t) - bx_{i3}(t), \end{cases} \tag{14}$$

where $a = 10, b = 8/3, c = 28$. The assumption (8) is clearly satisfied since chaotic attractors are bounded in a certain region [20]. Consider a weighted complex dynamical network (15) composed of six chaotic nodes.

$$\dot{x}_i(t) = Ax_i(t) + f\,(x_i(t)) + \varepsilon \sum_{j=1}^{6} c_{ij}\Gamma x_j(t), \tag{15}$$

where $A = \begin{bmatrix} -10 & 10 & 0 \\ 28 & -1 & 0 \\ 0 & 0 & -8/3 \end{bmatrix}$, $\Gamma = \begin{bmatrix} 1 & 0 & 0 \\ 0 & 1 & 0 \\ 0 & 0 & 1 \end{bmatrix}$, $\varepsilon = 0.1$. The topological structure of the network (15) represented by the matrix C is shown in Fig. 2.

$$C = (c_{ij})_{6\times6} = \begin{bmatrix} -2 & 0 & 0 & 2 & 0 & 0 \\ 0 & -2 & 1 & 1 & 0 & 0 \\ 0 & 0 & -1 & 0 & 0 & 1 \\ 0 & 0 & 0 & 0 & 0 & 0 \\ 1 & 0 & 2 & 0 & -5 & 2 \\ 0 & 0 & 0 & 0 & 1 & -1 \end{bmatrix}.$$

Consider another weighted complex dynamical network (16), whose initial values of state variables are chosen differently from (15).

Fig. 2 Topological structure of complex dynamical networks (15)

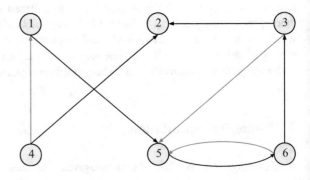

$$\begin{cases} \dot{y}_i(t) = A y_i(t) + f\left(y_i(t)\right) + \varepsilon \sum_{j=1}^{6} c_{ij} \Gamma y_j(t) + u_i(t), \\ u_i(t) = -k_i \left(y_i(t) - \bar{x}_i(t)\right), \\ x_i^\alpha(t) = \alpha_i(t) x_i(t). \\ \bar{x}_i(t) = x_i^\alpha(t) + (1 - \alpha_i(t)) y_i(t), \\ e_i(t) = y_i(t) - x_i(t), \\ \dot{k}_i = \delta_i \bar{\alpha} e_i^T(t) e_i(t). \end{cases} \tag{16}$$

The mathematical expectations of random variables $\{\alpha_i(t) \mid i = 1, 2, \ldots, 6\}$ are assumed to be the same as $\bar{\alpha} = 0.7$, and $\{\delta_i \mid i = 1, 2, \ldots, 6\}$ are set as $\delta = 2$ for simplicity. The initial values of state variables in the networks (15) and (16) are set randomly in the interval $(0, 1)$. The outer synchronization process of the networks (15) and (16) is shown in Fig. 3.

From Fig. 3, it is easy to find that the dynamical error variables of the corresponding nodes in the networks (15) and (16) converge quickly to the origin with random incomplete measurements of transmitted state variables. Figure 4 shows the

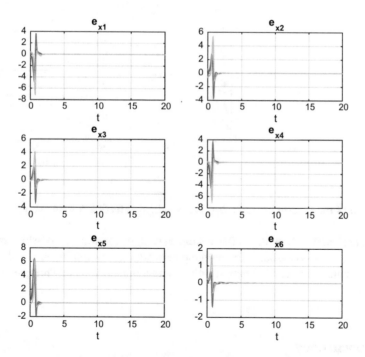

Fig. 3 Error variables of corresponding nodes $\{x_i \mid i = 1, 2, \ldots, 6\}$ in the outer synchronization process

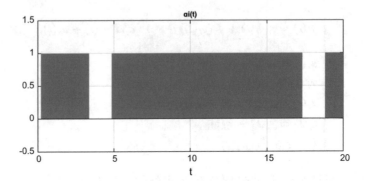

Fig. 4 Diagram of the random variable $\alpha_i(t)$ versus time t

Fig. 5 Adaptive feedback
gains $\{k_i\}$

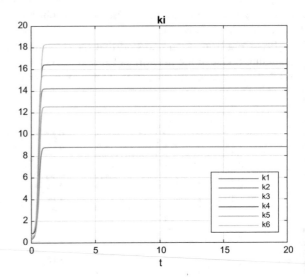

evolution of the random variable $\alpha_i(t)$. The adaptive feedback gains $\{k_i\}$ are shown
in Fig. 5.

Figures 3 and 4 show that, after a short period, the network (16) achieves completely the outer synchronization with the network (15) by employing the designed
adaptive controllers which function well with random incompletely measured information.

5   Conclusions

In this paper, we have investigated the outer synchronization for general weighted
complex dynamical networks with incomplete measurements of transmitted state
variables. We have proposed a novel method that performs well to balance the overly

deviated controllers affected by the incomplete measurements. Moreover, compared to most previous methods, it has no special restriction on the node dynamics. Sufficient criteria are deduced rigorously for designing the adaptive controllers to obtain the outer synchronization. Numerical simulations are given to verify the effectiveness of our proposed method dealing with the incomplete measurements of transmitted information.

Acknowledgements This work is supported by the National Natural Science Foundation of China (Grant Nos. 61374180, 61373136).

References

1. Strogatz, S.H.: Exploring complex networks. Nature **410**(6825), 268–276 (2001)
2. Watts, D.J., Strogatz, S.H.: Collective dynamics of 'small-world' networks. Nature **393**(6684), 440–442 (1998)
3. Barabási, A.L., Albert, R.: Emergence of scaling in random networks. Science **286**(5439), 509–512 (1999)
4. Barabási, A.L., Jeong, H., Néda, Z., Ravasz, E., Schubert, A., Vicsek, T.: Evolution of the social network of scientific collaborations. Phys. A Stat. Mech. Appl. **311**(3), 590–614 (2002)
5. Albert, R., Barabási, A.L.: Statistical mechanics of complex networks. Rev. Mod. Phys. **74**(1), 47 (2002)
6. Simonsen, I., Buzna, L., Peters, K., Bornholdt, S., Helbing, D.: Transient dynamics increasing network vulnerability to cascading failures. Phys. Rev. Lett. **100**(21), 218701 (2008)
7. Boccaletti, S., Latora, V., Moreno, Y., Chavez, M., Hwang, D.U.: Complex networks: structure and dynamics. Phys. Rep. **424**(4), 175–308 (2006)
8. Krioukov, D., Papadopoulos, F., Kitsak, M., Vahdat, A., Boguná, M.: Hyperbolic geometry of complex networks. Phys. Rev. E **82**(3), 036106 (2010)
9. Arenas, A., Daz-Guilera, A., Kurths, J., Moreno, Y., Zhou, C.: Synchronization in complex networks. Phys. Rep. **469**(3), 93–153 (2008)
10. Pecora, L.M., Carroll, T.L.: Master stability functions for synchronized coupled systems. Phys. Rev. Lett. **80**(10), 2109–2112 (1998)
11. Wang, X.F., Chen, G.: Synchronization in scale-free dynamical networks: robustness and fragility. IEEE Trans. Circuits Syst. I Fundam. Theory Appl. **49**(1), 54–62 (2002)
12. Wang, X.F., Chen, G.: Synchronization in small-world dynamical networks. Int. J. Bifurc. Chaos **12**(01), 187–192 (2002)
13. Chen, T., Liu, X., Lu, W.: Pinning complex networks by a single controller. IEEE Trans. Circuits Syst. I Regul. Pap. **54**(6), 1317–1326 (2007)
14. Nuno, E., Ortega, R., Basanez, L., Hill, D.: Synchronization of networks of nonidentical Euler-Lagrange systems with uncertain parameters and communication delays. IEEE Trans. Autom. Control. **56**(4), 935–941 (2011)
15. Korniss, G.: Synchronization in weighted uncorrelated complex networks in a noisy environment: optimization and connections with transport efficiency. Phys. Rev. E **75**(5), 051121 (2007)
16. Du, D., Fei, M., Jia, T.: Modelling and stability analysis of MIMO networked control systems with multi-channel random packet losses. Trans. Inst. Meas. Control. **35**(1), 66–74 (2013)
17. Li, J.N., Bao, W.D., Li, S.B., Wen, C.L., Li, L.S.: Exponential synchronization of discrete-time mixed delay neural networks with actuator constraints and stochastic missing data. Neurocomputing **207**, 700–707 (2016)
18. Han, F., Wei, G., Ding, D., Song, Y.: Finite-horizon bounded $H\infty$ synchronisation and state estimation for discrete-time complex networks: local performance analysis. IET Control. Theory Appl. **11**(6), 827–837 (2017)

19. Liu, M., Chen, H.: $H\infty$ state estimation for discrete-time delayed systems of the neural network type with multiple missing measurements. IEEE Trans. Neural Netw. Learn. Syst. **26**(12), 2987–2998 (2015)
20. Li, D., Wu, X., Lu, J.A.: Estimating the ultimate bound and positively invariant set for the Lorenz system and a unified chaotic system. J. Math. Anal. Appl. **323**(2), 844–853 (2006)

Diffusive Phenomena in Dynamic Networks: A Data-Driven Study

Letizia Milli, Giulio Rossetti, Dino Pedreschi and Fosca Giannotti

Abstract Everyday, ideas, information as well as viruses spread over complex social tissues described by our interpersonal relations. So far, the network contexts upon which diffusive phenomena unfold have usually been considered static, composed by a fixed set of nodes and edges. Recent studies describe social networks as rapidly changing topologies. In this work — following a data-driven approach — we compare the behaviors of classical spreading models when used to analyze a given social network whose topological dynamics are observed at different temporal granularities. Our goal is to shed some light on the impacts that the adoption of a static topology has on spreading simulations as well as to provide an alternative formulation of two classical diffusion models.

1 Introduction

Since the last decade, we are living two lives at the same time: one offline and one online. One of the facilities the WWW has granted us is the dismantling of physical distances, thus impacting the way diffusive phenomena evolve.

In the real world, we are discussing the spread of viruses such as *passive* contagion processes that do not require active agents to unfold. The diffusion of ideas, conversely, is an example of *active* process: Each can choose to adopt/advertise

L. Milli (✉) · D. Pedreschi
University of Pisa, Largo Bruno Pontecorvo, 2 Pisa, Italy
e-mail: letizia.milli@di.unipi.it, letizia.milli@isti.cnr.it

D. Pedreschi
e-mail: dino.pedreschi@di.unipi.it

L. Milli · G. Rossetti · F. Giannotti
KDD Lab. ISTI-CNR, via G. Moruzzi, 1 Pisa, Italy
e-mail: giulio.rossetti@isti.cnr.it

F. Giannotti
e-mail: fosca.giannotti@isti.cnr.it

© Springer International Publishing AG 2018
S. Cornelius et al. (eds.), *Complex Networks IX*, Springer Proceedings
in Complexity, https://doi.org/10.1007/978-3-319-73198-8_13

a new idea or not. When we move to the online world, we can experience both *passive* and *active* diffusion. These processes occur on top of social structures that have often been considered static. However, both passive and active processes require a direct contact with a content to spread from an already *infected* person to a susceptible one. Social interactions have a limited duration so that they dynamically shape the topology of our social graph.

In this work, we tackle the problem of understanding if, and how, dynamic network topology affects the diffusion of information. Is a static social network representation enough to simulate information spreading? Must topology dynamics be taken into account to understand the real diffusive phenomena better?

2   Related Works

Two different, yet related topics need to be reviewed and discussed: information spreading and dynamic social networks analysis.

Information Spreading. When we use the word *"spreading"* we think contagious diseases caused by biological pathogens. However, a plethora of phenomena can be linked to the concept of the epidemic: such as the spread of computer viruses [1], mobile phone virus [2], or the diffusion of knowledge in an online social network [3]. Here, we focus on the diffusion of innovations/idea. Rogers developed the diffusion of innovation theory in 1962 [4]: It aims to explain how an idea or product diffuses through a specific population or social system.

Dynamic Social Networks. With the explosion of human-generated data, the time has started representing a non-negligible entity. During the last decade, several works have provided novel interpretations of known problems, porting them from static to temporal networks: Motifs mining [5], link prediction [6], community discovery [7] are only a few examples. Indeed, [8] showed that it is mandatory to consider different granularity of temporal abstraction. Once understood the importance of ties dynamics for the overall network topology it becomes natural to study how they affect spreading phenomena.

Spreading on Dynamic Networks. Recently, the analysis of diffusive processes in dynamic networks has started to capture the attention of the research community, such as in [9] or [10] where the authors used the SI and SIR model, respectively, in dynamic contests. [11] and [12] are some of the few investigations of how dynamic networks affect the spread of information. Finally, in [13] a data-driven study similar to ours was performed. However, the authors were forced to synthesize network topology evolution, thus making impossible to observe the impact of characteristic phenomenon events on the diffusive process.

3 Problem Definition

Our analysis will be focused on answering the following questions:

Q1 : Can analyzing spreading phenomena on a static social graph lead to an over-estimate of the real volume of its diffusion?

Q2 : Do the choices made to keep track of topology dynamics impact the speed of diffusive processes?

Q3 : Is it safe to assume that spreading phenomena on a dynamic network topology unfold at a constant rate? Do the variations, as the diffusion progresses, of the number of nodes/edges impact the overall diffusion process?

To address such questions, we define three different scenarios. We model a network as an undirected graph denoted as $G = (V, E)$, where V is the set of the nodes and E is a set of interactions (edges), i.e., a triplet (u, v, t) where $u, v \in V$ and $t \in \mathbb{N}$ identify the time at which an *interaction* occurs between nodes u to v. We allow the presence of multiple interactions among the same pair of nodes. In the following, we will denote with E_{t_j} the set of interactions that appears in the graph at time t_j. We can formalize the problem in the following way:

Definition 1 (*Spreading problem*) Given a network $G = (V, E)$ observed for k consecutive snapshots, a diffusion model $\mathcal{D}$, and a set $I_{t_0} = \{n_1, n_2, ..., n_j\} \subseteq V$ identifying the initial infected nodes, we define the result of $\mathcal{D}(G, I_{t_0})$ as the ordered sequence $\mathcal{I} = \{I_{t_1}, ..., I_{t_k}\}$ of the nodes infected during each snapshot.

The scenarios we will analyze in our data-driven investigation are:

–S1 — Static topology. For each time t_i with $i = 1, ..., k$, we applied $\mathcal{D}$ to the full network $G = (V, E)$ using as infected node set at time t_i the result of $\mathcal{D}(G, I_{t_{i-1}})$. The set of edges will be $E = E_{t_1} \cup E_{t_2} \cup ... \cup E_{t_k}$.

–S2 — Snapshot evolution. For each time t_i with $i = 1, ..., k$, we compute $\mathcal{D}(G_{t_i}, I_{t_{i-1}})$ where $G_{t_i} = (V, E_{t_i})$.

–S3 — Interaction dynamics. For each time t_i with $i = 1, ..., k$, we apply $\mathcal{D}$ incrementally to the ordered stream of interaction in E_{t_i}.

In **S1**, a network will be built flattening all the interactions occurred in a single one, thus describing dynamic phenomena with a static structure. In **S2**, a network will be built for each snapshot and the spreading process computed on each one of them starting, incrementally, from the previous infection status. Finally, in **S3** all the interactions among nodes that occur during each snapshot will be analyzed in their temporal ordering: No network will be explicitly built.

Table 1 Base statistics of the analyzed interaction graphs

Network	Nodes	Interactions	Edges	#Observation
WEIBO	1 656 615	6 759 012	3 394 566	90 days
FB07	19 561	304 392	67 077	365 days

4 Data-Driven Study

To address our research questions, we used the following datasets:

WEIBO[1]: It is based on data from the popular Chinese micro-blog service WEIBO[2]. An interaction represents a direct message from two users. We selected the first 90 days of the year 2011.

Facebook: The FB07 network is a sample of the WOSN2009 [14] dataset and describes online interactions between Facebook users during 2007.

In Table 1 are reported the main statistics of the networks.

On such datasets, we simulated two classical compartmental models SI and SIR detailed in Sect. 4.1. For each scenario, in Sect. 4.2 we compared the diffusion trends obtained while varying network dynamic and the model's parameters; in Sect. 5, we discuss our results and underline their relations with the topology dynamic.

4.1 Diffusion Models

We chose SI and SIR to describe two different information diffusion scenarios:

D1 — Continuous advertising: After having adopted an idea/innovation, an agent continues to advertise it to its neighbors during each interaction;

D2 — Diminishing advertising: After having adopted an idea/innovation, an agent can decide to stop advertising it to its neighbors.

Since both models have been described for complete networks and static graphs, we will describe the modifications to apply them to the S2 and S3 scenarios.

SI: This model was introduced in 1927 by Kermack [15]. During the epidemics, an individual can belong to two states, *infected* (I) and *susceptible* (S); we adopt SI to simulate diffusion scenario D1. SI assumes that if a susceptible node comes into contact with an infected one, it becomes infected with probability β.

---

[1]http://www.wise2012.cs.ucy.ac.cy/challenge.html
[2]http://weibo.com

Algorithm 1 Interaction-based SI

Require: I_{t_0}: set of initial infected node
1: **for each** t_i in $\{1,...,k\}$ **do**
2:   $I_{t_i} = I_{t_{i-1}}$
3:   **for each** interaction (u, v, t_i) in E_{t_i} **do**
4:     **if** v in $I_{t_{i-1}}$ **then**
5:       $p = rand(0, 1)$ ▷ Random value in $[0,1]$
6:       **if** $\beta > p$ **then**
7:         add u to I_{t_i}
8:       **end if**
9:     **end if**
10:   **end for**
11:   **yield** I_{t_i} ▷ Return daily status
12: **end for**

S1: Static network. For every day t_i, each node $u \in V$ having at least an infected neighbor is evaluated to decide if it will become infected or not. SI sets the probability of infection for a node having n infected neighbors as $n\beta$: The more the infected neighbors a node has the higher its chance to join the I set.

S2: Snapshot-based evolution. The model applied at day t_i will use $I_{t_{i-1}}$ and $E_{t_{i-1}}$. Therefore, the node sets and the interactions of consecutive snapshot could vary. Naturally, the nodes not present during t_i do not take part in the diffusion process at time t_i. The probability of infection for a node u is $n_{t_i}\beta$ with $n_{t_i} \leq n$ restricting the set of infected neighbors to the ones that are present at time t_i.

S3: Interaction-based evolution. We can imagine such scenario as word of mouth spreading phenomena in which an idea or behavior can be shared/adopted only through a direct contact. We implement streaming SI as shown in Algorithm 1. In this model, an actor u involved into m interactions with infected nodes during the day t_i has a probability of infection equal to $\sum_{i=1}^{m} \beta$.

SIR: This model represents a variation of the previous one. Each node belongs to three states during the epidemics: the state *infected I*, *susceptible S*, and *removed R*. We adopt SIR to simulate diffusion scenario D2.

S1: Static network. We applied the classical formulation of the model on the flattened static graph. In SIR the idea/innovation is adopted with a $n\beta$ probability. Moreover, during each iteration, the probability that an infected node decides to stop advertising to its neighbors — thus joining the R set — is γ.

S2–S3: Dynamic networks. To comply with the topology dynamics described by S2 and S3, we adopted the SIR model with the same rationales used for SI. We omit the pseudocode for the interaction-based version of SIR since it differs from the one reported in Algorithm 1 solely for the evaluation of the removal probability γ.

4.2 Diffusion Analysis

We organized our simulations as follows:

-i: For each dataset, we randomly selected 10 sets of nodes each one covering 5% of the V: Such sets identify I_{t_0};

-ii: For each dataset, scenario, and initially infected status, we executed the SI and SIR models while setting their parameters;

-iii: We build the infection trend as the iteration-wise average of the runs over the 10 executions performed varying the initially infected nodes.

D1 — Continuous advertising. Figure 1 shows the results obtained by the simulation of the SI model on the two datasets.

Scenario S1. In WEIBO, Fig. 1a, setting $\beta = 0.01$ leads to an epidemic state covering almost 70% of the nodes. Increasing the values, a significant speedup in the diffusion process allows reaching almost the 80% of the nodes, after only 15–20 iterations. In FB07, Fig. 1d, the impact of β is more evident: A slight increase doubles the number of nodes infected after the first 50 iterations.

Scenario S2. This scenario leads to a significant reduction of the diffusion speed; in both WEIBO and FB07, the infection trends do not reach saturation. Observing the FB07 trend, Fig. 1e, only for $\beta = 0.5$ we can reach a final percentage of infected nodes "comparable" to the lowest one obtained by the same model on S1. In WEIBO, Fig. 1b, the pattern is similar.

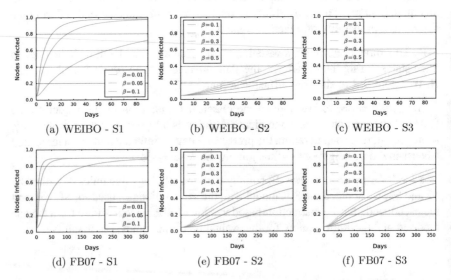

(a) WEIBO - S1   (b) WEIBO - S2   (c) WEIBO - S3

(d) FB07 - S1   (e) FB07 - S2   (f) FB07 - S3

Fig. 1 Simulation of SI models on both WEIBO and FB07: The curves represent the average percentage of infected nodes over time while varying the model parameter

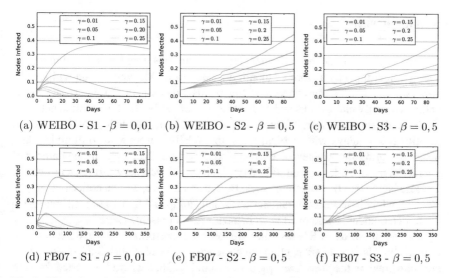

Fig. 2 Simulation of SIR models on both WEIBO and FB07: The curves represent the average percentage of infected nodes over time while varying the model parameters

Scenario S3. We observe a behavior similar to the one identified in S2; however, in this scenario the infection trends grow always faster than the ones in S2. Such speedup is due to the different way the probability of infection is calculated: In S2 a node having n infected neighbors is subject to a $n\beta$ probability of being infected, and in S3 the probability equals to $\sum_{i=1}^{m} \beta$ (where $m \geq n$ since during the same day multiple interactions can occur among the same pair of nodes).

D2 — Diminishing advertising. Figure 2 shows the results obtained by the simulation of the SIR model.

Scenario S1. In the simulation with SI, the diffusion reached in S1 with $\beta = 0.01$ is reachable in S2 and S3 when $\beta = 0.5$; so, we instantiate SIR fixing $\beta = 0.01$. In both datasets, we observe, Fig. 2a, d, the classic decay experienced by the infection trend in a SIR model. With lower values of γ ($\gamma = 0.01$), we found a rapid growth in the first period followed by a period where it rapidly decreased. For $\gamma >> \beta$, the growing phase is not present since all the initial infected nodes are more likely to being removed than to spread the infection.

Scenarios S2–S3. In Fig. 2b, c, e, f, we report for S2 and S3 the infection trends for $\beta = 0.5$. Similar to what happened in S1, for values of γ comparable to the β ones the trend curves steadily die out. However, the velocity of both infection and recovery diffusions is extremely lower w.r.t. the ones in S1.

5 Discussion

Our results suggest that the particular characteristics possessed by a dynamic system deeply affect the way a word of mouth diffusion of an idea/innovation will spread. Now, we concentrate our analysis on the WEIBO dataset. In Fig. 3a are shown the patterns of daily interactions and node presences of the WEIBO interaction network. Such trends show an overall increase in the number of interactions and nodes. We identify the Sundays with vertical lines; the WEIBO users tend to diminish their presence during the weekends. We can also observe a sharp peak in the number of interactions and nodes on the 34th day: Such day, 3 February 2011, identified the Chinese New Year. If we examine Figs. 1a, b, c and 2a, b, c, we can notice that in both S2 and S3, for all the tested parameters, a "small" jump highlights a sudden increase in the infected nodes, while in S1 such behavior is not present. Therefore, by adopting a flattened graph as in S1, not only we get an overestimate of the percentage of infected but also we do not capture the presence of special events. Such observations are confirmed by the *prevalence* plots shown in Fig. 3b, c where is reported for each day the number of novel infected nodes for SI and SIR respectively.

Once compared the diffusion trends in the three identified scenarios we can now provide answers to the research questions raised in Sect. 3:

A1: Yes, using an aggregate, static graph leads to an overestimate of the real network connectivity and, as a consequence, of all the diffusion processes.

A2: Yes, different temporal granularities for topology dynamics aggregation (e.g., snapshots and interactions) cause different spreading velocities.

A3: No, peculiar topology evolution patterns or the chosen diffusion model affects the rate of infection. In particular, cyclic patterns (weekend/weekdays) or special events (the Chinese New Year) characterize the rate at which diffusion occurs in SI, while the former loses its relevance with a SIR model.

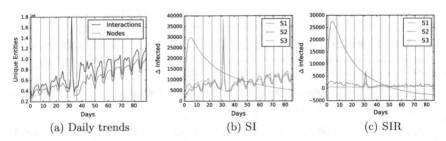

| (a) Daily trends | (b) SI | (c) SIR |

Fig. 3 **a** Daily trends in the WEIBO. Vertical lines identify Sundays. **b, c** Delta infection trend in SI (a) and SIR (b). The trends compare models having the following parameter settings — SI: S1 $\beta = 0.01$, S2–S3 $\beta = 0.5$; SIR: S1 $\beta = 0.01$ $\gamma = 0.01$, S2–S3: $\beta = 0.5$ $\gamma = 0.01$

6 Conclusions

In this work, we analyzed diffusive phenomena on dynamic social interaction graphs. We performed a data-driven study aimed to underline the real impact of network dynamics. After having modeled three different scenarios, we studied their impact on the outcome produced by classical compartmental models that we redefined to handle topology dynamics[3]. Our results show that analyzing diffusive phenomena without considering topology dynamic lead to relevant over estimate of the real speed and not capture the presence of special events.

As future work, we plan to study the other side of the problem; namely, the impact diffusive processes have on network topology.

Acknowledgments This work is funded by the EU's H2020 Program under the funding scheme "FETPROACT-1-2014: Global Systems Science (GSS)," grant agreement # 641191 CIMPLEX and under the scheme "INFRAIA-1-2014-2015: Research Infrastructures," grant agreement # 654024 *"SoBigData"*.

References

1. Szor, P.: Fighting computer virus attacks. In: USENIX (2004)
2. Havlin, S.: Phone infections. Science **324**, 1023–1024 (2009)
3. Burt, R.S.: Social contagion and innovation: cohesion versus structural equivalence. AJS **92**, 1287–1335 (1987)
4. Rogers, E.M.: Diffusion of Innovations. (1962)
5. Kovanen, L., Karsai, M., Kaski, K., Kertész, J., Saramäki, J.: Temporal motifs in time-dependent networks. J. Stat. Mech. Theory Exp. **2011**(11), (2011)
6. Tabourier, L., Libert, A.S., Lambiotte, R.: Predicting links in ego-networks using temporal information. EPJ Data Sci. **5**(1), (2016)
7. Rossetti, G., Pappalardo, L., Pedreschi, D., Giannotti, F.: Tiles: an online algorithm for community discovery in dynamic social networks. JMLR (2016)
8. Holme, P., Saramäki, J.: Temporal networks. Phys. Rep. **519**(3), (2012)
9. Gulyás, L., Kampis, G.: Spreading processes on dynamically changing contact networks. EPJ ST **222**(6), (2013)
10. Liu, C., Zhang, Z.K.: Information spreading on dynamic social networks. Commun. Nonlinear Sci. Numer. Simul. **19**(4), (2014)
11. Miritello, G., Moro, E., Lara, R.: Dynamical strength of social ties in information spreading. Phys. Rev. E **83**(4), (2011)
12. Weng, L., Ratkiewicz, J., Perra, N., Gonçalves, B., Castillo, C., Bonchi, F., Schifanella, R., Menczer, F., Flammini, A.: The role of information diffusion in the evolution of social networks. In: SIGKDD KDD (2013)
13. Stehlé, J., Voirin, N., Barrat, A., Cattuto, C., Colizza, V., Isella, L., Régis, C., Pinton, J.F., et al.: Simulation of an seir infectious disease model on the dynamic contact network of conference attendees. BMC Med. **9**(1), 87 (2011)
14. Viswanath, B., Mislove, A., Cha, M., Gummadi, K.P.: On the evolution of user interaction in facebook. In: WOSN (2009)
15. Kermack, W.O., McKendrick, A.: A contribution to the mathematical theory of epidemics. Proc. R. Soc. A **115**(772), (1927)

---

[3]All methods were made available within the NDLib library: https://goo.gl/1tstvG.

Fractal Analyses of Networks of Integrate-and-Fire Stochastic Spiking Neurons

Ariadne A. Costa, Mary Jean Amon, Olaf Sporns and Luis H. Favela

Abstract Although there is increasing evidence of criticality in the brain, the processes that guide neuronal networks to reach or maintain criticality remain unclear. The present research examines the role of neuronal gain plasticity in time-series of simulated neuronal networks composed of integrate-and-fire stochastic spiking neurons and the utility of fractal methods in assessing network criticality. Simulated time-series were derived from a network model of fully connected discrete-time stochastic excitable neurons. Monofractal and multifractal analyses were applied to neuronal gain time-series. Fractal scaling was greatest in networks with a mid-range of neuronal plasticity, versus extremely high or low levels of plasticity. Peak fractal scaling corresponded closely to additional indices of criticality, including average branching ratio. Networks exhibited multifractal structure, or multiple scaling relationships. Multifractal spectra around peak criticality exhibited elongated right tails, suggesting that the fractal structure is relatively insensitive to high-amplitude local fluctuations. Networks near critical states exhibited mid-range multifractal spectra width and tail length, which is consistent with the literature suggesting that networks poised at quasi-critical states must be stable enough to maintain organization but unstable enough to be adaptable. Lastly, fractal analyses may offer additional information about critical state dynamics of networks by indicating scales of influence as networks approach critical states.

1 Introduction

The last two decades have seen increasing discussion about the prevalence and role of criticality in the brain [1–4]. Criticality is a property of systems organized near phase transitions. Loosely speaking, a critical state is stable enough to maintain

A. A. Costa (✉) · M. J. Amon · O. Sporns
Department of Psychological and Brain Sciences, Indiana University,
Bloomington, IN 47405, USA

L. H. Favela
Department of Philosophy and Cognitive Sciences Program,
University of Central Florida, Orlando, FL 32816, USA

© Springer International Publishing AG 2018
S. Cornelius et al. (eds.), *Complex Networks IX*, Springer Proceedings
in Complexity, https://doi.org/10.1007/978-3-319-73198-8_14

organization, but is unstable enough to be adaptable so as to facilitate switches among states. Criticality has been experimentally demonstrated in neuronal systems in a variety of ways, for example, *in vivo* [5–8], *in vitro* [9, 10], and *in silico* (e.g., [11–14]). A natural question to ask is, "Why do brains exhibit criticality?" Though there is no agreed upon answer, a number of possibilities have been offered; for example, criticality maximizes the range of inputs that can be processed by neurons [15, 16], optimizes information processing [17, 18], and is a signature of brain dynamics in healthy nervous systems [19].

A primary indicator of criticality is power-law distributions. As an example, both size and duration of neuronal avalanches—cascades of neurons spiking consecutively —in cortical circuits are distributed according to power laws [9]. In regard to criticality, it is claimed that power-law distributions exhibit fractal scaling. Fractals are scale-free and self-similar spatial or temporal patterns, whereby the global pattern is maintained at various scales of observation [20, 21]. Mathematical fractals are perfectly self-similar across scale (e.g., Koch snowflakes and Sierpinski triangles). Natural fractals are statistically self-similar across scales (e.g., the coastline of Britain and tree branching). In this sense, power-law distributions can represent the scale-free self-similarity of fractals, although precise relationship between criticality, fractals, and power laws remains controversial (for review see [22]). While there is growing evidence of criticality in brains, an explanation of how brains reach and maintain such states remains elusive (cf. [23]).

Considering that power-law distributions can be indicators of criticality, and given that fractal scaling can be represented in terms of power-law distributions, fractal analyses may be useful methods for assessing the presence of criticality and quantifying how far a system is from criticality. In the current work, we test this hypothesis and attempt to understand more about the processes that guide a neuronal network to reach and/or maintain itself around criticality. To do this, we determine the scaling behaviors of temporal series of simulations of neuronal networks composed of stochastic spiking neurons [24] with gain plasticity [25] using monofractal detrended fluctuation analysis (DFA) and multifractal DFA (MFDFA). This neuronal model was first proposed as an explanation for self-organizing criticality (SOC) in cortical circuits in the brain, suggesting that neural circuits operate slightly above criticality (self-organizing supercriticality; SOSC).

In a classic paper, Bak, Tang, and Wiesenfeld showed that systems with spatial degrees of freedom self-organize into critical states reflected by $1/f$ (fractal) noise spectra, presenting minimally stable clusters at all length scales [21]. In accordance with Bak and colleagues, our main finding is that $1/f$ noise in time-series of neuronal gain plasticity corresponds to systems in (quasi-)critical states.

In the following section, we detail the mathematical neuronal model implemented and also the methods used for the analyses. Next, we present the fractal analysis results and conclude with a discussion of the results that may further illuminate the nature of neuronal criticality.

2 Method

2.1 Neuronal Model

The model considered here was first proposed by Costa et al. [25]. It consists of a network of $i = 1, \ldots, N$ discrete-time stochastic excitable neurons [14, 24–27]. Each neuron is connected to all other neurons j (i.e., it is a fully connected network). The presynaptic neuron j transmits signals to the postsynaptic neuron i proportionally to the synaptic strength W_{ij}.

This is an adaptation of the Galves–Löcherbach (GL) model [24] by adding neuronal gain plasticity. As in the GL model, each neuron has a membrane potential V_i that is evolved at each timestep t. In the special case of GL model with the filter function $g(t - t_s) = \mu^{t-t_s}$, where t_s is the time of the last firing of neuron i, the membrane potential at $t + 1$ can be discretized as:

$$V_i[t + 1] = \begin{cases} 0 & \text{if } X_i[t] = 1, \\ \mu V_i[t] + I_{ext} + \dfrac{1}{N} \displaystyle\sum_{j=1}^{N} W_{ij} X_j[t] & \text{if } X_i[t] = 0, \end{cases} \tag{1}$$

where I_{ext} represents external stimuli arriving at the postsynaptic neuron, while X_j is the state of the presynaptic neuron between the timesteps t and $t + 1$ ($X_j = 1$ when j spikes and $X_j = 0$ otherwise). The leakage factor $\mu \in [0, 1]$ reflects the diffusion of ions through the membrane.

Unlike the classic leaky integrate-and-fire (LIF) model [28], the neuron does not fire deterministically when $V_i[t + 1]$ exceeds a threshold. That fixed threshold potential is substituted by a probability of firing $\Phi(V_i[t])$, according to a *firing function* [24, 25, 29–32]. The *rational* firing function [14, 25, 29] is used to calculate the firing probability $0 \le \Phi(V_i[t]) \le 1$ for each neuron at each timestep:

$$\Phi_i[t](V_i[t]) = \frac{\Gamma_i[t](V_i[t] - V_T)}{1 + \Gamma_i[t] - (V_i[t] - V_T)} \, \Theta(V_i[t] - V_T) . \tag{2}$$

The neuronal gain Γ_i causes an amplification to the signal received by the neuron and is evolved as it follows:

$$\Gamma_i[t + 1] = \Gamma_i[t] + \frac{1}{\tau} \Gamma_i[t] - \Gamma_i[t] X_i[t] , \tag{3}$$

where τ is gain recovery time. The neuronal gain recovers each timestep and is decreased just after the neuron spikes. These dynamics are biologically plausible, corresponding to the reduction and recovery of sodium channels at the axon initial segment after spiking, as described in [33].

An advantage of studying neuronal gain plasticity instead of synaptic plasticity—as in previous works [12, 13, 34]—is the reduction in the number of equations evolved at each timestep: N equations for neuronal gains rather than $N(N - 1)$ equations for corresponding synapses [14, 25].

The results presented here come from simulations of networks with $N = 160,000$ neurons without external stimulus ($I_{ext} = 0$). After spiking, neurons do not have memory of previous timesteps ($\mu = 0$). The threshold potential is $V_T = 0$. Simulations were conducted in Fortran90.

Analyses were performed in time-series of average gains (computed over all neurons at each timestep) for different values of neuronal gain recovery times (τ). A long transient of 5 million timesteps was removed (see [25]), so that only the last 50,000 timesteps were analyzed.

2.2 Fractal Analyses

Detrended fluctuation analysis (DFA) is a type of monofractal analysis that removes local linear trends within specified windows of time in the data and then looks for statistical self-similarity in what remains. After linear detrending, the residual represents fluctuations around the global trend. For each window size, the log-log plot of the transformed frequency as a function of the transformed amplitude fluctuations reveals a linear relation indicating the degree of self-similarity across scaling, given by the Hurst exponent (H).

Hurst exponents approaching one ($H \approx 1$) represent $1/f$ noise or $1/f$ scaling. $1/f$ scaling indicates the presence of fractal structure within a signal or self-similar temporal or spatial patterns across scales [35–37]. $1/f$ noise contrasts with white noise, which represents relatively random or independent timesteps ($H \approx 0.5$). Hurst exponents close to 1.5 or higher ($H \approx 1.5$) represent Brownian motion. Brownian noise describes patterns of variability that exhibit a random walk pattern, with global structure and local independence [38, 39]. Brownian noise often can be used to describe the movement of natural systems, where it is not easy to predict a specific movement trajectory, but the trajectory is always dependent on the system's previous position. Lastly, blue noise is indicative of anti-persistence ($H \approx 0.0$), where positive data points tend to be followed by negative ones and vice versa such that this signal tends toward its mean. Anti-persistence is an indicator of a signal with short memory [39, 40].

Monofractal scaling identifies one scaling relationship that best characterizes a signal and assumes invariance across temporal or spatial scales. However, variance often occurs across scales. Multifractal analysis indicates the degree to which power-law structure and self-similarity across scales are heterogeneous across a signal. Unlike monofractal analysis, multifractal analysis is capable of characterizing different local scaling properties across a signal [41]. Multifractal detrended fluctuation analysis (MFDFA) adds an additional q parameter to DFA, which weights the influence of small and large fluctuations or root-mean-square (RMS). RMS of variation around local trends is successively raised to the value of each q parameter [8]. The more negative the q value, the more strongly it is influenced by segments with small

RMS. Conversely, more positive q values are influenced by segments with large RMS and q's of 0 are neutral to the influence of relatively small or large RMS [41]. The variation of Hurst exponents ($H_{max} - H_{min}$) based on q provides an index of multifractality or the degree to which $1/f$ scaling varies across the signal [41]. The multifractal spectrum can be plotted to represent the variation, as well as the relative length of the right and left tails of the multifractal spectra. Multifractal spectra with long right tails indicate that the fractal structure is relatively insensitive to local fluctuations with large magnitudes, and long left tails suggest that spectra are relatively insensitive to local fluctuations with small magnitudes [41].

Prior to analyses, data were normalized and outliers ±4 standard deviations were trimmed [39]. Minimum (min = 4) and maximum (max = 4,096) window sizes were selected to accommodate the sample size of the time-series ($S = 50,000$). The minimum window size was chosen to avoid local RMS fluctuation errors, and a maximum window size was selected to represent a significant portion of the time-series while allowing for multiple windows across the time-series [42]. Windows were overlapped by 50% to provide better estimates of within-window variability [8, 43, 44]. A linear detrending procedure was utilized to examine power-law scaling in the residuals (e.g., [42]). Preliminary analyses indicated that data were fractional Brownian motion, suggesting that data did not need to be integrated prior to analysis [45, 46].

3   Results

DFA was used to determine the extent to which a dynamical sequence of average neuronal gain (Γ^*) exhibited $1/f$ scaling. Hurst exponents were derived for average neuronal gain time-series for different τ values (Fig. 1). Hurst exponents (H) peak in the range of $1/f$ (fractal) noise $\tau \approx 1,920$ ($H = 0.91$). Hurst exponents were positively associated with average gain ranging between $\tau = 160$ and $\tau = 1,920$. After peaking at $\tau \approx 1,920$, Hurst exponents began to decline with increasing τ values ($\tau = 1,920 - 12,320$). Hurst exponents were in the range between white and $1/f$ noise ($M = 0.81 \pm 0.09$, min = 0.67, max = 0.91), with the greatest degree of $1/f$ scaling $\tau \approx 1,920$.

For the same systems, the average branching ratio (σ^*)—a common measure for criticality characterization [13, 47–49] —was also computed (see Fig. 1a). In theory, $\sigma^* = 1$ corresponds to a stable (critical) system, while lower values of σ^* represent the activity decreasing in a subcritical system, and $\sigma^* > 1$ indicates the activity increasing in supercritical ones. Figure 1b reveals how H changes due to variations in the average branching ratio corresponding to the networks with different τ values.

According to analytical calculations (see [25]), we can have the critical values $\Gamma_C = 1/W_C$ in the stationary state for the model with neither synaptic nor gain plasticity, i.e., fix Γ and W values for all neurons. In the simulations, we have fix $W_C = 1$, which implies in $\Gamma_C = 1$. However, for the model containing neuronal gain plasticity, as presented here, we can also assert due to analytical calculations that Γ^*

Fig. 1 Relationship between characteristic neuronal gain recovery time (τ) and fractal scaling (Hurst exponent), average branching ratio of active neurons (σ^*) and branching ratio. **a** As fractal scaling increases $\tau = 160$ to $1{,}920$ and peaks $\tau = 1{,}920$, the branching ratio approaches criticality ($\sigma^* \approx 1$). After this point, the fractal scaling reduces, suggesting that greater fractal scaling of gain is associated with gain criticality. **b** As characteristic neuronal gain recovery time (τ) increases, the branching ratio moves away from supercriticality ($\sigma^* > 1$) to subcriticality ($\sigma^* < 1$). The Hurst exponents peak for $\tau = 1{,}920$

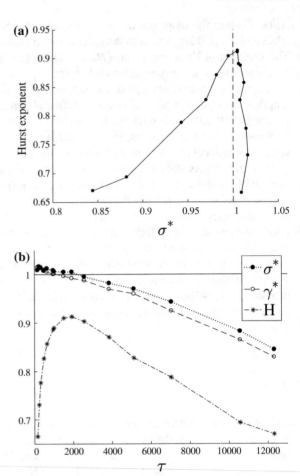

depends on τ [25]. In this sense, we expect variations of the average gain in time-series of networks with different τ values.

Figure 1b compares three different measures to estimate criticality in neuronal networks with different characteristic times of recovery τ. The three measures are the Hurst exponent (H), branching ratio (σ^*), and average gain (Γ^*). Peak $1/f$ noise, as well $\Gamma^* \approx 1$ (for this specific case of the model with $\mu = 0$, $V_T = 0$, $I_{ext} = 0$, $W_{ij} = W = 1$) and $\sigma \approx 1$, is indicative of criticality.

MFDFA applied an additional parameter to the DFA analysis, such that minimum and maximum q-orders were selected to examine the influence of segments with large and small fluctuations on the degree of neuronal gain fractal scaling. The q-orders ranged from $q_{min} = -5$ to $q_{max} = 5$ with a step size of $q_{step} = 2$ [41]. Neuronal gain time-series exhibited multifractality, with average minimum Hurst exponents $M = 0.56 \pm 0.12$ and maximum Hurst exponents $M = 1.70 \pm 0.12$ having an average multifractal spectra width $M = 1.06 \pm 0.33$. Multifractal spectra width

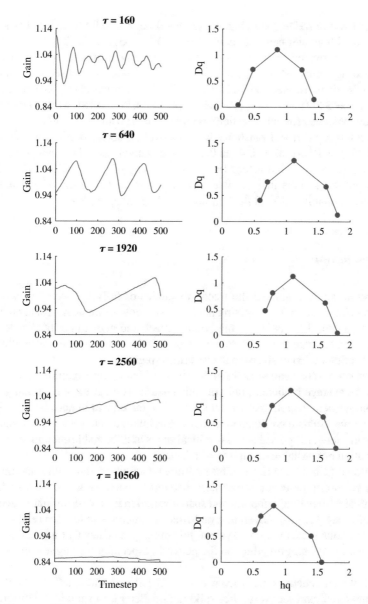

Fig. 2 First, $t = 500$ timesteps of neuronal gain time-series (left column) and multifractal spectra of corresponding time-series (50, 000 timesteps; right column) for select τ values ranging $\tau = 160$–10, 560. Gain time-series with lower characteristic time of neuronal gain recovery (τ) demonstrate higher frequency fluctuations, as compared to gain time-series for higher τ. Fast timescale fluctuations are reflected in multifractal spectra with extended left tails. As τ increases, slow-frequency and low-amplitude timescale fluctuations of neuronal gain become more pronounced, such that left multispectra tails are truncated compared to right tails

was negatively correlated with higher τ, $r = -0.66$, $p = 0.02$, such that $1/f$ scaling stabilized with greater neuronal gain $M = 1.06 \pm 0.33$; see Fig. 2.

In terms of overall multifractal spectra tail length, multifractal spectra exhibited elongated right tails ($M_{left-right} = 0.33 \pm 0.23$). Differences between left and right tail lengths demonstrated extended right tails for neuronal gain plasticity ranging between $\tau = 320$ and $\tau = 12,320$ ($M_{left-right} = 0.39 \pm 0.18$), while extended left tails were observed for gain plasticity between $\tau = 160$ and $\tau = 240$ ($M_{left-right} = -0.08 \pm 0.04$). Neuronal plasticity was positively associated with left tail values ($r = 0.75$, $p = 0.001$, $M = 0.43 \pm 0.17$) and negatively associated with differences between left and right tail values ($r = -0.73$, $p = 0.02$, $M = 0.33 \pm 0.23$). Elongated right tail lengths indicate that networks with higher levels of plasticity are relatively insensitive to local fluctuations with large magnitudes.

4 Discussion

The present study evaluated the fractal dynamics of fully connected networks of stochastic integrate-and-fire spiking neurons in order to examine neuronal gain plasticity as a factor that may drive the brain to reach and maintain critical states. Our findings demonstrate the appropriateness of utilizing monofractal and multifractal detrended fluctuation analyses to assess critical regimes.

We verify fractal scaling as an indicator of criticality by examining its relationship to the average branching ratio of active neurons and average neuronal gain. In addition, we demonstrate that fractal scaling is reduced within networks exhibiting more extreme (sub/supercritical systems) versus intermediate values of average neuronal gain. Networks poised at quasi-critical states must be stable enough to maintain organization but unstable enough to be adaptable.

Neuronal gain time-series exhibit multifractal structure, indicating that multiple scaling relationships occur across each network's signal (Fig. 2). This finding is compatible with natural temporal and spatial variation in scale-invariant structure of biomedical signals [50]. Given that multifractal structures reflect the relative influence of various scales within a system, this finding indicates that a broad range of scales exert a meaningful effect on the network, especially for larger average gains (lower τ values).

Multifractal spectra shifted from elongated left and shortened right tails with faster neuronal gain recovery after spiking (smaller τ), to shortened left tails and elongated right tails with slower recovery time (larger τ; Fig. 2). Specifically, networks with small τ values exhibited local fluctuations with relatively high-amplitude fluctuations, such that multifractal spectra were relatively sensitive to local fluctuations with larger amplitudes than to less-pronounced fluctuations. Along these lines, networks with greater τ exhibited low-frequency amplitude fluctuations, such that multifractal spectra were relatively sensitive to smaller fluctuations, as compared to larger amplitude fluctuations.

In summary, the present study identifies fractal scaling in neuronal networks as a viable measure for identifying critical states. In addition, the present study indicates that neuronal gain plasticity may play a significant role in modulating system criticality. Future work will address whether the results presented here are consistent for smaller or larger network sizes. Additionally, fractal analyses may also be fruitfully applied to time-series of other signals related to neuronal activity, including empirical recordings of *in vivo* and *in vitro* neuronal networks.

Acknowledgements This article was produced as part of the activities of FAPESP Research, Innovation and Dissemination Center for Neuromathematics (grant #2013/07699-0, S.Paulo Research Foundation). AAC also thanks grants #2016/00430-3 and #2016/20945-8 São Paulo Research Foundation (FAPESP).

References

1. Chialvo, D.R.: Critical brain networks. Phys. A **340**, 756–765 (2004)
2. Beggs, J., Timme, N.: Being critical of criticality in the brain. Front Psychol. **3**, 163 (2012)
3. Favela, L.H.: Radical embodied cognitive neuroscience: addressing "grand challenges" of the mind sciences. Front Hum Neurosci. **8**, 796 (2014)
4. Hesse, J., Gross, T.: Self-organized criticality as a fundamental property of neural systems. Front Syst. Neurosci. **8**, 166 (2014)
5. Poil, S.S., van Ooyen, A., Linkenkaer-Hansen, K.: Avalanche dynamics of human brain oscillations: relation to critical branching processes and temporal correlations. Hum. Brain Mapp **29**, 770–777 (2008)
6. Petermann, T., Thiagarajan, T.C., Lebedev, M.A., Nicolelis, M.A.L., Chialvo, D.R., Plenz, D.: Spontaneous cortical activity in awake monkeys composed of neuronal avalanches. Proc. Natl. Acad. Sci. **106**, 15921–15926 (2009)
7. Hahn, G., Petermann, T., Havenith, M.N., Yu, S., Singer, W., Plenz, D., Nikolić, D.: Neuronal avalanches in spontaneous activity in vivo. J. Neurophysiol. **104**, 3312–3322 (2010)
8. Favela, L.H., Coey, C.A., Griff, E.R., Richardson, M.J.: Fractal analysis reveals subclasses of neurons and suggests an explanation of their spontaneous activity. Neurosci. Lett. **626**, 54–58 (2016)
9. Beggs, J.M., Plenz, D.: Neuronal avalanches in neocortical circuits. J. Neurosci. **23**, 11167–11177 (2003)
10. Beggs, J.M., Plenz, D.: Neuronal avalanches are diverse and precise activity patterns that are stable for many hours in cortical slice cultures. J. Neurosci **24**(22), 5216–29 (2004)
11. de Arcangelis, L., Perrone-Capano, C., Herrmann, H.J.: Self-organized criticality model for brain plasticity. Phys. Rev. Lett. **96**, 028107 (2006)
12. Levina, A., Herrmann, J.M., Geisel, T.: Dynamical synapses causing self-organized criticality in neural networks. Nat. Phys. **3**, 857–860 (2007)
13. Costa, A.A., Copelli, M., Kinouchi, O.: Can dynamical synapses produce true self-organized criticality? J. Stat. Mech. Theory Exp. **2015**, P06004 (2015)
14. Brochini, L., Costa, A.A., Abadi, M., Roque, A.C., Stolfi, J., Kinouchi, O.: Phase transitions and self-organized criticality in networks of stochastic spiking neurons. Sci. Rep. **6** (2016)
15. Kinouchi, O., Copelli, M.: Optimal dynamical range of excitable networks at criticality. Nat. Phys. **2**, 348–351 (2006)
16. Shew, W.L., Yang, H., Petermann, T., Roy, R., Plenz, D.: Neuronal avalanches imply maximum dynamic range in cortical networks at criticality. J. Neurosci. **29**, 15595–15600 (2009)
17. Beggs, J.M.: The criticality hypothesis: how local cortical networks might optimize information processing. Philos. Trans. R. Soc. A **366**, 329–343 (2008)

18. Shew, W.L., Plenz, D.: The functional benefits of criticality in the cortex. Neuroscientist **19**, 88–100 (2013). PMID: 22627091
19. Massobrio, P., de Arcangelis, L., Pasquale, V., Jensen, H.J., Plenz, D.: Criticality as a signature of healthy neural systems. Front. Syst. Neurosci. **9**, (2015)
20. Mandelbrot, B.B.: The fractal geometry of nature, Updated edn. W. H. Freeman and Company, New York (1982)
21. Bak, P., Tang, C., Wiesenfeld, K.: Self-organized criticality: an explanation of the 1/f noise. Phys. Rev. Lett. **59**, 381–384 (1987)
22. Watkins, N.W., Pruessner, G., Chapman, S.C., Crosby, N.B., Jensen, H.J.: 25 years of self-organized criticality: concepts and controversies. Space Sci Rev. **198**, 3–44 (2016)
23. Tetzlaff, C., Okujeni, S., Egert, U., Wörgötter, F., Butz, M.: Self-organized criticality in developing neuronal networks. PLoS Comput. Biol. **6**, e1001013 (2010)
24. Galves, A., Löcherbach, E.: Infinite systems of interacting chains with memory of variable length - a stochastic model for biological neural nets. J. Stat. Phys. **151**, 896–921 (2013)
25. Costa, A.A., Brochini, L., Kinouchi, O.: Self-organized supercriticality and oscillations in networks of stochastic spiking neurons. Entropy. **19**, 399 (2017)
26. Gerstner, W., van Hemmen, J.L.: Associative memory in a network of 'spiking' neurons. Netw. Comput. Neural **3**, 139–164 (1992)
27. Gerstner, W., Kistler, W.M.: Spiking Neuron Models: Single Neurons, Populations. Plasticity. Cambridge University Press, Cambridge (2002)
28. Lapicque, L.: Recherches quantitatives sur l'excitation électrique des nerfs traitée comme une polarisation. J. Physiol. Pathol. Gen. **9**, 620–635 (1907): Translation: Brunel, N., van Rossum, M.C.: Quantitative investigations of electrical nerve excitation treated as polarization. Biol. Cybern. **97**, 341–349 (2007)
29. Larremore, D.B., Shew, W.L., Ott, E., Sorrentino, F., Restrepo, J.G.: Inhibition causes ceaseless dynamics in networks of excitable nodes. Phys. Rev. Lett. **112**, 138103 (2014)
30. Duarte, A., Ost, G.: A model for neural activity in the absence of external stimuli. Markov Process. Relat. Fields **22**, 37–52 (2016)
31. De Masi, A., Galves, A., Löcherbach, E., Presutti, E.: Hydrodynamic limit for interacting neurons. J. Stat. Phys. **158**, 866–902 (2015)
32. Galves, A., Löcherbach, E.: Modeling networks of spiking neurons as interacting processes with memory of variable length. J. Soc. Franc. Stat. **157**, 17–32 (2016)
33. Kole, M.H., Stuart, G.J.: Signal processing in the axon initial segment. Neuron **73**, 235–247 (2012)
34. Campos, J.G.F., Costa, A.A., Copelli, M., Kinouchi, O.: Correlations induced by depressing synapses in critically self-organized networks with quenched dynamics. Phys. Rev. E **95**, 042303 (2017)
35. Kello, C.T., Beltz, B.C., Holden, J.G., Van Orden, G.C.: The emergent coordination of cognitive function. J. Exp. Psychol. Gen **136**, 551 (2007)
36. Holden, J.G., Van Orden, G.C., Turvey, M.T.: Dispersion of response times reveals cognitive dynamics. Psychol. Rev **116**, 318 (2009)
37. Van Orden, G.C., Kloos, H., Wallot, S.: Living in the pink: Intentionality, wellbeing, and complexity. In: Philosophy of Complex Systems, Handbook of the philosophy of science, vol. 10. (2011)
38. Gilden, D.L.: Cognitive emissions of 1/f noise. Psychol. Rev **108**, 33 (2001)
39. Holden, J.G.: Gauging the fractal dimension of response times from cognitive tasks, pp. 267–318. A Webbook Tutorial, Contemporary Nonlinear Methods for Behavioral Scientists (2005)
40. Delignieres, D., Ramdani, S., Lemoine, L., Torre, K., Fortes, M., Ninot, G.: Fractal analyses for 'short' time-series: a re-assessment of classical methods. J. Math. Psychol **50**, 525–544 (2006)
41. Ihlen, E.A.: Introduction to multifractal detrended fluctuation analysis in matlab. Front Psychol. **3** (2012)
42. Botcharova, M., Farmer, S.F., Berthouze, L.: Markers of criticality in phase synchronization. Front. Syst. Neurosci. **8** (2014)

43. Hardstone, R., Poil, S.S., Schiavone, G., Jansen, R., Nikulin, V.V., Mansvelder, H.D., Linkenkaer-Hansen, K.: Detrended fluctuation analysis: a scale-free view on neuronal oscillations. Front Psychol. **3** (2012)
44. Linkenkaer-Hansen, K., Nikouline, V.V., Palva, J.M., Ilmoniemi, R.J.: Long-range temporal correlations and scaling behavior in human brain oscillations. J. Neurosci. **21**, 1370–1377 (2001)
45. Eke, A., Herman, P., Bassingthwaighte, J., Raymond, G., Percival, D., Cannon, M., Balla, I., Ikrényi, C.: Physiological time-series: distinguishing fractal noises from motions. Pflügers Arch. **439**, 403–415 (2000)
46. Delignières, D., Marmelat, V.: Theoretical and methodological issues in serial correlation analysis. In: Progress in Motor Control, pp. 127–148. Springer, Berlin (2013)
47. Haldeman, C., Beggs, J.M.: Critical branching captures activity in living neural networks and maximizes the number of metastable states. Phys. Rev. Lett. **94**, 058101 (2005)
48. Timme, N.M., Marshall, N.J., Bennett, N., Ripp, M., Lautzenhiser, E., Beggs, J.M.: Criticality maximizes complexity in neural tissue. Front. Psychol. **7**, (2016)
49. Wilting, J., Priesemann, V.: Branching into the unknown: inferring collective dynamical states from subsampled systems (2016). arXiv preprint arXiv:1608.07035
50. Lopes, R., Ayache, A.: Tenets, methods, and applications of multifractal analysis in neurosciences. In: The Fractal Geometry of the Brain, pp. 65–79. Springer, Berlin (2016)

Part IV
Network Science Applications

Cultivating Tipping Points: Network Science in Teaching

Catherine Cramer, Ralucca Gera, Michaela Labriole, Hiroki Sayama,
Lori Sheetz, Emma Towlson and Stephen Uzzo

Abstract Current education systems continue to be based predominantly on reduc-
tionist mindsets in which teaching is conducted on a subject-by-subject and module-
by-module basis. Improvement is planned and implemented using a linear, causal,
independent-problem-to-solution approach, with very little consideration given to
the interconnectedness among the various components and ideas involved in these
complex knowledge systems. This situation presents a need to think about how
understanding these connections can improve the learning of complex ideas. It also
constitutes an opportunity to provide a multifaceted intervention for communities
of learners, which would, itself, be a coordinated network of collaborative efforts to
develop a network literate populace. In this paper, the authors describe addressing
these issues through a multi-phase, multi-year approach to professional development
with formal and informal educators; the outcomes of this work; and next steps.

1 Introduction

While network science provides opportunities to develop many of the skills, habits
of mind, and core ideas that are essential for today's interconnected world [1, 2], they

C. Cramer (✉) · M. Labriole · S. Uzzo
New York Hall of Science, Queens, NY, USA
e-mail: ccramer@nysci.org

R. Gera
Naval Postgraduate School, Monterey, CA, USA

H. Sayama
Center for Collective Dynamics of Complex Systems, Binghamton University, Binghamton, NY,
USA

H. Sayama · E. Towlson
Center for Complex Network Research, Northeastern University, Boston, MA, USA

L. Sheetz
Center for Leadership and Diversity in STEM, U.S. Military Academy at West Point, West Point,
NY, USA

© Springer International Publishing AG 2018   175
S. Cornelius et al. (eds.), *Complex Networks IX*, Springer Proceedings
in Complexity, https://doi.org/10.1007/978-3-319-73198-8_15

are neither addressed nor utilized in extant elementary/secondary education curricula nor teaching practice. However, theories of complex systems inform us of the possibility of a "phase transition" or "tipping point" induced by collective actions taking place on a dynamical network of interconnected components [3]. Through leveraging the connections in the network science research community toward the goal of developing and framing the needs of twenty-first-century learners, the authors are cultivating such collective action. The plan is to strive for a tipping point through increased attention to, and involvement in, identifying the needs of these learners and the potential value of bringing network science to teaching. Specific to the purpose of this paper is efforts to facilitate the development of practices and resources with educators, and promote network thinking among K-12 students, teachers, and administrators through developing curriculum materials, lesson plans, and practical learning resources for K-12 classrooms across all domains of knowledge; providing rigorous professional development opportunities for both formal (school) and informal (cultural institutions, camps, after-school and other community-based programs) educators; and increasing the awareness of the demand for network science education among researchers.

This paper describes the trajectory of the development of models bringing researchers together with K-12 formal educators in order to introduce the educators to network science tools and concepts and the researchers to classroom challenges; to collaborate on using network thinking in K-12 classroom settings; to create network approaches to curriculum development; and to develop a path for mapping *Network Literacy Essential Concepts and Core Ideas* [4] to learning standards.

2 Model I: Small Teams, Co-learning

This work began in 2010 with the NSF-funded *Network Science for the Next Generation,* also known as *NetSci High*, designed to address a skills gap between current teaching and learning and STEM practice [5]. The kinds of advanced skills needed by the twenty-first century workforce include: the ability to interact with and derive meaning from large amounts of data, and facility with visual representations, metaphors and reference systems for abstract large-scale spatial and dynamic data streams (necessary to see and make sense of patterns in complex data); and the ability to create and understand more sophisticated scientific models. (Higher-order thinking is needed to develop and interpret probabilistic and stochastic models to allow exploratory and inductive skills to be used to identify patterns and characterize behaviors across a wide range of differing environments and processes.)

Students in the STEM "pipeline" need to be prepared for this new reality as they enter tertiary education and the modern day workforce. However, exposure to these data-driven science skills is largely unavailable to most primary and secondary school students, particularly students in underserved communities. Such lack of access sends students down a path that misses important opportunities to fully participate in advances in modern society. NetSci High was developed to address this skills gap through a rigorous program of network science training and research. It is a regional educational outreach program designed to empower high school students and teachers to harness the power of network modeling and analysis, resulting in a more holistic, dynamic understanding of the "interdependence" among components and the evolution of relationships among various things around us. NetSci High provides interventions in STEM teaching and learning that directly address the need for twenty-first-century skills while targeting female, minority, and economically disadvantaged students. It provides an alternative and advanced pathway to develop rigorous skill-based curricula, resources, and programs that utilize the rapidly growing science of complex networks as a vehicle through which students can learn computational and analytical skills for network-oriented data analysis, as well as how these skills can lead to breakthroughs in solving real-world problems. NetSci High explores innovative approaches that, as this work demonstrates, can capture the interest and imagination of underrepresented populations to explore science research problems using computational tools and methods [6–8].

The goal of NetSci High is to prepare and mentor teams of high school students from underserved communities to do yearlong original network science research projects. In the original design, the role of the high school teachers was intended to be as co-learners and facilitators with the students. The teachers would then transition to a mentorship role during the academic year. This structure was piloted in an immersive two-week summer workshop at Boston University in 2012. Feedback from participants in the workshop revealed that the comfort level of teachers was very low, as they were accustomed to being the authority in the classroom and found the experience of co-learning with students to be a barrier to their learning. As a result, it was modified for the 2013 project year, with a three-day teacher workshop designed and executed to precede the student workshop in order to provide a preparatory experience for teachers that would allow them to more effectively mentor students during their workshop experience [9]. It became clear as an outcome of this project that approaches for training teachers to mentor network science research differed from those used with students (Figs. 1, 2, 3, and 4).

Summative project evaluation at the end of the 2015 project year used a mixed method, post hoc model suited to the generative nature of the project [10]. It included

Fig. 1 Teacher and student participants at NetSci High intensive summer workshop at Boston University, July 2013

Fig. 2 Ninth Grade faculty
at teacher Professional
Development January 2014

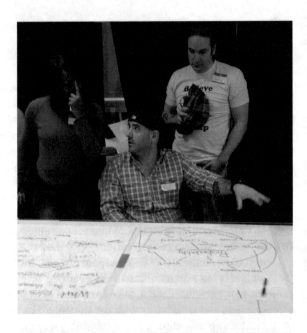

Fig. 3 Educators mapping
learning standards to
Network Literacy Essential
Concepts, July 2015

three open-ended questions intended to elicit teacher perspectives on the effect of
participation in the program. The teachers were asked to reflect on their own learning,
network skills development, and the potential for the experience of participation
having sustained effects on their own practice ($n = 16$). They indicated that the
professional development and workshops themselves well prepared them to mentor
students during the research phase of the program. They also indicated that the
network perspective was likely to continue to influence their approach to certain

Fig. 4 NiCE workshop
attendees doing hands-on
network exercise

content in their teaching practice, and they were highly appreciative of the new technical skills, with these skills clearly being seen as having a distinct educational value.

Further, the teachers tended to cite the value of the more theoretical learning, with a special emphasis on the innovative qualities of the NetSci High process. What the teachers learned through participation in this project was decidedly not something they were likely to have otherwise gathered (concepts such as nodes, edges, betweenness, clustering coefficients, centrality, and eigenvectors), and they also learned from each other during the intensive summer workshops. The teachers expressed appreciation of the new technical skills, tools, and approaches, with these skills clearly seen as having a distinct educational benefit, and they stated that their knowledge of network science had grown "exponentially" over the three years of the project and had given them a new perspective on this new and emerging field of science. The teachers also felt that the project helped them develop a comfort level in regards to supervising student research outside of their area of expertise.

3   Model II: Cross-Disciplinary Curriculum

The next phase for bringing network science to teaching was curriculum development, based on the belief that once instructors had a grasp of the network science paradigm, they would be empowered to identify, develop, and test potentially effective interventions in instructional practice on their own. Further, it was believed that introducing network science in this manner would highlight its potential as a helpful tool for the teachers in their existing challenges rather than as an add-on to their workload, and specifically to explore whether connecting subjects thematically across multi-disciplinary classrooms could make difficult academic topics more accessible to students.

The first workshop was for a multi-disciplinary team consisting of the entire ninth-grade faculty from one of the Title 1 (87% low income, 97% minority) urban schools participating in NetSci High (Chelsea Career & Technical Education High School,

New York City Public Schools). They were brought together to test the interdisciplinary approach in a one-day workshop at the New York Hall of Science in 2014. Teachers represented: English, Algebra, Global Studies, Special Education, Living Environment, Career Management, Graphic Arts, and Technology departments. The Assistant Principal also attended. Teachers received an introduction to network science through participation in a data-gathering exercise that resulted in a network visualization, plus several presentations of case studies–illustrating how all of their teaching disciplines can use the network science approach–as well as an introduction to *Gephi*, network analysis and visualization software. During the second half of the day, the teachers mapped their curriculum to network science concepts, and drew connections among their curricular themes in order to choose multi-disciplinary topics for lesson planning. Their task upon returning to their home school was to develop a team to teach a lesson or unit of instruction.

A short survey given at the end of the workshop showed a majority of the teachers to have a high degree of confidence in their grasp of network science concepts. The teachers ranged from minimally to somewhat confidant in their understanding of network modeling, the use of Gephi, network research, and kinds of networks. They gave higher scores of somewhat to very confidant in their abilities to work in a group on the day's projects. A majority of teachers saw a need for more specificity in how to use network science to address curricular topics across disciplines. This pointed to what we understood to be the next step in the trajectory of this work: deeper, more interdisciplinary learning and practice. Topics the teachers surfaced that they considered good possibilities for using a multi-disciplinary approach using network thinking included: Medieval Europe; feudal Japan; the Islamic world; the Renaissance; African Kingdoms; the Reformation; absolute monarchs; early Latin American civilizations; world religion; systems of equations; inequalities; exponential functions; quadratic functions; abuse of power; gender roles; benefits of free republics; attaining power and influence; body cells and homeostasis; body systems and homeostasis; and cell function as it relates to body system function. One of the participating math teachers was able to use these new concepts to deliver a lesson plan to his students based on rumor profusion in Twitter in order to introduce the topic of comparing exponential to linear functions.

4 Model III: Mapping Network Literacy to Learning Standards

In July 2015, seven teachers from New York and Vista, California attended a two-day workshop at the New York Hall of Science focused on mapping the Network Literacy Essential Concepts to learning standards. The Network Literacy Essential Concepts were developed over a yearlong period (2014–2015) as a response to the question: What should every citizen know about networks [11]? The teacher workshop was designed to test whether state-mandated learning standards, such as the

Next Generation Science Standards (NGSS) [12], could be more easily applied in the classroom through a network lens. This required an intensive period of review of the standards, including the cross-cutting concepts, a recent addition to learning standards, which lends itself to the application of network science. (Examples of cross-cutting concepts include systems and system models; patterns; and cause and effect.) Educators looked for specific standards that could be taught using Network Literacy concepts. Educators then developed lesson plans that both used network concepts and addressed specific standards. Examples of their lesson plans include a network of counties based on energy usage; Hudson River food webs; and evidence of the common ancestry of diversity. However, it became apparent to workshop facilitators that there was a significant gap in teacher understanding of network applications in the classroom, necessitating further work.

5   Model IV: Interdisciplinary Learning

The Networks in Classroom Education (NiCE) teacher workshop was held at the US Military Academy at West Point as a four-day workshop in July 2017, supported by the US Army Research Office [13]. Its goal was to educate teachers and administrators across K-12 instruction and nationally about network science and to enable those teachers and administrators to bring network science thinking and ideas to their students, schools, and districts. During the workshop, network thinking was not only presented as concepts to be taught to students, it was also actively utilized as a tool to make curriculum development and delivery easier and more successful, and to explore and explicate school-wide challenges. The participating teachers and administrators developed presentations and concrete lesson plans that utilized network science and network thinking in classroom practice. These lesson plans collectively demonstrate a tremendous opportunity to improve education by quantitatively identifying curricular elements central to interdisciplinary learning and sequencing the implementation of curriculum so that these central topics may be accessible to a greater range of students.

    During NiCE, we exposed 21 primary and secondary educators from around the US to network science concepts and tools (10 from Central and Southern California, 7 from New York State, and 4 from North Carolina). The educators represented a wide range of disciplines—from STEM fields to the humanities—including specialty fields from reading, to music, to English as a second language, as well as age ranges from pre-K to high school. In small groups, these educators worked together using network science concepts were learning to develop network-inspired interdisciplinary modules and lesson plans that they brought back to their schools–public, private, traditional, and charter–with the goal to enhance interdisciplinary learning that builds upon connections and better aligns education with the complex world it serves. Lesson plans developed during the workshop include: *Balance Networks in Music Ensembles*; *Interdependent Relationships in Ecosystems: 3rd–5th Grade*; *Sharks, A Slice of NiCE*; *K-2 Network Science*; *Using Networks to Explore Nanotechnology*; *How Things Are Made (resource mapping)*; *Networking in a Calculus*

Environment; *Networks in World History*; *Networking Interdisciplinary Grade Level Standards*; A *World Without Fish*, and *Community Networking Within Two Schools*.

Results from a survey conducted with attendees indicated that participating teachers: believe that what they learned in the workshop is applicable to implementing science standards; can apply network science as an interdisciplinary approach to connecting concepts across curricula; understand that network thinking can be empowering to students by providing relevant skills to solving a variety of real-world problems and making connections; can both teach directly about networks and apply network science in unit planning for social studies, science, language, music, art, and math; network science provides more opportunities to both collaborate with other teachers as well as to cultivate collaboration among students; network science can provide opportunities for students to visualize and play with data; and teachers are eager and willing to learn more about network science to improve practice.

The output of these teachers in the NiCE workshop constitutes a nexus of what the authors learned from in the previous workshops. It was by far the most successful in terms of how deeply the teachers demonstrated a capacity to understand and apply the network paradigm to developing classroom resources, align them with standards, and engage in discourse around the synthesis of their prior knowledge and the new knowledge of networks, the science of connections.

6   Conclusions

The creation and execution of teacher professional development in network science, described herein, constitutes an evolution in both the understanding of the needs of teachers, and a deeper understanding of the degree to which network science is a radical shift in thinking about science and learning in general. The transdisciplinary nature of network science does not fit the reductive paradigm normally used to design curriculum, to chunk knowledge into pieces and deliver them as instruction, and to test student knowledge based on this reductive and disconnected approach. The value of bringing network science to teaching helps to introduce process skills, knowledge of complexity, transdisciplinarity, and analytical methods to science and math, all of which are now demanded of teaching with the Next Generation Science Standards. But as the NiCE workshop demonstrates, it also requires a tipping point in seeing connections and applying network science across disciplines and throughout grades.

7   Next Steps

As this work advances, network science might not only function as an effective way into instruction and learning of advanced skills and knowledge in data-driven science, but may also act as a bridge to new process-oriented standards, and advance teaching practice overall. We will continue to build on these ideas to create more effective learning settings for network science and opportunities for teachers to use network science to improve practice. And we will continue to test and validate the alignment of the *Network Literacy: Essential Concepts and Core Ideas* to NGSS and other emerging learning standards to build bridges between network literacy and teaching and learning communities, and expand teacher training, curriculum and resource development across learning settings.

Acknowledgements The authors would like to acknowledge the National Science Foundation (BCS Award #1027752 and DRL Award #1139478) and the US Army Research Office for supporting this important work.

References

1. Harrington, H.A., Beguerisse-Díaz, M., Rombach, M.P., Keating, L.M., Porter, M.A.: Commentary: teach network science to teenagers. Netw. Sci. **1**(2), 226–247 (2013)
2. Sánchez, A., Brändle, C.: More network science for teenagers (2014). arXiv:1403.3618
3. Dorogovtsev, S.N., Goltsev, A.V., Mendes, J.F.: Critical phenomena in complex networks. Rev. Mod. Phys. **80**(4), 1275 (2008)
4. NetSciEd: Network Literacy Essential Concepts and Core Ideas (2017). http://tinyurl.com/networkliteracy. Accessed 22 Nov 17
5. Cramer, C., Sheetz, L., Sayama, H., Trunfio, P., Stanley, H. E., Uzzo, S.: NetSci High: bringing network science research to high schools. In: Complex Networks, vol. VI, pp. 209–218. Springer (2015)
6. Buldyrev, S., Parshani, R., Paul, G., Stanley, H.E., Havlin, S.: Catastrophic cascade of failures in coupled networks. Nature **464**(7291). Springer Nature, Berlin (2010)
7. Cohen, R., Erez, K., Ben Avraham, D., Havlin, S.: Resilience of the internet to random breakdowns. Phys. Rev. Lett. **85**(4646). American Physical Society, College Park, MD (2000)
8. Trunfio, P., Hoffman, M., Shann, M.: Partnerships between graduate fellows and Boston area high school teachers. Presented at Annual meeting of American Chemical Society, New York, September 7 (2003)
9. Faux, R.: Summer Workshop Evaluation Report. Davis Square Research Associates, Boston, MA (2014)
10. Faux, R.: Evaluation of the NetSci High ITEST Project: Summative Report. Davis Square Research Associates, Boston, MA (2015)
11. Cramer, C., Porter, M., Sayama, H., Sheetz, L., Uzzo, S.: What are essential concepts about networks? J. Complex Netw. **3**(4). Oxford University Press, Oxford (2015)
12. NGSS Lead States: Next Generation Science Standards: For States, By States. The National Academies Press, Washington, DC (2013)
13. US Military Academy Network Science Center: Networks in Classroom Education Teacher Workshop. United States Military Academy at West Point (2017). https://sites.google.com/a/binghamton.edu/nice-teacher-workshop/. Accessed 22 Nov 17

Terrorist Network Analyzed with an Influence Spreading Model

Vesa Kuikka

Abstract Al Qaeda's network structure before the tragic events of 9/11/2001 is studied using a method of social network analysis. The method is based on a modeling framework to assess the influence of a node in a complex network with respect to spreading information via different paths between source and target nodes. The same framework is used consistently to compute closeness and betweenness centrality measures as well as to detect subcommunities. Centrality measures taking into account all possible paths between source and target nodes, not just the shortest paths, are useful in modeling resilience of covert networks. Along these lines, new versions of node and link betweenness centrality measures are proposed.

1 Introduction

A method of social network analysis is applied to an Al Qaeda terrorist network. The proposed algorithm for detecting subcommunities uses influence measures where influence of nodes is considered in both directions; a node has influence on others and a node is influenced by others. Network topology, and possible directed connections, and unequal weights of nodes and links are essential features of the model. All the parameters of the model have probabilistic interpretations giving flexibility in various applications. Different measures, such as closeness centrality and betweenness centrality, and community detection are computed with the same modeling framework.

The modeling methodology is used in analyzing Al Qaeda's network structure before the tragic events of 9/11/2001. The purpose of this study is to demonstrate possibilities of discovering new phenomena and the usefulness of the method with real-world social network data. Terrorist network data centered on the 19 dead hijackers have been collected from public sources after the terrorist attacks of September 11, 2001 [10, 11, 18]. The empirical data used in this study has been published

V. Kuikka (✉)
Finnish Defence Research Agency, PO BOX 10, Tykkikentäntie 1,
11311 Riihimäki, Finland
e-mail: Vesa.Kuikka@mil.fi

© Springer International Publishing AG 2018
S. Cornelius et al. (eds.), *Complex Networks IX*, Springer Proceedings
in Complexity, https://doi.org/10.1007/978-3-319-73198-8_16

in Krebs [10, 11]. The aim of studying covert, illegal, and terrorist networks is to uncover common and specific characteristics of these networks [2]. Social network analysis provides a number of analytical methods to help in preventing or investigation of terrorist and criminal activity. The two aspects of prevention and prosecution have been discussed in more detail in [10, 11].

Observations of the study in [11] are such that many of the hijackers were distant from each other in the network structure. Many hijackers on the same flight were more than two steps away from each other. Krebs [11] describes the network structure of the hijackers' connections like the shape of a serpent. Keeping cell members distant from each other and from other cells minimizes damage to the network if a cell member is captured or otherwise compromised. Krebs [11] has also studied the phenomenon of transitory shortcuts in the network structure to accomplish its goals. There has been a balance between security and resilience. Trusted prior contacts were a strength in coordinating tasks. The 19 hijackers had other accomplices that were conduits for money and also provided needed skills and knowledge.

The terrorist network has 62 nodes and 306 links (ties, connections) between the nodes. The 62 members of the network were connected by a number of ways: attended the same college, took flight classes together, bought flight tickets using the same address, bought flight tickets together, were known to be together in week before the attacks, and had last known address in the same area. Krebs [11] determined three categories for strength of ties between the members of the network.

Different measures for evaluating, comparing, and categorization network members and subgroups have been presented in the literature [17]. Network topology is an essential factor in modeling complex networks. Node degree is a basic measure for node centrality in a general network. Different network metrics can be given interpretations even though their meanings largely overlap. Degrees reveal activity between neighbors in the network. Closeness centrality shows ability to access others in the network and monitor what is happening. Betweenness centrality measures control or a role as a broker over the flow in the network.

A normalized version of reciprocal closeness centrality [14] is defined by

$$C_c(i) = \frac{\sum_{j \neq i} \left(g_{ij}^{-1} \right)}{N - 1} \tag{1}$$

where the geodesic distance g_{ij} is the shortest distance between ego and all its others. N is the total number of nodes in the network. A generalized version of the measure has been defined in [1]. (Freeman) betweenness centrality measure has been defined in [6] where the betweenness of a node n is defined as the number of shortest paths between other pairs of nodes that pass through n.

Various network measures have been developed using, for example, local structural characteristics [3, 4], geodesic distances [1], and random walks [13]. Geodesic distances and random walks are based on considering paths of different lengths.

Different methods and algorithms have been presented for detecting communities in social networks [7, 8, 15–17]. Algorithms for detecting communities in network structures are minimum-cut method, hierarchical clustering, Girvan–Newman algorithm, modularity maximization, statistical inference, and clique-based methods [5, 12]. One of the community detection methods, a diffusion community in a complex network, described in review article [5], is a set of nodes that are grouped together by diffusion or percolation of the same property in the network. Many of the algorithms are based on the Kernighan–Lin algorithm [9]. The algorithm repeatedly moves, starting from some initial division, the vertices that most increase or least decrease a particular measure used for optimizing the division in the network.

In this paper, we use a theoretical framework to assess the influence of a node in a complex network with respect to spreading of information. On the basis of the method, consistent network measures, such as closeness centrality and betweenness centrality, are defined. The same methodology is used to detect communities in the network. In the following sections, after presenting the mathematical background, we examine the network structures of Al Qaeda terrorists involved in September 11, 2001, attacks.

2 Modeling Influence Measures in Social Networks

The closeness and betweenness centrality measures and the community detection algorithm of this study are based on computing the probabilities of influence $P_{s,t}$ from source node s to target node t. These quantities are calculated by computer algorithm because of a high number of possible paths in network structures. The quantities are functions of time T which we usually omit in notations.

Information spreading via different paths to a target node is calculated probabilistically taking into account the not mutually exclusive events when the same influence has been initiated from the same source node. The spreading is modeled as a replication process. The probability of a replication event is assumed to not dependent on nodes' states.

If two paths of lengths L_1 and L_2 have L_3 common links at the beginning of their paths, the conditional probabilities via $L_1 - L_3$ and $L_1 - L_3$ links given that the spreading has occurred via L_3 links are denoted by $p_{(L_1-L_3)}$ and $p_{(L_2-L_3)}$. We denote the unconditional probability by p_{L_3}. The probability for spreading via the two routes is

$$
\begin{aligned}
G_{s,t,d} &= p_{L_3}\left(p_{(L_1-L_3)} + p_{(L_2-L_3)} - p_{(L_1-L_3)}p_{(L_2-L_3)}\right) \\
&= p_{L_3}p_{(L_1-L_3)} + p_{L_3}p_{(L_2-L_3)} - \frac{p_{L_3}p_{(L_1-L_3)}p_{L_3}p_{(L_2-L_3)}}{p_{L_3}} \\
&= p_{L_1} + p_{L_2} - \frac{p_{L_1}p_{L_2}}{p_{L_3}}
\end{aligned}
\tag{2}
$$

where p_{L_1} and p_{L_2} are the unconditional probabilities of spreading via the links of path lengths L_1 and L_2. Weighting factors for nodes and links are incorporated by multiplying the unconditional probabilities in (2) with node and link weighting factors associated with the paths corresponding these terms. The weighting factor for a source node is assumed to have the value of 1.0. In the Poisson distribution model, the unconditional probabilities describing temporal spreading at time T on a path of length L is

$$p_L = 1 - \sum_{l=0}^{L-1} e^{-\lambda T} \frac{(\lambda T)^l}{l!} \tag{3}$$

where λ is the intensity parameter of the Poisson distribution.

In the algorithm, paths are combined iteratively in the descending order of common path lengths in a set of possible paths between source node and target node. In the final phase of the algorithm, all the common links have been considered ending up with D- independent terms. The probability for influence of source node s on target node t is

$$P_{s,t} = 1 - \prod_{d=1}^{D} \left(1 - G_{s,t,d}\right), \tag{4}$$

where D is the degree of source node s. Usually $P_{s,t} \neq P_{t,s}$ because of network topology, directional links, and different weighting factors of nodes and links. We introduce two closeness centrality measures as

$$P_{s,\bullet} = \frac{1}{N} \sum_{t=1}^{N} P_{s,t}, s = 1, \ldots, N \tag{5}$$

and

$$P_{\bullet,t} = \frac{1}{N} P_{s,t}, t = 1, \ldots, N \tag{6}$$

The number of nodes in the network is denoted by N. Equation (5) describes the role of node s as source of influence and (6) describes the role of node t as target of influence spreading. Again, usually the values of the quantities are not equal for a node s ($P_{s,\bullet} \neq P_{\bullet,s}$). We introduce a quantity describing betweenness centrality as

$$B_n = \frac{1}{N^2} \sum_{\substack{i,j=1 \\ i,j \neq n}}^{N} \left(1 - \prod_{d-1}^{D} \left(1 - G_{i,j,d}\right)\right) \tag{7}$$

In (7), node n is removed from the network and the effects on all the other nodes in the network are computed. The normalization factor in (7) is chosen to be N^2 (in the literature $N(N-1)$ has been used in some definitions).

The quantity in (7) is not very illustrative because nodes with high betweenness centrality values have low values of B_n. A better measure for betweenness centrality is

$$b_n = \frac{C - B_n}{C} \tag{8}$$

where C is a cohesion measure for the whole network.

$$C = \frac{1}{N^2} \sum_{i,j=1}^{N} \left(1 - \prod_{d=1}^{D} (1 - G_{i,j,d}) \right) \tag{9}$$

The algorithm for community detection uses the centrality measures of (5) and (6). The idea in modeling community detection is based on the concept of node's role in the network as source and target of influence. Both of these aspects have a role in community formation. Two subcommunities in a social network are detected by searching local maxima $\arg\max_{V,\overline{V}} (P)$ of (10):

$$P = \sum_{s=i_V}^{i_{N_V}} P_{s,\bullet} (V) + \sum_{s=i_V}^{i_{N_V}} P_{\bullet,t} (V) + \sum_{s=i_V}^{i_{N_V}} P_{s,\bullet} (\overline{V}) + \sum_{s=i_V}^{i_{N_V}} P_{\bullet,t} (\overline{V}) \tag{10}$$

where V and $\overline{V}$ is split into two factions of the network of N nodes with $N = N_V + N_{\overline{V}}$. An analog for community detection of the Kernighan–Lin algorithm [9] is used for searching the local maxima. Equation (10) could be generalized to more than two factions with the Kernighan–Lin algorithm moving vertices, one at a time, that most increase the measure used for optimizing the division, between these factions. In this paper, only divisions to two factions are searched.

Typically, social networks with weak interactions between nodes or social networks in their early development phases have several local maxima with different compositions. These factions can overlap with each other. In many cases, unions and intersections of the divisions are also local maxima of (10) with some parameters of the model. If a union or intersection is not identified as a local maximum, these sets of nodes could still be considered as possible subgroups of the network. In dynamic community, building processes sets of nodes divided by different community boundaries may be left as outsiders. This is more probable if the measure of (10) has a low value, or several divisions have almost equal numerical values.

3 Summary of the Results

Table 1 summarizes the four groups of hijackers listed in the same order as in [10]. The first column shows node numbers of the network in Fig. 1 Columns 2 and 3 show results of the models (betweenness centrality and closeness centrality) to be discussed in the following sections. The numerical results and the rankings obtained with

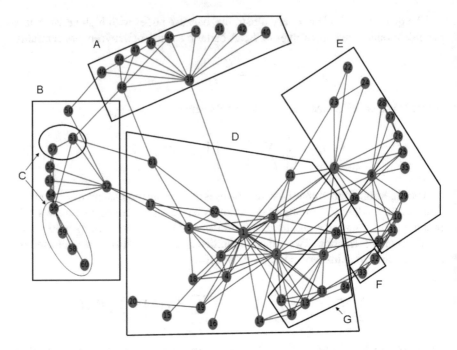

Fig. 1 Subcommunities identified with time value $T = 0.5$ and model parameters $\lambda = 0.5$, $w_N = w_L = 1.0$, $L = 6$

these new measures can be compared with the traditional closeness and betweenness centrality values for the terrorist network in [10, 11]. The proposed method takes into account more interactions in the network, not just the shortest paths between the nodes. This kind of detailed modeling can lead to more accurate results.

Table 2 shows the corresponding data for the most important nodes of the network. Figure 1 shows the network topology (structure) of the terrorist network. Results of the community detection algorithm are indicated in the figure. Also, these results will be discussed later in this paper. We argue that subcommunity structures similar to those of Fig. 1 are characteristic to covert social networks. Subcommunity D has a specific structure of central nodes and a subgroup G inside the subcommunity. Subcommunities A and B are connected to the core structures of subcommunity D only with a few links. Members of subcommunity E are connected with more links to D due to their actual roles in the hijackings. The network topology of connections between the terrorists is partly self-organized, but the main operational principles have been planned and implemented systematically before the attacks [11].

4 Closeness Centrality

Closeness centrality is one of the most important measures in social network analysis. In this paper, we use the definition of (4) because it is consistent with other measures used in this study. Figure 2 shows node degrees and the values of closeness centrality measures for two time values T = 1 and T = 4. Degree values and T = 4 values are multiplied by 0.3653 and 0.1333, respectively, for scaling the sum of measures of the 62 nodes. Weighting values for nodes $w_N = 0.5$ and links $w_L = 1.0, 0.75$, and 0.5 for strong, medium, and weak connections published in [10, 11] are used.

The highest values of closeness centrality at T = 1 are for nodes {1, 2, 7, 3, 39, . . .} in this order. The highest values of degrees are for nodes {1, 2, 7, 39, . . .}. Results for time values T = 1 and T = 4 describe different temporal development phases of influence spreading in the network. Results for low time values describe young social networks or new ideas in social networks. These both aspects can exist simultaneously. Nodes {1, 2, 39, 48, 56} show different patterns when compared with most of the other nodes {4, 6, 10, 12, 14, 18, 19, 21, 25, 36, 37, 38, 61, 62, . . .}. The five nodes are in better positions to spread new ideas.

The values and rankings of reciprocal closeness centrality of (1) [10, 11] are not equal to the results of (5). Equation (1), and the closeness centrality measure of this paper have different definitions. The reason is that (1) takes into account only the shortest distances between ego and others. Degree of nodes considers only connections to the nearest neighbors. The measures of this paper take into account influence spreading to all nodes in the network, but at early phases of influence spreading nodes far away from a source node have not been reached. Node degree values and the results for closeness centrality at times T = 1 and T = 4 agree with this. Based on this discussion, we assume that low values of time T are more suitable for analysis of the terrorist network as we assume that the network has self-organized to change plans and act in a short notice. On the other hand, friendly relations of the network have been developing during a long period of time before the hijackings.

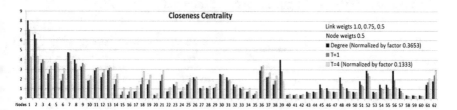

Fig. 2 Closeness centrality for two values of time T and degrees of nodes

5 (Node)Betweenness Centrality

The second measure usually used in social network analysis is (node) betweenness centrality. This measure has a classical definition as the number of shortest paths between other pairs of nodes that pass through node [6]. Values and rankings of the Freeman betweenness and the betweenness centrality of this paper are not the same. The reason is similar to the case with closeness centrality discussed before.

The first data series in Fig. 3 shows betweenness centrality values at T = 1 with node weighting values $w_N = 0.5$ for all nodes, and the second data series shows the results with $w_N = 1.0$ for the hijacker nodes (gray bars). Link weighting factor values are the same as in Fig. 2 The highest values of betweenness centrality are for nodes $\{1, 2, 7, 3, 8, 5, 9, 4, 39, 36, 13, 11, 21, 52, 6, \ldots\}$. These can be compared with the highest values of closeness centrality for nodes $\{1, 2, 7, 3, 5, 9, 8, 36, 11, 13, 4, 39, 12, 52, 6, \ldots\}$. The rankings of the first four nodes are the same. Nodes $\{8, 4, 39, \ldots\}$ have higher betweenness rankings than closeness rankings. Their positions in the network are favorable as brokers of information between other nodes as can be seen in Fig. 1.

Next, we examine the second data series in Fig. 3. Obviously, higher node weighting factors are observed on hijacker nodes. Different patterns for nodes $\{14, 21, 61, \ldots\}$ and $\{39, 52, 56, \ldots\}$ uncover their special positions with respect to the hijacker nodes. They are closely connected with the hijacker nodes, and as a consequence, their betweenness centrality measures are sensitive to hijacker node weighting factors.

Figure 4 shows ratios of closeness centrality to betweenness centrality as calculated from (5) and (8). The first 19 bars are for hijackers, and in addition, some important nodes are selected for comparison. A clear pattern is observed as nodes $\{1, 2, 3, 7, 8\}$ have a low ratio, meaning that their skills as brokers of information are emphasized with respect to their leader skills (they can be good leaders as well) when compared with other hijackers or other members of the terrorist network. On the other hand, nodes $\{14, 61, 62\}$ have high ratios, meaning that their leader skills are emphasized (they can be good brokers of information as well).

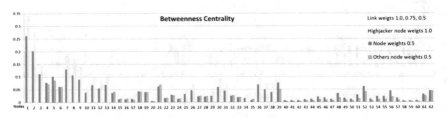

Fig. 3 Betweenness centrality for two choices of node weighting factors, T = 1 Betweenness centrality for two choices of node weighting factors, T = 1

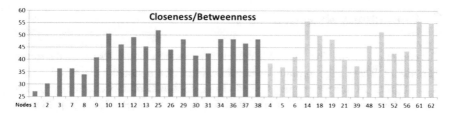

Fig. 4 Ratios of closeness to betweenness values of Figs. 2 and 3, $w_N = 0.5$, $T = 1$

6 Link Betweenness Centrality

Similar to the node betweenness centrality, a classical link betweenness centrality has been defined that counts the number of geodesic paths that run along links [17]. A new version of the link betweenness centrality is proposed in this section. The link betweenness centrality is computed analogously to the node betweenness centrality of (8) where instead of removing a node, one link is removed from the network. Figure 5 shows the results of 50 most important links $\{1 - 2, 2 - 1, 1 - 39, 39 - 1, 13, 1 - 5, 1 - 9, 1 - 8, 1 - 4, 9 - 1, \ldots\}$ of the terrorist network of Fig. 1. The results demonstrate the central role of node 1. Connections from nodes 2 and 39 to node 1 are the most important. The order of connections can be used for analyzing the social network structure.

A member of the social network could utilize the link betweenness values in prioritizing the order of sending or forwarding information in order to optimize the effects. This measure could also be helpful in allocating countermeasures against terrorist and criminal connections.

The alternative of adopting the new link betweenness centrality measure, instead of the closeness measure, can also be used for detecting communities. This idea has been earlier implemented in the Girvan–Newman algorithm [7, 16] which detects communities by progressively removing links from the original network.

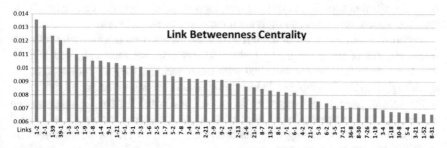

Fig. 5 Link Betweenness centrality of 50 highest values. Parameters $w_N = 0.5$, $w_L = 1.0, 0.75, 0.5$, and $T = 1$ (same as in Figs. 2 and 3 first data series)

7 Community Detection

The community detection algorithm detects seven subcommunities inside the terrorist network. These subcommunities are shown in Fig. 1 denoted by letters A – G. C is a subgroup of subcommunity B, and G is a subgroup of subcommunity D. The first airplane crashed to WTC north with five terrorists {1, 9, 11, 12, 37}. In curly brackets, the nodes are listed in the order of closeness centrality. Node 1 is in central role in the whole operation. Node 1 is a member of subcommunity D, and the other nodes are members of subgroup G. Terrorists {7, 8, 36, 26, 25} crashed to Pentagon, and they are all members of subcommunity E. Terrorists {2, 13, 30, 10, 34} crashed WTC south, and they are in subcommunities D and E with nodes {13, 34} in subgroup G. Four terrorists {3, 38, 31, 29} crashed in Pennsylvania, and they are in subcommunities D and E with node {38} in subgroup G.

In Fig. 1, no link weighting values are used. Using link weighting factors 1.0, 0.75, and 0.5 for the three categories of connection strengths published in [10, 11], almost identical results are obtained. A new overlapping division composed of subcommunity E plus nodes {9, 11, 12, 13, 32, 33, 34, 37} appears. It is possible that more weak structures are discovered with other choices of node or link weighting factors. Typically, lower values of weighting factors and lower time values result in more subcommunities. This occurs because networks with weaker connections more easily split up. After all, the network topology information with appropriate values of time T usually provides the main results of the analysis. If node and link weighting factors are available, they should be used to get more accurate results.

Some conclusions can be made on the results. All the hijackers are in subcommunities D and E. All the nodes in subgroup G are hijackers on different planes that could be interpreted as an inner circle among hijackers. The central nodes in each of the four planes {1, 2, 3, 7} are in subcommunity D with one exception of node {7} in subcommunity E. Other nodes in subcommunity D were not on the plains. This kind of structures can be characteristic to covert networks. Subcommunities A and B are detected as clear-cut divisions of the terrorist network. Members of subcommunities A and B did not get on the planes. They were conduits for money and provided needed skills and services. In this manner, factions of the network can be helpful in profiling subcommunities and individual members of the network.

Figure 6 shows divisions of the terrorist network into two factions with four different choices of node weighting factors. The two factions are indicated with 0s and 1s (if more factions are searched with the algorithm, they are indicated correspondingly). In the first table node weighting factors are $w_N = 0.5$ for all the nodes. In the second table, node weighting factors are $w_N = 0.5$ for the 19 hijackers and $w_N = 0.25$ for the others. In the third table, node weighting factors are $w_N = 1.0$ and $w_N = 0.5$ correspondingly. In the fourth table, $w_N = 1.0$ for all the nodes. In all the tables, the time value of spreading process is $T = 0.5$, the Poisson intensity parameter $\lambda = 0.5$ and link weighting factors are $w_L = 1.0$. The first two columns show the number of nodes in two factions. The third column in the first table shows

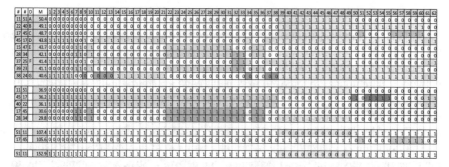

Fig. 6 Identified subcommunities with different node weights. In the following, the first number gives the weight factor used for the 19 dead hijackers and the second number gives the weighting factor for the rest of the nodes: $w_N = 0.5/0.5$, $w_N = 0.5/0.25$, $w_N = 1.0/0.5$, and $w_N = 1.0/1.0$. Model parameters $T = 0.5$ and $\lambda = 0$.

the label of the divisions shown in Fig. 1. The fourth column shows the numerical value of the measure in (10).

The results are analyzed starting from lines with the highest values of the measure in (10). The division between nodes denoted by A and the rest of the network is the most optimal. This division appears in all the four tables of Fig. 6. Seven different subcommunities or subgroups can be discovered form the first table.

Interestingly, when different node weighting factors ($w_N = 0.5$ and $w_N = 0.25$) for the hijackers and the rest of the network are used in the model, the second table shows a more clear picture and less divisions. The second line suggests that faction A+C is a subcommunity. Consistently, the third line also indicates that the overlapping A+B is a subcommunity. The fourth line discovers subcommunity E. Because more information is used in the second table compared to the first table, these results may be more accurate.

The third table with stronger connections also discovers division C. No other divisions are found because of high cohesion of the network. In the fourth table with very high equal node weighting factors, only one division A is discovered.

8 Conclusions

The method is based on a modeling framework to assess the influence of a node in a complex network with respect to replicating information via paths of different lengths between source and target nodes. The same framework is used consistently to compute closeness centrality and betweenness centrality measures as well as to detect subcommunities in the social network. Time is an important parameter in dynamic network spreading models. In the same network structure, fast and slow processes can exist at the same time.

These centrality measures are different from the classical reciprocal closeness centrality [14] and Freeman betweenness measures [6] which consider only the shortest paths between source and target nodes. In terrorist or criminal networks, possibilities of using all the paths of social network are important to maintain resilience of the network in case nodes are removed from paths between source and target nodes. The node betweenness measure of (8) in this paper and the Freeman betweenness measures have different values and rankings in the network [10, 11]. The difference can be explained by different usage and interpretations of the two betweenness measures.

The community detection algorithm can be used to analyze possible subcommunities or closely connected members of the network. Structures characteristic to covert social networks are discovered in the terrorist network of Al Qaeda. Members in the subcommunities have different roles in planning, financing, and carrying out the terrorist attacks. Subcommunities having only supporting functions are loosely connected to the more tightly connected core structures of the network. The subcommunities detected by the algorithm have no overlapping members. In the terrorist network, the subcommunity of seven terrorists is discovered who were all on three plains used in the crashes. All the five terrorists on the first plane, which crashed WTC north, except the most central node of the whole operation, are in this subcommunity. The five terrorists on the fourth plane, which crashed in Pentagon, are all in another subcommunity discovered by the algorithm. In this subcommunity are also ten other members of the terrorist network who were not on planes.

A new community detection method and a new node closeness centrality measure are proposed in this paper. The community detection method can also be used with many other closeness centrality measures. A new version of link betweenness centrality is also defined in this paper. The measure is a new tool for analyzing undirected or directed and weighted or non-weighted connections in social network structures.

References

1. Agneessens, F., Borgatti, S.P., Everett, M.G.: Geodesic based centrality: unifying the local and the global. Soc. Netw. **49**, 12–26 (2017)
2. Baker, W.E., Faulkner, R.: The social organization of conspiracy: illegal networks in the heavy electrical equipment industry. Am. Sociol. Rev. **58**(6), 837–860 (1993)
3. Borgatti, S.P.: Identifying sets of key players in a social network. Comput. Math. Organ. Theor. **12**, 21–34 (2006)
4. Borgatti, S.P., Everett, M.: A graph-theoretic perspective on centrality. Soc. Netw. **28**, 466–484 (2006)
5. Coscia, M., Giannotti, F., Pedreschi, D.: A classification for community discovery methods in complex networks. Stat. Anal. Data Min. **4**(5), 512–546 (2011)
6. Freeman, L.C.: Centrality in social networks: conceptual clarification. Soc. Netw. **1**, 215–239 (1979)
7. Girvan, M., Newman, M.E.J.: Community structure in social and biological networks. Proc. Natl. Acad. Sci. U S A **99**(12), 7821–7826 (2002)
8. Karrer, B., Newman, M.E.J.: Stochastic blockmodels and community structure in networks. Phys. Rev. E **83**(1), 016107 (2011)

9. Kernighan, B.W., Lin, S.: An efficient heuristic procedure for partitioning graphs. Bell. Syst. Tech. J. **49**, 291–307 (1970)
10. Krebs, V.E.: Mapping networks of terrorist cells. Connections **24**(3), 43–52 (2001)
11. Krebs, V.E.: Uncloaking terrorist networks. First Monday **7**(4) (2002)
12. Lancichinetti, A., Fortunato, S.: Community detection algorithms: a comparative analysis. Phys. Rev. E **80**, 056117 (2009)
13. Newman, M.E.J.: A measure of betweenness centrality based on random walks. Soc. Netw. **27**(1), 39–54 (2003)
14. Newman, M.E.J.: The structure and function of complex networks. SIAM Rev. **45**, 167–256 (2003)
15. Newman, M.E.J.: Detecting community structure in networks. Eur. Phys. J **B38**(2), 321 (2004)
16. Newman, M.E.J., Girvan, M.: Finding and evaluating community structure in networks. Phys. Rev. E **69**, 026113 (2004)
17. Newman, M.E.J.: Networks. An Introduction. Oxford University Press, Oxford (2010)
18. Ouellet, M., Bouchard, M., Hart, M.: Criminal collaboration and risk: the drivers of Al Qaeda's network structure before and after 9/11. Soc. Netw. **51**, 171 (2016)

Author Attribution Using Network Motifs

Younis Al Rozz and Ronaldo Menezes

Abstract The problem of recognizing the author of unknown text has concerned linguistics and scientists for a long period of time. The authorship of the famous Federalist Papers remained unknown until Mosteller and Wallace solved the mystery in 1964 using the frequency of functional words. After that, many statistical and computational studies were published in the fields of authorship attribution and stylistic analysis. Complex networks, gaining much popularity in recent years, may have a role to play in this field. Furthermore, several studies show that network motifs, defined as statistically significant subgraphs within a network, have the ability to distinguish networks from distinctive disciplines. In this paper, we succeed in the utilization of network motifs to distinguish the writing style of 10 famous authors. Using statistical learning algorithms, we achieved an accuracy of 77% in classifying 100 books written by ten different authors, which outperformed the results from other works. We believe that our method proved the importance of network motifs in author attribution.

1 Introduction

An author's writing style can be considered as an example of a behavioral biometric. The words used by people and the way they structure their sentences are unique and can frequently be used to identify the author of a certain work. The task of author attribution is gained attention among researchers in the fields of statistical physics, natural language processing, and data and information science. A thorough survey of the techniques used in authorship attribution can be found in [18]. Applications of authorship attribution are not only limited to the literature stylometry [4] but also expanded to other fields such as social media forensic [16] and e-mail fraud

Y. Al Rozz (✉) · R. Menezes
BioComplex Laboratory, Computer Science, Florida Institute of Technology,
Melbourne, FL, USA
e-mail: yyounis2013@my.fit.edu

R. Menezes
e-mail: rmenezes@cs.fit.edu

© Springer International Publishing AG 2018
S. Cornelius et al. (eds.), *Complex Networks IX*, Springer Proceedings
in Complexity, https://doi.org/10.1007/978-3-319-73198-8_17

detection [8]. As researchers find complex networks a promising field in linguistic studies [2], more and more authorship attribution works based on text networks saw the light of day. Measurement from word co-occurrence network topology combined with traditional statistical methods like frequency of functional words and intermittency were used to attribute authors [1, 3].

Network motif defined by Milo et al. [11] as a statically significant subgraphs pattern occurred in real-world networks compared to random ones has gained a lot of attention because of its ability in discriminating networks from different discipline [19]. In this work, we utilized network motifs as a fingerprint to attribute authors by their writing style. More precisely, we extract network motifs from directed co-occurrence networks of 100 books by 10 well-known authors and then we use five machine learning algorithms to classify the authors by their network motif signature. We show that four-node directed network motifs alone can be utilized to attribute authors of different books.

The paper is organized as follows. Section 2 is an overview of the efforts spent by other researchers on the subject of author attribution. In Sect. 3, we describe our dataset and steps taken place in order to extract the network motif from the text networks. The classification methods and results are explained in Sect. 4. Finally, we conclude our work in Sect. 5 with a road map for future work.

2 Motivation and Related Work

Several studies exist that deal with the importance of network motifs in natural language networks. The first attempt to classify different networks including word co-occurrence using network motifs was made by Milo et al. [12]. Li et al. [9] extracted and studied three and four-node directed motif structure of 72,923 two-character Chinese words network. They found that feed-forward loop (FFL) motif structure is significant in their network. Rizvić et al. [15] examined three-node (triads) network motifs extracted from directed co-occurrence networks of five Croatian texts and compared their results with other languages. They realized that there is a similarity between the Croatian language networks triad significance profiles and other previously studied languages. Cabatbat et al. [7] compared five-node network motifs among other network measures of the Bible and the Universal Declaration of Human Rights (UDHR) translations in eight languages. Pearson correlation coefficient and mutual information were used to compare the metrics of real texts with random texts from other sources. Their finding is that the distribution of network motif frequency is beneficial in recognizing similar texts. Biemann et al. [6] realized that motif signatures serve to discriminate co-occurrence networks of natural language from artificially generated ones. To assist their finding, they present additional results on peer-to-peer streaming, co-authorship, and mailing networks. The directed motif of size 3 and undirected motif of size 4 was used in their work.

All the previous works showed the ability of various size network motifs of discriminating text from different languages and genre. They did not utilize machine learning algorithms to support their findings. On the other hand, Marinho et al. [10] achieved 57.5% accuracy in their best scenario of attributing eight authors of 40 novels with three-node directed network motifs. An important aspect of author attribution task is the feature frequency [18]. To capture an author style more preciously, the feature should be more frequent. This motivates us to use the frequency of the 199 four-node directed network motif in an attempt to attribute the authors under study.

3 Datasets and Methodology

3.1 Data Collection and Network Creation

The dataset used in this work comprised of 100 literature books authored by 10 different authors, 10 books for each individual author. The books are listed in Table 1 and were collected from the Project Gutenberg Web site.[1] Each book was limited to 20 thousand words which are the length of the shortest book in the set. Text preprocessing steps were applied to remove punctuation, numbers and non-Latin alphabets, and all letters were converted to lowercase. We preserved functional words (stop words) in the text as their frequency has been proven to reflect stylistic aspects of the text and improve authorship attribution task [5, 13, 17]. A sample text from Charles Dickens's "A Christmas Carol" novel and the resulted preprocessed text are shown in (Fig. 1a, b, respectively) to illustrate this process.

Next, we created the directed co-occurrence networks from the result of the preprocessed text of the 100 books. Co-occurrence networks can be constructed based on the sentence, paragraph, or the whole text boundary. We chose the sentence boundary as it produces less dense network hence reduces the amount of time required to extract network motifs. Sentence boundary is defined by period, exclamation point, and question mark [14]. The network constructed from the preprocessed text is depicted in (Fig. 1c).

3.2 Feature Extraction

A plethora of network motif extraction tools exist, and each one has its pros and cons related to the number of motif's nodes count and the algorithm speed. We chose the iGraph[2] implementation for its flexibility and fast execution time. Tran et al. [19] suggested that small undirected network motifs cannot reveal differences among networks from different disciplines, while large ones do. Based on this argument and

---

[1] http://www.gutenberg.org.
[2] http://igraph.org.

Table 1 Authors used in our experiments and their book titles

Authors	Book titles
Bernard Shaw 1856–1950	Man and Superman, Candida, Arms and the Man, The Philanderer, Caesar and Cleopatra, Pygmalion, Major Barbara, Heartbreak House, The Devil's Disciple, Cashel Byron's Profession
Charles Dickens1812–1870	A Christmas Carol, A Tale of Two Cities, The Pickwick Papers, Oliver Twist, Great Expectations, David Copperfield, Little Dorrit, Our Mutual Friend, The Life and Adventures of Nicholas Nickleby, Dombey and Son
George Eliot 1819–1880	The Essays of George Eliot, Impressions of Theophrastus Such, Silas Marner, Scenes of Clerical Life, The Mill on the Floss, Adam Bede, Romola, Daniel Deronda, Felix Holt The Radical, MiddleMarch
Herbert George Wells 1866–1946	Tales of Space and Time, The Food of the Gods and How It Came to Earth, The Country of the Blind and Other Stories, The Invisible Man, The First Men in the Moon, The Island of Doctor Moreau, The War of the Worlds, The Time Machine, In the Days of the Comet, Ann Veronica
Jack London 1876–1916	The Call of the Wild, White Fang, The Iron Heel, Before Adam, Martin Eden, The People of the Abyss, The Night-Born, The Sea Wolf, South Sea Tales, The Valley of the Moon
Mark Twain 1835–1910	The Adventures of Tom Sawyer, Adventures of Huckleberry Finn, Life on the Mississippi, The Mysterious Stranger and Other Stories, A Tramp Abroad, Following the Equator, The Innocents Abroad, Roughing It, The Prince and The Pauper, A Connecticut Yankee in King Arthur's Court
Oscar Wilde 1854–1900	A House of Pomegranates, The Duchess of Padua, Vera, Lady Windermere's Fan, A Woman of No Importance, Intentions, An Ideal Husband, Lord Arthur Savile's Crime and Other Stories, The Importance of Being Earnest, The Picture of Dorian Gray
Sir Arthur Conan Doyle 1859–1930	Rodney Stone, The Adventures of Sherlock Holmes, A Duet, The Tragedy of The Korosko, The Refugees, Uncle Bernac, The Valley of Fear, The Hound of the Baskervilles, Sir Nigel, The Lost World
William Henry Giles Kingston 1814–1880	Hendricks the Hunter, The Three Lieutenants, The Three Midshipmen, The Three Commanders, Peter the Whaler, Ben Burton, The Three Admirals, Adventures in Africa, In the Wilds of Florida, Peter Trawl
William Shakespeare 1564–1616	Hamlet, Prince of Denmark, The Life of Henry the Fifth, The Merchant of Venice, The Tragedy of Antony and Cleopatra, The Tragedy of Coriolanus, The Tragedy of Julius Caesar, The Tragedy of King Lear, The Tragedy of Othello, Moor of Venice, The Tragedy of Romeo and Juliet, The Winter's Tale

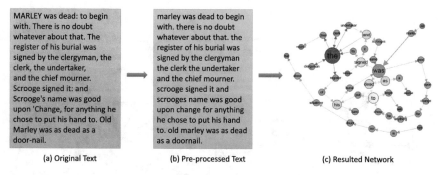

(a) Original Text   (b) Pre-processed Text   (c) Resulted Network

Fig. 1 Sample text from Charles Dickens's "A Christmas Carol" novel showing the stages of text preprocessing and the co-occurrence network created from the text

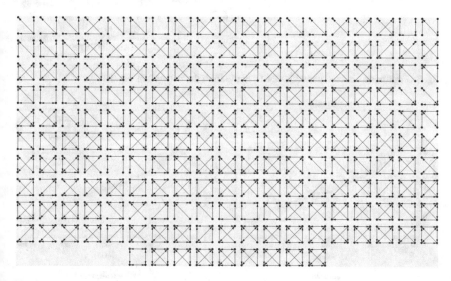

Fig. 2 199 different orientation of the directed four-node network motif

the importance of feature frequency explained at the end of Sect. 2, we chose the directed four-node network motifs shown in Fig. 2. For each book in the dataset, we extracted the 199 motifs from their directed network and then a data frame contains the motifs frequencies was created. Figure 3 illustrates a sample four-node directed motif extracted from the example network of Fig. 1c. The frequency distribution of the extracted four-node motifs from the books of Bernard Shaw, H.G. Wells, Jack London, and William Shakespeare is shown in Fig. 4.

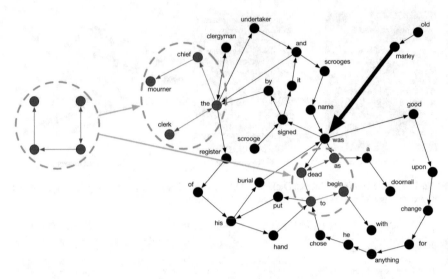

Fig. 3 Four-node directed network motif sample from the network of Fig. 1

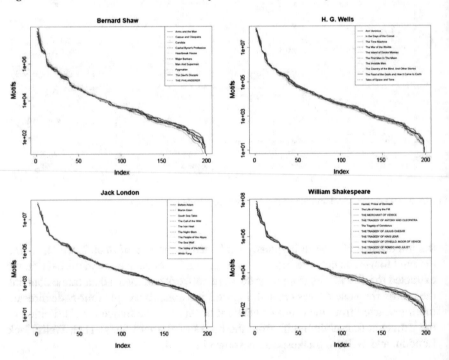

Fig. 4 Four-node network motif sorted frequency of the networks created from the books by Bernard Shaw, H.G. Wells, Jack London, and William Shakespeare

Table 2 Average classification accuracy results for the four-node directed motifs when splitting the dataset into 75% for training set and 25% for testing set with 100 times random shuffling

	Complete set	75%	50%	25%	10%
KNN	42	42	48	53	52
Decision tree	41	45	46	45	50
Random forest	56	58	58	59	64
SVM	53	56	62	63	67
MLP	66	68	70	70	68

4 Motif-Based Classification

For this part of the work, we utilized five supervised machine learning classification algorithms, namely K-nearest neighbors (KNN), decision trees, random forests, support vector machines (SVM), and multilayer perceptrons (MLP). They are all part of the scikit-learn[3] machine learning package for Python. As we try to attribute 10 authors, we have a multi-class classification problem with the number of samples ($N = 100$) which represents the number of books and the dimension of the feature set ($D = 199$) was relatively high. We used two cross-validation methods, the first one is to split our dataset into 75% training set and 25% testing set and then shuffle the dataset and repeat the operation for 100 times. The second method was leave-one-out, where the dataset is split into 99 sample for training and one sample for testing then iterates through the remaining samples. The average classification accuracy was calculated with both methods for all the algorithms used in the work. All the datasets were standardized by scaling to unit variance and removing the mean. The classification was performed on all the feature sets, that is the whole 199 four-node directed motifs and then recursive feature elimination (RFE) feature selection method is used to find the best 75, 50, 25, and 10% features, respectively. An alternative method mostly used in the literature is to choose significant motifs based on the highest Z-scores, but we preferred to collect the whole set of motifs and then use feature selection methods to choose the best set.

The results of classification using the first cross-validation method of shuffling and splitting the dataset are listed in Table 2, while Table 3 lists the results of the leave-one-out cross-validation method. As can be seen from both tables, the two basic classification algorithms KNN and the decision trees did not perform well compared to more sophisticated algorithms. Although KNN gives us an average accuracy of 60% when using 25% of the dataset and the leave-one-out cross-validation method, it is still lower than the accuracy obtained from the other classification methods. The best classification accuracy of 77% was obtained when the MLP classifier used with leave-one-out validation method.

---

[3]http://scikit-learn.org.

Table 3 Average classification accuracy results for the four-node directed motifs with leave-one-out cross-validation method

	Complete set	75%	50%	25%	10%
KNN	41	42	51	60	54
Decision tree	44	55	47	55	57
Random forest	61	56	62	60	65
SVM	61	66	70	66	71
MLP	72	73	75	77	72

5   Conclusion and Future Work

Throughout this work, we attempted to attribute 10 authors of 100 books using four-node directed network motifs. Functional words (stop words) were kept during text preprocessing as they proven by many previous works to reflect author style and increase the accuracy of attributing authors. The results we obtained herein outperformed other works when network motifs were the only feature used in attributing authors. Also, the number of 100 books used in this work is much higher than other works, which statistically means if we used the same smaller dataset, we will get better classification accuracy. This proves the importance of network motifs in recognizing the variety of writing styles among different authors. This opens the door for future work to generalize this method in attributing text from a different genre and translation assessment. Other possibilities are to study the effect of extracting higher motif order on the accuracy of classification.

References

1. Akimushkin, C., Amancio, D.R., Oliveira Jr., O.N.: Text authorship identified using the dynamics of word co-occurrence networks. PloS one **12**(1), e0170527 (2017)
2. Al Rozz, Y., Hamoodat, H., Menezes, R.: Characterization of written languages using structural features from common corpora. In: Workshop on Complex Networks CompleNet, pp. 161–173. Springer, Berlin (2017)
3. Amancio, D.R.: A complex network approach to stylometry. PloS one **10**(8), e0136076 (2015)
4. Arefin, A.S., Vimieiro, R., Riveros, C., Craig, H., Moscato, P.: An information theoretic clustering approach for unveiling authorship affinities in Shakespearean era plays and poems. PloS one **9**(10), e111445 (2014)
5. Biber, D.: Variation Across Speech and Writing. Cambridge University Press, Cambridge (1991)
6. Biemann, C., Krumov, L., Roos, S., Weihe, K.: Network motifs are a powerful tool for semantic distinction. Towards a Theoretical Framework for Analyzing Complex Linguistic Networks, pp. 83–105. Springer, Berlin (2016)
7. Cabatbat, J.J.T., Monsanto, J.P., Tapang, G.A.: Preserved network metrics across translated texts. Int. J. Mod. Phys. C **25**(02), 1350092 (2014)

8. Chen, X., Hao, P., Chandramouli, R., Subbalakshmi, K.P.: Authorship similarity detection from email messages. In: International Workshop on Machine Learning and Data Mining in Pattern Recognition, pp. 375–386. Springer, Berlin (2011)
9. Li, J., Xiao, F., Zhou, J., Yang, Z.: Motifs and motif generalization in Chinese word networks. Procedia Comput. Sci. **9**, 550–556 (2012)
10. Marinho, V.Q., Hirst, G., Amancio, D.R.: Authorship attribution via network motifs identification. In: 2016 5th Brazilian Conference on Intelligent Systems (BRACIS), pp. 355–360. IEEE (2016)
11. Milo, R., Shen-Orr, S., Itzkovitz, S., Kashtan, N., Chklovskii, D., Alon, U.: Network motifs: simple building blocks of complex networks. Science **298**(5594), 824–827 (2002)
12. Milo, R., Itzkovitz, S., Kashtan, N., Levitt, R., Shen-Orr, S., Ayzenshtat, I., Sheffer, M., Alon, U.: Superfamilies of evolved and designed networks. Science **303**(5663), 1538–1542 (2004)
13. Mosteller, F., Wallace, D.L.: Inference in an authorship problem: a comparative study of discrimination methods applied to the authorship of the disputed federalist papers. J. Am. Stat. Assoc. **58**(302), 275–309 (1963)
14. Nunberg, G.: The Linguistics of Punctuation. CSLI Lecture Notes. Cambridge University Press, Cambridge (1990)
15. Rizvić, H., Martinčić-Ipšić, S., Meštrović, A.: Network motifs analysis of croatian literature. arXiv:1411.4960 (2014)
16. Rocha, A., Scheirer, W.J., Forstall, C.W., Cavalcante, T., Theophilo, A., Shen, B., Carvalho, A.R.B., Stamatatos, E.: Authorship attribution for social media forensics. IEEE Trans. Inf. Forensic Secur. **12**(1), 5–33 (2017)
17. Segarra, S., Eisen, M., Ribeiro, A.: Authorship attribution through function word adjacency networks. IEEE Trans. Sig. Process. **63**(20), 5464–5478 (2015)
18. Stamatatos, E.: A survey of modern authorship attribution methods. J. Assoc. Inf. Sci. Technol. **60**(3), 538–556 (2009)
19. Tran, N.T.L., DeLuccia, L., McDonald, A.F., Huang, C.-H.: Cross-disciplinary detection and analysis of network motifs. Bioinform. Biol. Insights **9**, 49 (2015)

Complex Networks Reveal a Glottochronological Classification of Natural Languages

Harith Hamoodat, Younis Al Rozz and Ronaldo Menezes

Abstract The success of humans cannot be attributed to language, but it is certainly true that language and modern humans are inseparable. This work focuses on revealing the structure of 20 Indo-European languages belonging to three sub-families (Romance, Germanic, and Slavic) from a chronological perspective. In order to find the chronological characteristic features of these languages, we use (1) Heaps' law, which describes the growth of vocabulary (distinct words) in a corpora for each language to the total number of words in the same corpora and (2) structural properties of networks created from word co-occurrence in corpora of 20 written languages. Using clustering approaches and entanglement, we show that in spite of differences from years of being used separately and differences in alphabets, one can find language characteristics that lead to cluster of languages resembling the organization according to historical sub-families and chronological relations.

1 Introduction

The development of societies leads to the use of different tones and words creating different dialects for the same language. Over time, those dialects change by adding or removing words until they are considered as a new language. Moreover, the migration of human populations groups contributed to the formation of languages because the geographical separation of populations acts as a catalyst for changes in vocabulary. In fact, this analogy is similar to how different species emerged as a result of geographical separation. This evolution of language formation means that today there are thousand of different languages currently being used [17]. Due to

H. Hamoodat (✉) · Y. Al Rozz · R. Menezes
BioComplex Laboratory, Computer Science, Florida Institute of Technology,
Melbourne, FL, USA
e-mail: hhamdon2013@my.fit.edu

Y. Al Rozz
e-mail: yyounis2013@my.fit.edu

R. Menezes
e-mail: rmenezes@cs.fit.edu

© Springer International Publishing AG 2018
S. Cornelius et al. (eds.), *Complex Networks IX*, Springer Proceedings
in Complexity, https://doi.org/10.1007/978-3-319-73198-8_18

209

the nature of their formation, many of these languages can be grouped together into a language family. The languages in each family are related through descent to a common ancestral language. Parental languages transfer some of its characteristics to derived languages; thus, we can say that the derived languages within a language family are "genetically" related [23].

There are about 100 language families in the world, e.g., Indo-European, Afro-Asiatic, and Niger-Congo. The Indo-European family has the largest number of speakers among all families known (more than 40% of the human population). It contains about 445 languages many of them widely spoken such as Spanish, English, Russian, and Portuguese [10]. According to Linguists, the Indo-European family can be divided into several sub-families such as Germanic, Italic-Romance, Slavic, and Baltic [7].

The availability of large volumes of data today encourages researchers to study the relation between languages using regularities extracted from corpora of text. In this work, we show that even without lexical distance analysis or word-pair relations, and focusing merely on the structure built from syntax, we can detect useful structure of language families.

2 Related Work

Although a number of studies have been done in the history of languages and how they derived from each other, there is no unanimity on the origin of human languages because of the lack of direct evidence and empirical data [4]. Due to the difficulty to determine the specific date of language separation, the researchers try to study the relationship between languages and convert the result into an estimate for when a pair diverged. However, the calculation of the distance between pair of language is one of the most efficient methods to use it for chronological estimation. Linguistic distance—how different one language or dialect is from another [22]—can be computed by the lexical distance of the language vocabulary [12, 21].

There are several distance measure algorithms that can be applied on text like Hamming distance, Levenshtein distance, and Jaro–Winkler distance [25]. Levenshtein is commonly used, and it is a metric for measuring the difference between two string sequences. The Levenshtein distance between two words is the minimum number of single-character edits (insertions, deletions, or substitutions) required to change one word into the other.

Petroni and Serva [21] created a chronological family tree for Indo-European and the Austronesian group. They used fifty different languages for both cases depending on two Swadesh list dataset, one for Indo-European and one for Austronesian. The authors created matrices of the lexical distances between languages for the two families. Each matrix contains 1225 elements to describe all pairs in a group. Then, they calculated the absolute timescale for those pairs. In order to calculate the distance between each language pair, one takes the average of the distances between the word pairs. They used a modification of the Levenshtein distance and normalized it by the number of characters for longer of the two words, which is reasonable if two words differ by one character this is much more important for short words than it is for long words. They found that the result from the method above is relatively similar to those found by glottochronologists.

The use of a cognate set of words to study the time of language divergence is not new. In fact, Gray et al. [11] studied the time separation between 87 Indo-European languages from a dataset of 2,449 cognate sets coded as discrete binary characters. They applied likelihood models of lexical evolution to solve the problem of accuracy of tree topology and branch length estimation. A Bayesian inference of phylogeny was used to enhance the estimation of tree topology and branch lengths. Also, they used rate-smoothing algorithms to reduce the rate variation across the tree. Last, they tried to examine subsets of languages using split decomposition, and the result showed a strong identity for the tree when comparing a subset result with complete one. They found the results are in agreement with the Anatolian theory for the origin time of Indo-European languages. Furthermore, a number of studies have been done for the classification of languages using text characteristics without looking to the time divergence [2, 15].

3 Methodology

3.1 Data Curation and Model

In this work, we utilize a large amount of textual data called the Leipzig corpora collection [9]. The languages chosen for this work were Romanian (Ron), French (Fra), Catalan (Cat), Italian (Ita), Spanish (Spa), Portuguese (Por), German (Ger), Dutch (Dut), Danish (Dan), Norwegian (Nor), Swedish (Swe), English (Eng), Slovenian (Slv), Bulgarian (Bul), Polish (Pol), Russian (Rus), Ukrainian (Ukr), Croatian (Cro), Czech (Ces), and Slovak (Slk). These languages were chosen because they are good representations of three large sub-families of the Indo-European family, which are Italic, Germanic, and Slavic. The text corpus for each language was constructed from Wikipedia and news pages to ensure vocabulary diversity. We made the size of the corpus for each language consistent; each language corpus is composed of 1 million sentences. After the entire text was converted to lower case, and the punctuation and special characters were removed, we used 100,000 words from each corpus for the work developed in this paper. The second kind of data we used relates to the languages, tree topology, branches length, and divergence period between languages (year the languages separate), which we reconstructed from several works [11, 12, 21] in linguistics. This hierarchy was done for the 20 languages we deal with in this paper and is used as the ground truth (see Fig. 1).

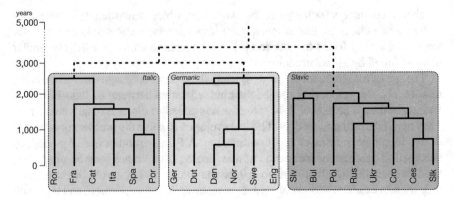

Fig. 1 A dated phylogenetic tree of 20 Indo-European languages with three sub-families, Italic, Germanic, and Slavic. The dates on the y-axis are approximations for when these languages split from a common language

3.2 Feature Extraction

We extracted a set of 19 features for each language; we want to demonstrate that one could use these features (or some of them) to unveil a structure similar to the ground truth. The first two features represent the vocabulary richness of the language as expressed by Heaps' law [13]. The parameters k and β describe the vocabulary growth (distinct words) in texts as a function the total number of words seen [2, 16]. More formally, $V_R(n) = kn^{\beta}$ where V_R is the number of vocabulary words in the text of size n, k and β are parameters determined experimentally from the fitting of Heaps' law.

The other 17 features were obtained from the word co-occurrence network for each language. The network is simple and built having words as nodes and linking words if they appear in the corpus consecutively. The edges' weights represent the frequency in which the two words appear next to each other. The networks follow a power-law distribution and have community structures (we used Louvain modularity [3]); the number of communities is an important feature (*com*). The features α_d and α_s represent, respectively, the scaling of the degree distribution and the distribution of community sizes. The size of the network is given by the number of nodes n and number of edges m.

There are many other structural characteristics that can be computed from the networks. For this work, we exhaustively added many features without too much concern for an exact number. The purpose is to make sure we are capturing as many uncorrelated metrics as possible. Later we worked reducing the dimensions and identifying the most significant parameters. The degree k of a node is the number of edges connected to it. The higher average degree $\langle k \rangle$ the network has, the more density it is [6]. From Table 1, we can clearly see that the Slavic languages have a

Table 1 Each line in this table represents 19-dimension feature vector for the language shown in the first column

Languages	κ	β	α_d	α_s	n	m	$\langle k \rangle$	C_4	C	$\langle C_d \rangle$	$\langle C_b \rangle$	$\langle C_c \rangle$	D	$trans$	η_∇	ℓ	r	Q	com
Portuguese	6.40	0.702	2.302	1.343	20,641	70,816	6.86	0.044	0.186	0.00033	0.00010	0.305	11	0.0103	11.729	3.331	-0.135	0.392	47
Spanish	7.63	0.694	2.323	1.462	22,258	73,026	6.56	0.059	0.241	0.00030	0.00010	0.315	14	0.0088	12.972	3.217	-0.227	0.351	111
Italian	8.28	0.689	2.291	1.399	22,885	77,693	6.79	0.035	0.170	0.00030	0.00010	0.302	13	0.0113	11.721	3.357	-0.223	0.363	55
Catalan	7.69	0.686	2.324	1.335	20,856	68,005	6.52	0.073	0.277	0.00030	0.00010	0.322	10	0.0084	13.551	3.151	-0.210	0.364	44
French	7.41	0.690	2.289	1.324	20,700	73,241	7.08	0.051	0.257	0.00030	0.00010	0.322	09	0.0109	16.628	3.146	-0.245	0.336	58
Romanian	8.91	0.683	2.307	1.252	22,821	75,361	6.60	0.043	0.175	0.00028	0.00010	0.305	10	0.0106	11.306	3.325	-0.185	0.371	33
Dutch	6.54	0.700	2.175	3.529	20,485	72,745	7.10	0.081	0.320	0.00030	0.00010	0.326	11	0.0157	26.030	3.102	-0.219	0.304	31
German	0.23	1.008	2.214	1.427	24,296	73,841	6.08	0.088	0.260	0.00020	0.00009	0.317	10	0.0120	16.121	3.200	-0.195	0.352	112
Danish	5.70	0.720	2.217	4.804	22,234	71,612	6.44	0.066	0.246	0.00020	0.00010	0.311	10	0.0130	16.535	3.259	-0.183	0.358	34
Norwegian	6.13	0.706	2.231	4.456	20,571	63,997	6.22	0.090	0.298	0.00030	0.00010	0.322	10	0.0108	15.349	3.143	-0.210	0.364	30
Swedish	4.65	0.743	2.186	1.330	24,071	70,887	5.89	0.081	0.278	0.00020	0.00010	0.316	11	0.0086	11.808	3.209	-0.199	0.386	44
English	9.88	0.650	2.368	1.404	17,448	68,762	7.88	0.074	0.318	0.00040	0.00010	0.339	09	0.0107	22.913	2.994	-0.193	0.291	47
Bulgarian	5.41	0.728	2.449	1.854	23,655	58,746	4.97	0.061	0.185	0.00020	0.00009	0.306	17	0.0034	5.091	3.323	-0.189	0.503	496
Slovenian	7.58	0.716	2.343	1.791	28,669	83,470	5.82	0.037	0.122	0.00020	0.00008	0.286	11	0.0105	8.593	3.558	-0.117	0.396	62
Russian	7.51	0.719	2.334	4.502	29,333	81,405	5.55	0.045	0.123	0.00010	0.00008	0.285	10	0.0057	5.415	3.576	-0.112	0.428	57
Ukrainian	4.41	0.765	2.345	2.629	29,363	78,155	5.32	0.054	0.147	0.00018	0.00008	0.289	15	0.0066	5.654	3.543	-0.159	0.438	36
Czech	4.71	0.765	2.387	1.878	31,486	83,320	5.29	0.041	0.101	0.00016	0.00008	0.274	12	0.0057	4.298	3.726	-0.086	0.438	64
Slovak	7.07	0.733	2.288	2.305	32,542	87,625	5.39	0.029	0.086	0.00016	0.00008	0.270	13	0.0075	4.896	3.775	-0.081	0.431	65
Croatian	7.31	0.716	2.317	2.003	27,369	63,826	4.66	0.039	0.144	0.00017	0.00010	0.267	14	0.0040	2.693	3.819	-0.134	0.550	132
Polish	5.92	0.734	2.390	3.155	27,390	72,721	5.31	0.048	0.122	0.00019	0.00009	0.277	16	0.0082	5.123	3.678	-0.130	0.470	70

lower $\langle k \rangle$ compared to all other languages in the dataset, while the English language has the higher one.

In addition to the network clustering coefficient (C), a measure of the degree to which nodes in a graph tend to cluster together, we calculate the square clustering (C_4) which is the quotient between the number of squares and the total number of possible squares [14].

Similar to the concept of clustering (C) is the concept of transitivity ($trans$) [24] of the network. Moreover, both C and $trans$ depend on the number of triangles (cliques of three nodes) in the network, so we have also included these features ($trans$ and $\eta_\triangledown$, respectively) as part of our set of metrics. Another important feature of networks is the average path length (ℓ) between two nodes which is also included in our list. Croatian has the longest value for $\ell = 3.81$ steps, while the shortest one was English with $\ell = 2.99$. This is likely because morphological languages like most of Slavic languages tend to have long sentences than analytic languages like English and Dutch [1]. The diameter of the network D is the largest shortest path and another important feature we included here. Note that at this point, the idea is to have an exhaustive list of features that could represent a language.

Related to community detection algorithms is the modularity of the network given by Q which is designed to measure the strength of a division of a network into groups; a measure the community structure [8]. The value of Q for all 20 networks was calculated using the approach proposed by Newman [19]. Based on this metric, Croatian has the largest modularity value of 0.550, while the lowest value was 0.291 scored by English.

Centrality measures are used to identify the important nodes within a network; here we used degree centrality (C_d) which is highly correlated to $\langle k \rangle$, betweenness (C_b), and closeness (C_c) as defined by Borgatti [5]. However since we want a network feature, we represent the average of all these values given by $\langle C_d \rangle$, $\langle C_b \rangle$, and $\langle C_c \rangle$. Last, we compute the degree assortativity of the network which is given by r [18].

After all the analysis, we had a 19-dimension feature vector for each language as depicted in Table 1. This vector is used in clustering the networks, but we will also try to identify the significant features and reduce the dimension.

4 Results and Discussion

In order to compare the tree resulted from the hierarchical clustering with the ground truth tree (Fig. 1), we measured the quality of the alignment of the two trees by calculating the entanglement function. Entanglement is a measure between 1 (full entanglement) and 0 (no entanglement) which corresponds to a good alignment. We took all the possible combination of the 19 parameters in the matrix, for each combination, we constructed a tree and compared it with the ground truth in order to find the entanglement. Table 2 contains the best 10 entanglement from all combinations. Furthermore, we can evoke which features have high impact on the results like

Table 2 Best 10 Entanglement with its combinations

Entanglement	κ	β	α_d	α_s	n	m	$\langle k \rangle$	C_4	C	$\langle C_d \rangle$	$\langle C_b \rangle$	$\langle C_c \rangle$	D	$trans$	$\eta_\triangledown$	ℓ	r	Q	com
0.0602616					✓		✓	✓				✓		✓		✓	✓		
0.0602616					✓		✓	✓				✓		✓		✓	✓		✓
0.0604673			✓			✓	✓							✓	✓	✓	✓	✓	✓
0.0604673			✓		✓	✓	✓					✓		✓	✓		✓	✓	✓
0.0604673			✓		✓	✓	✓							✓	✓	✓	✓	✓	✓
0.0604673			✓			✓	✓							✓		✓	✓	✓	✓
0.0604673			✓		✓	✓								✓	✓	✓	✓	✓	✓
0.0653795			✓						✓	✓		✓		✓		✓	✓		✓
0.0663400			✓				✓		✓		✓	✓			✓		✓		
0.0687276					✓		✓	✓						✓		✓	✓		✓

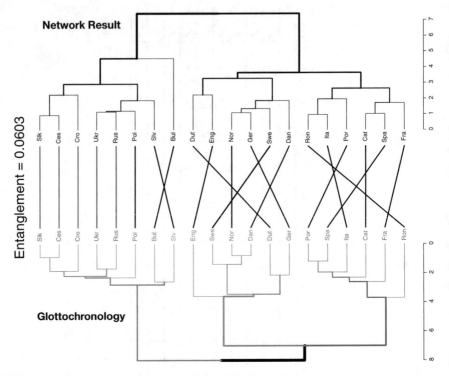

Fig. 2 Entanglement between two trees using the best entanglement case

$trans, r, \ell,$ and $\langle k \rangle$ which they appeared in the most cases. In contrast, there are some parameters useless for this work like Heaps' law parameters and $\langle C_b \rangle$ (Table 2).

The best combination between all the cases has the entanglement value of 0.06 (first case in Table 2), this case has only seven parameters, which are the smallest combination parameters that give better values (Fig. 2). The hierarchical clustering was not only able to distinguish the Slavic languages from the non-Slavic language but also to capture the branches relation and distances for this sub-family with one exception which is the Bulgarian language (discussed later). Moreover, it was ambidextrous to recognize the Germanic from Romance languages with some differences in the branches relation like Germany with Norwegian instead of the Dutch language.

In order to check the consistency of result, we tested the sensitivity of removing languages. First, we remove one language each time and calculate the average entanglement for all cases. Secondly, we remove two languages and calculate the average entanglement, and so on (Fig. 3b). The average entanglement increased until the sixth language removed and then started to be constant at a high level, which means that the topology of the tree is completely destroyed and the removal of more languages does not affect the result.

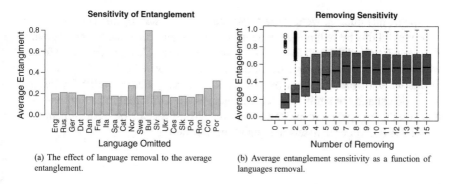

(a) The effect of language removal to the average entanglement.

(b) Average entanglement sensitivity as a function of languages removal.

Fig. 3 Entanglement sensitivity as a function of removing languages

To test for certain language impact on the average entanglement and tree topology, we removed one language each time and recalculated the average entanglement. The language with high average entanglement in Fig. 3a means the most effective language on the tree topology. In our languages set, when we removed the Bulgarian language which occupied a whole branch in the network result cluster, the average entanglement became very high (0.79) which means the branches relation is very tangled. The unpredictable behavior of the Bulgarian language may be due to several reasons; first, the number of unique words (nodes) is less than others Slavic languages. Also, words in the Bulgarian language are most likely to connect with another word several times which describes the reason why the language has a number of connections less than all other language networks in the dataset. On the other hand, several important dissimilarities exist between the Bulgarian language and other Slavic languages. For instance, Bulgarian is an analytic language and its unique morphological features tend toward the Balkan family of languages. The Bulgarian language roots back to the Proto-Slavic branch of the Indo-European language family which have common features with the Indo-Iranian languages, more specifically, the Germanic family, but it was much similar to the Baltic family of languages. Finally, a lot of the words in the Bulgarian language were borrowed from the Turkish and Greek languages [20].

5 Conclusion

In this study, we used the topological measurements extracted from word co-occurrence networks of 20 Indo-European languages along Heaps' law parameters to construct the hierarchical cluster that represents the chronological distance between those languages. The comparison that we made of our results with the glottochronological classification based on the lexical distance between word fluctuation among different languages shows a strong agreement between the two methods. In order to support this finding, we test the tolerance of the cluster against languages variation. We did this by removing one language a time and calculate the entanglement. Also, we extracted the best features that give the lowest entanglement; these features we believe they best describe the chronological difference between languages. The results we get from this work open the door for many future works; for instance, we could expand our study to include languages from different main families. Also, it is possible to apply our method to find the closest translation of document to the original text in order to assets the quality of translation.

References

1. Abramov, O., Mehler, A.: Automatic language classification by means of syntactic dependency networks. J. Quant. Linguist. **18**(4), 291–336 (2011)
2. Al Rozz, Y., Hamoodat, H., Menezes, R.: Characterization of written languages using structural features from common corpora. In: Workshop on Complex Networks CompleNet, pp. 161–173. Springer, Berlin (2017)
3. Blondel, V.D., Guillaume, J.-L., Lambiotte, R., Lefebvre, E.: Fast unfolding of communities in large networks. J. Stat. Mech. Theory Exp. **2008**(10), P10008 (2008)
4. Bolhuis, J.J., Tattersall, I., Chomsky, N., Berwick, R.C.: How could language have evolved? PLoS Biol. **12**(8), e1001934 (2014)
5. Borgatti, S.P.: Centrality and network flow. Soc. Netw. **27**(1), 55–71 (2005)
6. Bosu, A., Carver, J.C.: How do social interaction networks influence peer impressions formation? a case study. In: IFIP International Conference on Open Source Systems, pp. 31–40. Springer, Berlin (2014)
7. Campbell, L.: American Indian Languages: The Historical Linguistics of Native America. Oxford University Press, Oxford (2000)
8. de Arruda, H.F.: Costa, L.da F., Amancio, D.R.: Topic segmentation via community detection in complex networks. Chaos Interdiscip. J. Nonlinear Sci. **26**(6), 063120 (2016)
9. Goldhahn, D., Eckart, T., Quasthoff, U.: Building large monolingual dictionaries at the leipzig corpora collection: from 100 to 200 languages. In: LREC, pp. 759–765 (2012)
10. Gordon, R.G., Grimes, B.F., et al.: Ethnologue: Languages of the World, vol. 15. SIL International, Dallas (2005)
11. Gray, R.D., Atkinson, Q.D.: LangUage-Tree Divergence Times Support the Anatolian Theory of Indo-European Origin, vol. 426. Nature Publishing Group, London (2003)
12. Gray, R.D., Atkinson, Q.D., Greenhill, S.J.: Language evolution and human history: what a difference a date makes. Philos. Trans. R. Soc. Lond. B Biol. Sci. **366**(1567), 1090–1100 (2011)
13. Herdan, G.: Type-Token Mathematics, vol. 4. Mouton, Berlin (1960)
14. Lind, P.G., Gonzalez, M.C., Herrmann, H.J.: Cycles and clustering in bipartite networks. Phys. Rev. E **72**(5), 056127 (2005)
15. Liu, H., Xu, C.: Can syntactic networks indicate morphological complexity of a language? EPL (Europhys. Lett.) **93**(2), 28005 (2011)
16. Lü, L., Zhang, Z.-K., Zhou, T.: Zipf's law leads to heaps' law: analyzing their relation in finite-size systems. PloS One **5**(12), e14139 (2010)
17. McWhorter, J.H.: The Story of Human Language. Teaching Company (2004)
18. Newman, M.E.J.: Assortative mixing in networks. Phys. Rev. Lett. **89**(20), 208701 (2002)
19. Newman, M.E.J.: Modularity and community structure in networks. Proc. Natl. Acad. Sci. **103**(23), 8577–8582 (2006)
20. Osenova, P.: Bulgarian. Revue belge de philologie et d'histoire **88**(3), 643–668 (2010)
21. Petroni, F., Serva, M.: Language distance and tree reconstruction. J. Stat. Mech. Theory Exp. **2008**(08), P08012 (2008)

22. Renfrew, C., McMahon, A., Trask, R.L.: Time depth in historical linguistics. The Macdonald Institute for Archaelogical Research (2000)
23. Rowe, B.M., Levine, D.P.: A Concise Introduction to Linguistics. Routledge, Abingdon-on-Thames (2015)
24. Schank, T., Wagner, D.: Approximating clustering-coefficient and transitivity. Universität Karlsruhe, Fakultät für Informatik (2004)
25. Van der Loo, M.P.J.: The stringdist package for approximate string matching. R J. 2 (2014)

A Percolation-Based Thresholding Method with Applications in Functional Connectivity Analysis

Farnaz Zamani Esfahlani and Hiroki Sayama

Abstract Despite the recent advances in developing more effective thresholding methods to convert weighted networks to unweighted counterparts, there are still several limitations that need to be addressed. One such limitation is the inability of the most existing thresholding methods to take into account the topological properties of the original weighted networks during the binarization process, which could ultimately result in unweighted networks that have drastically different topological properties than the original weighted networks. In this study, we propose a new thresholding method based on the percolation theory to address this limitation. The performance of the proposed method was validated and compared to the existing thresholding methods using simulated and real-world functional connectivity networks in the brain. Comparison of macroscopic and microscopic properties of the resulted unweighted networks to the original weighted networks suggests that the proposed thresholding method can successfully maintain the topological properties of the original weighted networks.

1 Introduction

Network science has become an integral part of analyzing complex systems whose aggregate behavior cannot be explained by the summation of their parts [1]. This is in part due to the rapid advancement of data acquisition techniques that enable empirical measurement from components of complex systems, where these measurements can be used to estimate the links (similarities) between the system components. For example, neuroimaging data such as functional magnetic resonance imaging (fMRI) and electroencephalogram (EEG) are extensively used in computational neuroscience

F. Z. Esfahlani (✉) · H. Sayama
Department of Systems Science and Industrial Engineering, Center for
Collective Dynamics of Complex Systems, Binghamton University, P.O. Box 6000,
Binghamton, NY 13902-6000, USA
e-mail: fzamani1@binghamton.edu

H. Sayama
e-mail: sayama@binghamton.edu

© Springer International Publishing AG 2018   221
S. Cornelius et al. (eds.), *Complex Networks IX*, Springer Proceedings
in Complexity, https://doi.org/10.1007/978-3-319-73198-8_19

to study the connectivity between different regions of the brain [2]. However, since the empirical data often include measurement noise and the connection between components of the system (nodes of the network) are generally estimated using the statistical measurements, the resulted networks are dense graphs including many weak links. Analyzing such dense networks is often challenging due to the larger memory requirement, higher computational time complexity, and the limited number of measures to characterize the topology and properties of the weighted networks. Hence, usually, the weighted networks are mapped to an unweighted counterpart by binarizing the edge weights using a specific threshold where the connection weights smaller than the predefined threshold are discarded.

Various thresholding methods have been introduced in the literature, which can be categorized into two main categories of "absolute thresholds" and "proportional thresholds" [3]. The absolute thresholding methods use a fixed threshold value to binarize the weighted networks, whereas the proportional methods generally use some statistics of the connection weights (such as mean, median, or the p-th percentile) for thresholding the weighted networks. However, none of these methods take into account the topological integrity of the original weighted network, which becomes problematic when some key edges that are critical for maintaining the macroscopic and microscopic properties of the network are removed. This is especially important as it has been shown that such topological changes can significantly impact the derived network metrics such as centrality measures [4]. Furthermore, the majority of thresholding methods in the literature neglect the importance of weak links, which has been shown to provide useful information about the underlying properties of the network [5, 6].

According to our best knowledge, the only method that has partially addressed the previous challenges is the 99% connectedness method by Bassett et al. [7]. Even though this method takes into account the overall connectedness of the nodes, it assumes the original weighted network is dense. In other words, for sparser networks where the average node degree is very small, this method does not guarantee the topological integrity of the network. This might not be a problem for functional connectivity analysis, but it is problematic in effective connectivity analysis where the resulted weighted networks are generally sparse.

To tackle these issues, here we propose a new thresholding method based on the percolation theory which takes into account the whole network topology during the binarization process. More specifically, we use the maximum threshold that maintains the topological integrity of the original network, which in this case is defined as the size of the largest connected component in the network. The rationale behind this method is to retain a minimum number of edges that keep the network in the same level of global connectedness. The performance of the proposed method was compared to existing statistical thresholding methods using both simulated (based on a simple linear model) and real-world (Attention Deficit Hyperactivity Disorder-200 (ADHD-200) and the Center for Biomedical Research Excellence (COBRE)) datasets. According to the results, the proposed percolation-based thresholding method was capable of maintaining the topological properties of the original weighted network at both macroscopic and microscopic levels.

2 Materials and Methods

2.1 Percolation-Based Thresholding Method

The basic idea of the percolation-based thresholding method is to identify the minimum number of edges that maintain the giant component identified in the original weighted network. To achieve this, we start binarizing the network using an initial threshold (i.e., the maximum edge weight). Next, we characterize the size of the largest connected component in the binarized network as a function of edge weight threshold θ (we call this $n(\theta)$) and compare it with the size of the largest connected component in the original network (denoted by n_0), which is usually the same as the number of nodes for most weighted networks constructed using real-world datasets. This process is repeated by gradually decreasing θ, until the critical threshold θ_c is achieved, which is the first (largest) threshold value that satisfies

$$n(\theta_c) = \alpha n_0. \tag{1}$$

Here α is the level of the connectedness of the network. In this paper, we used $\alpha = 1$ so that we can guarantee the same connectedness of the network after binarization. Relaxing this criterion (using $\alpha < 1$) results in more sparse networks, which could be beneficial for specific applications. However, since we want to minimize the impact of thresholding on the topological integrity of the network, we keep α at one in this study. Figure 1 provides a visual summary of the proposed percolation-based thresholding method.

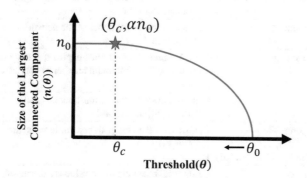

Fig. 1 A schematic illustration of the percolation-based thresholding method. First, the weighted network is binarized using an initial edge weight threshold (θ_0). Next, the largest connected component in the binarized network is characterized as a function of edge weight threshold θ and compared to the size of the largest connected component in the original network (n_0). The threshold θ is gradually decreased until the stopping criterion $n(\theta_c) = \alpha n_0$ is reached

2.2 Performance Evaluation

To evaluate the effectiveness of the thresholding method in maintaining the properties of the original weighted networks, a simple distance measure (d) was used to calculate the difference between the original and thresholded network properties:

$$d = |x_t - x_w|. \tag{2}$$

Here x_t represents the network property of the thresholded network and x_w represents the same network property calculated using the original weighted network. The distance value between different macroscopic and microscopic network properties from the thresholded networks using percolation-based thresholding were compared to several proportional thresholding methods including mean, p-th percentile of the edge weights ($p \in \{\%99, \%95, \%75, \%50(median), \%25, \%5, \%1\}$) [8], and methods proposed by Bassett et al. [7] including weighted average node degree that must be maintained at a minimum number of connected nodes, the connectedness of at least 99 % of the nodes, and a fixed thresholding method of average degree where the average degree must not be smaller than the $\ln(|nodes|) * 2$. Table 1 provides a summary of the methods examined in this study.

Table 1 Thresholding Methods. Label # refers to the corresponding label in the horizontal axis of Figs. 3, 5, and 6

Label	Method	Short description	Reference
1	Proposed percolation-based method	Identify the minimum number of edges that maintain the giant component in the original weighted network	
2	99% connectedness	At least 99% of the nodes of the network must be connected	[7]
3, 4	Average degree	(a) Weighted average node degree that must be maintained at a minimum number of connected nodes	[7]
		(b) The average degree must not be smaller than the $\ln(|nodes|) * 2$	
5, 6, 7 8, 9, 10, 11	p-th percentile $p \in$ {%99, %95, %75, %50 (median), %25, %5, %1}	Edges less than the p-th percentile value are discarded	[8]
12	Mean	Edges less than the mean value are discarded	[8]

2.3   Datasets

We have used three simulated datasets and two real-world datasets to test the performance of the proposed percolation-based thresholding method. The simulated datasets with node sizes of 64, 125, and 216 (i.e., the number of Regions of Interest (ROI)) were generated using a simple linear model with random design matrices as described in [9]. More specifically we have

$$y = MB + e, \tag{3}$$

where B is the weight matrix, M is the design matrix, and e is the random noise. Here, B corresponds to a 3D image with five blocks at the corners and one in the middle to simulate active brain regions (Fig. 2), M is random normal variables smoothed with Gaussian fields to imitate the observed fMRI data, and e is the Gaussian random noise chosen such that we have a signal-to-noise ratio of 10 dB.

For the real-world datasets, we used two major public fMRI datasets including ADHD-200 [10] and the COBRE dataset [11]. The ADHD-200 dataset includes fMRI scans from Attention Deficit Hyperactivity Disorder (ADHD) patients and Typically Developing (TD) Children, whereas the COBRE dataset includes scans from schizophrenia patients. The ADHD-200 dataset was preprocessed according to the Athena pipeline [12] using the Analysis of Functional NeuroImages (AFNI) and the FMRIB's Software Library (FSL) tools, whereas the COBRE dataset was preprocessed according to the CIVET pipeline [13] using the NeuroImaging Analysis Kit (NIAK). For each dataset, the weight of edges between the network nodes (in this case the number of ROI) was estimated using correlation, partial correlation, and tangent connectivity measures [14]. The summary of networks used in this study is shown in Table 2.

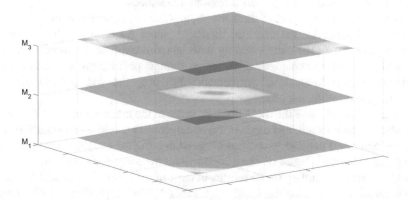

Fig. 2 Simulated active brain regions

Table 2 Datasets

Data	Sample size (# of networks)	Total # of nodes	Reference
Simulated	20	64, 125, 216	[9]
Attention Deficit Hyperactivity Disorder (ADHD)	20	114	[10]
Schizophrenia (SZ)	72	39	[11]
Typically Developed Children (TD)	20	114	[10]

3 Results

Figure 3 shows the mean threshold value calculated for different datasets using various thresholding methods, and Fig. 4 shows an example of thresholding a weighted functional connectivity network where edges below the identified threshold value using the percolation-based thresholding method are discarded. As seen in Fig. 3, the 99% connectedness method provides similar threshold values to the proposed percolation-based thresholding method when using the partial correlation and tangent values, but provides a slightly higher threshold value for the correlation-based connectivity. In general, average degree-based thresholding methods provided smaller threshold values except for networks obtained from the correlation measure for real-world datasets. Moreover, 99, 95, and 75 percentiles had higher threshold values, while 50, 25, 5, 1 and mean resulted in a lower threshold value than the percolation-based method. The 99% percentile method provided the largest thresholding values, which means the binarized networks using this method will be much sparser than the binarized networks using other thresholding methods.

To understand the impact of these threshold values on maintaining the properties of weighted networks, the obtained threshold values for each weighted network was used to extract the corresponding unweighted network and several measures including the macroscopic (density, average shortest path length, and modularity) and microscopic (closeness and degree centralities) properties were calculated before and after the thresholding process.

Figures 5 and 6 show the mean distance value of the macroscopic (density, average shortest path length, and modularity) and microscopic (closeness centrality and degree centrality) network properties between the original weighted networks and the thresholded ones based on three connectivity measures for simulated and real-world datasets. According to the results, in most of the cases, the distance between the original network properties and the binarized network properties was smaller when using thresholding methods based on the network topology (percolation-based method, 99% connectedness, and degree-based method). Having said that in some cases (e.g., mean distance of modularity calculated using tangent connectivity

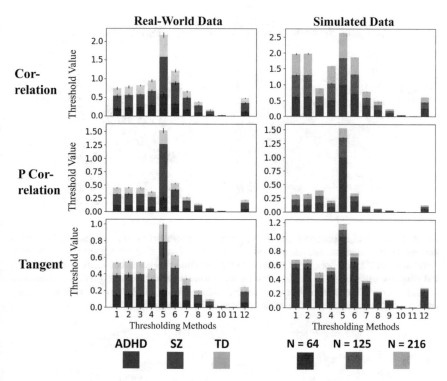

Fig. 3 Threshold values obtained from thresholding methods for three real-world (ADHD, SZ, and TD children) and three simulated datasets with node size $N \in \{64, 125, 216\}$. The horizontal axis in each plot shows the thresholding methods presented in Table 1. Each rows refers to a connectivity measure including: correlation which is the simplest connectivity method that quantifies the linear interdependency of two time series data, partial correlation which removes the effect of controlling random variables when calculating the interdependency [15], and the tangent that uses residuals of correlation matrices in the tangent space to estimate a covariance matrix [16]

measure), thresholding based on p-th percentile (i.e., 25, 5, and 1%) resulted in a lower distance value between weighted and thresholded network. However, this was mainly because of selecting a very small threshold value where only small number of links were removed from the original weighted network, and hence the resulted unweighted networks were dense. Taking into account this limitation of statistical methods, the performance of the percolation-based method was especially good for preserving the modularity of the correlation and partial correlation-based networks in real-world datasets. Interestingly, the thresholding results using three tested connectivity measures showed a similar pattern in maintaining degree centrality which would indicate the robustness of degree centrality measure to outliers and noise.

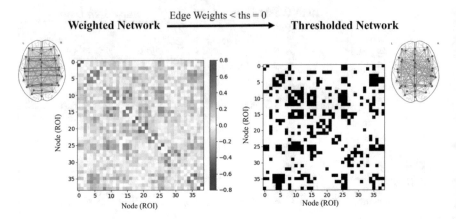

Fig. 4 Example of a functional connectivity matrix/network before and after thresholding with the percolation-based thresholding method. The functional connectivity matrix represents the mean correlation for 72 networks in SZ dataset. In the weighted functional connectivity matrix, rows and columns represent ROIs (nodes), and the cell colors represent the strength of the connection between nodes. In the thresholded functional connectivity matrix, black and white cells indicate the presence and absence of the connections between ROIs, respectively

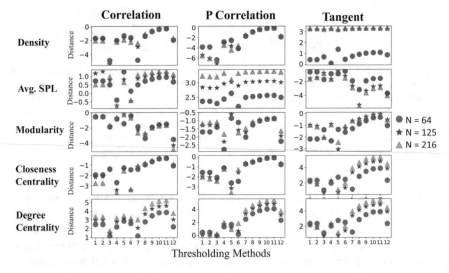

Fig. 5 Evaluation of the macroscopic and microscopic network properties after thresholding for three simulated datasets with node size $N = [64, 125, 216]$. Each column refers to a connectivity measures (correlation, partial correlation, and tangent) and each row refers to macroscopic/microscopic network properties. In each subplot, the horizontal axis represents the different thresholding methods, and the vertical axis represents log of the distance value

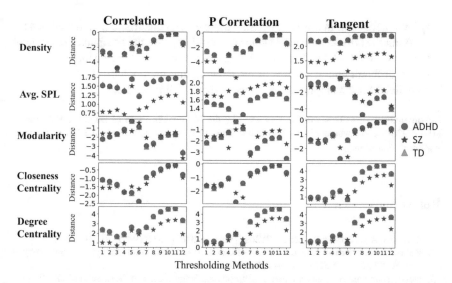

Fig. 6 Evaluation of the macroscopic and microscopic network properties after thresholding for three real-world datasets. Each column refers to a network connectivity measure of correlation, partial correlation, and tangent, and each row refers to a macroscopic/microscopic network properties. In each subplot, the horizontal axis represents the different thresholding methods, and the vertical axis represents log of the distance value

4 Conclusion

Choosing an appropriate threshold value to convert weighted networks to unweighted ones is a major challenge in complex system analysis as the imprecise selection of such threshold could significantly alter the original network topology, which could subsequently bias the derived network properties such as centrality. In this study, we proposed a percolation-based thresholding method that maintains the topological integrity and connectivity of the original weighted network by minimizing the number of isolated nodes after binarization. More specifically, the proposed method resulted in thresholded networks with a similar macroscopic and microscopic network properties (in particular modularity, average shortest path length, and degree centrality) to the original weighted network. This is especially important in the field of network neuroscience where changes of network properties could introduce a significant bias to the outcome of the study. Even though the current study focuses on the neuroscience applications, the percolation-based thresholding method could also be used in other domains such as analysis of social networks and genetic networks. Examples include scientific collaboration networks where nodes are scientists and edges represent co-authorships [17], or gene co-expression networks where nodes are genes and edges represent gene similarities [18].

The proposed method has limitations. Being an iterative procedure, the computational complexity of the proposed method is high, and it might not be suitable for very large networks. In this regard, using more efficient heuristic search methods for identifying the critical threshold value (θ_c) could improve the computational time of the proposed method. Furthermore, theoretical studies that describe the relationship between different characteristics of the network and its topological integrity could be beneficial for developing more effective thresholding methods.

Acknowledgements This work is supported by the National Institute of Health (NIH) grant # MH112925-01. We would like to also thank Gregory P. Strauss and Katherine Visser for their support of this work.

References

1. Barabási, A.L.: Network science. Philos. Trans. R. Soc. A **371**(1987), 20120, 375 (2013)
2. Rubinov, M., Sporns, O.: Complex network measures of brain connectivity: uses and interpretations. Neuroimage **52**(3), 1059–1069 (2010)
3. van den Heuvel, M.P., de Lange, S.C., Zalesky, A., Seguin, C., Yeo, B.T., Schmidt, R.: Proportional thresholding in resting-state fmri functional connectivity networks and consequences for patient-control connectome studies: Issues and recommendations. Neuroimage **152**, 437–449 (2017)
4. Van Wijk, B.C., Stam, C.J., Daffertshofer, A.: Comparing brain networks of different size and connectivity density using graph theory. PlOS one **5**(10), e13, 701 (2010)
5. Alexander-Bloch, A.F., Gogtay, N., Meunier, D., Birn, R., Clasen, L., Lalonde, F., Lenroot, R., Giedd, J., Bullmore, E.T.: Disrupted modularity and local connectivity of brain functional networks in childhood-onset schizophrenia. Front. Syst. Neurosci. **4** (2010)
6. Granovetter, M.: The strength of weak ties: a network theory revisited. Sociol. Theory. 201–233 (1983)
7. Bassett, D.S., Meyer-Lindenberg, A., Achard, S., Duke, T., Bullmore, E.: Adaptive reconfiguration of fractal small-world human brain functional networks. Proc. Natl. Acad. Sci. **103**(51), 19518–19523 (2006)
8. Cohen, M.X.: Analyzing Neural Time Series Data: Theory and Practice. MIT Press, USA (2014)
9. Michel, V., Gramfort, A., Varoquaux, G., Eger, E., Thirion, B.: Total variation regularization for fmri-based prediction of behavior. IEEE Trans. Med. Imaging **30**(7), 1328–1340 (2011)
10. Milham, M.P.: The adhd-200 dataset. http://fcon_1000.projects.nitrc.org/indi/adhd200/ (2011). Accessed 01 May 17
11. Aine, C.: The center for biomedical research excellence (cobre) dataset. http://fcon_1000. projects.nitrc.org/indi/retro/cobre.html (2011). Accessed 01 May 17
12. Bellec, P., Chu, C., Chouinard-Decorte, F., Benhajali, Y., Margulies, D.S., Craddock, R.C.: The neuro bureau adhd-200 preprocessed repository. Neuroimage **144**, 275–286 (2017)
13. Ad-Dabbagh, Y., Lyttelton, O., Muehlboeck, J., Lepage, C., Einarson, D., Mok, K., Ivanov, O., Vincent, R., Lerch, J., Fombonne, E., et al.: The civet image-processing environment: a fully automated comprehensive pipeline for anatomical neuroimaging research. In: Proceedings of the 12th Annual Meeting Of The Organization For Human Brain Mapping, p. 2266. Florence, Italy (2006)
14. Craddock, R.C., Jbabdi, S., Yan, C.G., Vogelstein, J.T., Castellanos, F.X., Di Martino, A., Kelly, C., Heberlein, K., Colcombe, S., Milham, M.P.: Imaging human connectomes at the macroscale. Nat. Methods **10**(6), 524–539 (2013)
15. Wang, Y., Kang, J., Kemmer, P.B., Guo, Y.: An efficient and reliable statistical method for estimating functional connectivity in large scale brain networks using partial correlation. Front. Neurosci. **10** (2016)

16. Varoquaux, G., Baronnet, F., Kleinschmidt, A., Fillard, P., Thirion, B.: Detection of brain functional-connectivity difference in post-stroke patients using group-level covariance modeling. In: International Conference on Medical Image Computing and Computer-Assisted Intervention, pp. 200–208. Springer, Berlin (2010)
17. Newman, M.E.: The structure of scientific collaboration networks. Proc. Natl. Acad. Sci. **98**(2), 404–409 (2001)
18. Spellman, P.T., Sherlock, G., Zhang, M.Q., Iyer, V.R., Anders, K., Eisen, M.B., Brown, P.O., Botstein, D., Futcher, B.: Comprehensive identification of cell cycle-regulated genes of the yeast saccharomyces cerevisiae by microarray hybridization. Mol. Biol. Cell **9**(12), 3273–3297 (1998)

Discovering Patterns of Interest in IP Traffic Using Cliques in Bipartite Link Streams

Tiphaine Viard, Raphaël Fournier-S'niehotta, Clémence Magnien and Matthieu Latapy

Abstract Studying IP traffic is crucial for many applications. We focus here on the detection of (structurally and temporally) dense sequences of interactions that may indicate botnets or coordinated network scans. More precisely, we model a MAWI capture of IP traffic as a link streams, i.e., a sequence of interactions (t_1, t_2, u, v) meaning that devices u and v exchanged packets from time t_1 to time t_2. This traffic is captured on a single router and so has a bipartite structure: Links occur only between nodes in two disjoint sets. We design a method for finding interesting bipartite cliques in such link streams, i.e., two sets of nodes and a time interval such that all nodes in the first set are linked to all nodes in the second set throughout the time interval. We then explore the bipartite cliques present in the considered trace. Comparison with the MAWILab classification of anomalous IP addresses shows that the found cliques succeed in detecting anomalous network activity.

1 Introduction

Attacks against online services, networks, and information systems, as well as identity thefts, have annual costs in billions of euros. Network traffic analysis and anomaly detection systems are of crucial help in fighting such attacks. In particular, much work is devoted to the detection of anomalous patterns in traffic. This work mostly relies on pattern search in (sequences of) graphs, which poorly captures the dynamics of traffic, or on signal analysis, which poorly captures its structure. Instead, we use here

T. Viard (✉) · R. Fournier-S'niehotta
CEDRIC CNAM, 75003 Paris, France
e-mail: tiphaine.viard@cnam.fr

R. Fournier-S'niehotta
e-mail: fournier@cnam.fr

C. Magnien · M. Latapy
CNRS, UMR 7606, LIP6, Sorbonne Universités, UPMC Univ Paris 06, 75005 Paris, France
e-mail: clemence.magnien@lip6.fr

M. Latapy
e-mail: matthieu.latapy@lip6.fr

© Springer International Publishing AG 2018
S. Cornelius et al. (eds.), *Complex Networks IX*, Springer Proceedings
in Complexity, https://doi.org/10.1007/978-3-319-73198-8_20

the *link stream* framework, which allows us to model both structural and temporal aspects of traffic in a consistent way. Within this framework, a clique is defined by a set of nodes and a time interval such that all the nodes in that set continuously interact with each other during this time interval. Such patterns may be the signature of various events of interest like botnets, DDoS, and others.

We present in Sect. 2 our *link stream* modeling of traffic captures, which calls for specific notions, as this traffic is bipartite. We present in Sect. 3 fast heuristics for finding cliques of interest in practice. We briefly present the MAWI dataset that we use and then detail the results of our experiments in Sect. 4. Related work is overviewed in Sect. 5, and we discuss the main perspectives in Sect. 6.

2   Traffic as a Link Stream

IP traffic is composed of packets, each with a source address and a destination address. Routers forward these packets toward their destination, and one may capture traffic traces by recording the headers of packets managed by a router, along with the time at which they manage them. Such traces are generally collected on firewalls, access points, or border routers. As a consequence, they often have a bipartite nature; they capture exchanges between two disjoint sets of devices (for instance, the ones in a company LAN and the outside Internet), as these routers are not in charge of traffic between devices within one of these sets. This leads to the definition of **bipartite link streams** that extend the classical definitions of bipartite graphs [1] and of link streams [2, 3].

A bipartite link stream $L = (T, \top, \bot, E)$ is defined by a time span T, a set of top nodes $\top$, a set of bottom nodes $\bot$, and a set of links $E \subseteq T \times \top \times \bot$. If $(t, u, v) \in E$ then we say that u and v are linked at time t. We say that $l = (b, e, u, v)$ is a link of L if $[b, e]$ is a maximal interval of T such that u and v are linked at all t in $[b, e]$. We call $\bar{l} = b - e$ the duration of l. See Fig. 1 for an illustration. We consider undirected links only: We make no distinction between (t, u, v) and (t, v, u) in E.

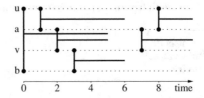

Fig. 1 A bipartite link stream $L = (T, \top, \bot, E)$ with $T = [0, 10]$, $\top = \{u, v\}$, $\bot = \{a, b\}$, and $E = ([1, 6] \cup [8, 10]) \times \{(u, a)\} \cup [0, 5] \times \{(u, b)\} \cup ([2, 5] \cup [7, 10] \times \{(v, a)\} \cup [3, 6] \times \{(v, b)\}$. In other words, the links of L are $(1, 6, u, a)$, $(8, 10, u, a)$, $(0, 5, u, b)$, $(2, 5, v, a)$, $(7, 10, v, a)$, and $(3, 6, v, b)$. We display nodes vertically and time horizontally, each link being represented by a vertical line at its beginning that indicates its extremities, and a horizontal line that represents its duration

To model traffic as a link stream, we transform packets exchanged at timestamps into continuous interactions; we consider that two IP addresses are continuously linked together from time b to time e if they exchange at least one packet every second within this time interval. This leads to the following definition of E: It is the set of all (t, u, v) such that u and v exchanged a packet at a time t' such that $|t' - t|$ is lower than half a second.

3 Finding Cliques of Interest

In a bipartite graph $G = (\top, \bot, E)$, a clique is a couple $(C_\top, C_\bot)$ with $C_\top \subseteq \top$ and $C_\bot \subseteq \bot$ such that $C_\top \times C_\bot \subseteq E$. In other words, there is a link between each element of $C_\top$ and each element of $C_\bot$.

We define similarly a clique in a bipartite link stream as a tuple $(C_\top, C_\bot, I)$ with $C_\top \subseteq \top$, $C_\bot \subseteq \bot$ and I an interval of T such that $I \times C_\top \times C_\bot \subseteq E$. In other words, each element of $C_\top$ is linked to each element of $C_\bot$ for the whole duration of I. We call $|C_\top \cup C_\bot|$ the size of the clique and $|I|$ its duration. A clique is maximal if there is no other clique $(C'_\top, C'_\bot, I')$ such that $C_\top \subseteq C'_\top$, $C_\bot \subseteq C'_\bot$ and $I \subseteq I'$. See Fig. 2 for an illustration.

We explore cliques as patterns of interest in IP traffic. It is easy to extend to the bipartite case the (non-bipartite) algorithm presented in [4]. However, we face two issues. First, clique computations are costly, and enumerating all cliques is out of reach in our case; we therefore resort to sampling. Second, not all maximal cliques are interesting: In particular, a node and its neighbors is a bipartite clique, but has little interest for us.

As a consequence, we focus on *balanced* cliques: A clique $(C_\top, C_\bot, I)$ is balanced if and only if $||C_\top| - |C_\bot|| \leq 1$. We then search for balanced cliques as follows. We iteratively build $(C_\top, C_\bot, I)$ from the empty clique of maximal duration, $(\emptyset, \emptyset, T)$. At each step, we choose a random node v in $\top$ or $\bot$ such that v is linked to all nodes in $C_\bot$ or $C_\top$, respectively, during an interval $I_v \subseteq I$, and we update the current clique into $(C_\top \cup \{v\}, C_\bot, I_v)$ or $(C_\top, C_\bot \cup \{v\}, I_v)$, respectively. In order to ensure

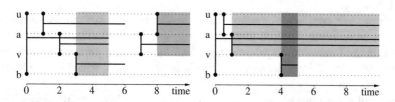

Fig. 2 Left: Two maximal cliques of the bipartite link stream of Fig. 1: $(\{u, v\}, \{a, b\}, [3, 5])$ (in blue), and $(\{u, v\}, \{a\}, [8, 10])$ (in green). **Right:** From the blue clique $(\{u, v\}, \{a\}, [1, 10])$, our algorithm builds the clique $(\{u, v\}, \{a, b\}, [4, 5])$ by adding b, which reduces the time span from $[1, 10]$ to $[4, 5]$

the obtained clique is balanced, we alternatively choose v in $\top$ and in $\perp$ at each iteration. We stop when no node can be added.

This process samples a maximal balanced clique, and we run it many times to obtain a large set of such cliques (see Sect. 4). It is clear that this set may be biased by the sampling process, which is not uniform. However, since our primary goal is to explore the relevance of *some* cliques in the context of IP traffic, studying this bias is out of our scope. Notice also that the found clique is maximal among the set of balanced cliques, but not necessarily maximal in the stream: It may be included in an unbalanced clique.

Our sampling process tries to add as many nodes to the clique as possible, which generally induces a reduction of its duration; see Fig. 2. As we are both interested in large *and* long balanced cliques, we include in the sampled set all the intermediary maximal balanced cliques built during the process.

4 Dataset and Results

We use an IP traffic capture from the MAWI archive, more precisely from the DITL[1] initiative, from June 24, 2013, 23:45 to June 25, 2013, 00:45. This trace lasts $3,600$ s, during which $88,266,534$ packets are sent involving $992,466$ nodes. $408,751$ nodes are part of WIDE, and $583,715$ are outside of WIDE. WIDE/Non-WIDE sets are nearly balanced, improving our chances of finding large balanced cliques.

We transform the data into a bipartite link stream $L = (T, \top, \perp, E)$, where $T = [0, 3600]$, $\top$ contains all observed WIDE IP addresses, and $\perp$ all other observed IP addresses. As said in Sect. 2, E is the set of links obtained by considering that nodes interact for one second every time they exchange a packet. E contains $6,206,295$ links.

In addition to this raw data, we use the MAWILab database [5], which gives labels locating traffic anomalies in the MAWI archive. These labels are obtained using an advanced graph-based methodology that compares and combines different and independent anomaly detectors. This database indicates that there is a total of 488 anomalous IP addresses in our dataset that we will use to interpret the results in the following.

We ran our sampling algorithm on the MAWI dataset. We ran 14 instances in parallel on a server[2] for 106 days (3 months and 17 days). This led to a total of 1,291,084,661 sampled cliques, among which there were a great number of duplicates: $198,718,323$ are distinct maximal balanced cliques, which we study in the rest of the paper.

We call *anomalous clique* a clique which contains at least one IP address that is flagged as anomalous in the MAWILab database.

---

[1] http://mawi.wide.ad.jp/mawi/ditl/ditl2013/.
[2] A Linux machine with 24 cores at 2.9 GHz and 256 GB of RAM.

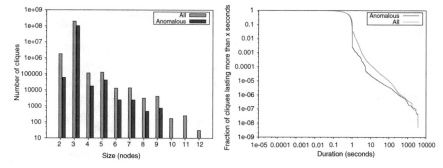

Fig. 3 Size distribution (top) and duration inverse cumulative distribution (bottom) of the balanced cliques found by our sampling. For instance, the fraction of cliques lasting more than 10 s is 0.0001

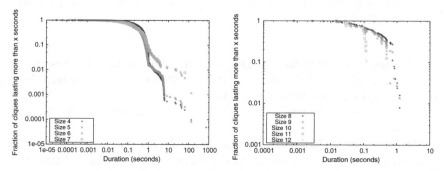

Fig. 4 Inverse cumulative distribution of durations of sampled balanced cliques of a given size. Left: For sizes from 4 to 7 nodes. Right: For sizes from 8 to 12 nodes

We show in Fig. 3 the distribution of sampled clique sizes and the inverse cumulative distribution of their durations. Many are of size 2 or 3, which has little interest: Cliques of size 2 are single links, and bipartite cliques of size 3 are just composed of a node linked to two other nodes. Since any node with k neighbors leads to $k(k-1)/2$ balanced 3-cliques, the large number of such cliques is unsurprising and brings no significant information. We therefore focus on cliques of size 4 or more. Our sample contains 275, 647 such cliques.

The sampled cliques of 4 nodes or more involve 29, 744 distinct nodes, 94 of which anomalous. The fraction of anomalous nodes in these cliques therefore is $3.1 \cdot 10^{-3}$, much larger than in the whole dataset, $5 \cdot 10^{-4}$. This indicates that maximal balanced cliques are related to anomalous activity, as suspected.

While most maximal balanced cliques have a duration close to 1 second, the duration distributions show that there are very long cliques. However, duration is highly influenced by size, as explained above, and so we display the duration distribution for each clique size separately in Fig. 4 (for readability, we show them in two plots). As expected, the duration of large cliques is in general shorter than for cliques involving

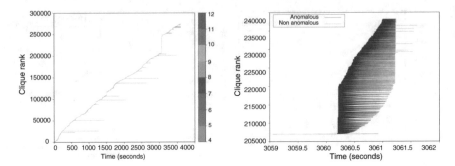

Fig. 5 Left: Representation of the temporal features of our balanced cliques of size 4 or more. Each clique $(C_T, C_\perp, I)$ is represented by a horizontal line of length $|I|$ with a color representing its size (color code on the right). These lines are ranked according to their beginning time. **Right:** Detailed view of the [3059; 3062] interval. A very large number of anomalous cliques start within the same second (in purple)

only few nodes. The 18 cliques spanning more than 500 s all have 4 or 5 nodes, the longest being of size 4 and duration 1045.75 s. Still, there are cliques involving 7 nodes for more than 10 s, up to almost 100 s. Larger cliques all have a duration close to 1 second or less. Importantly, among the 18 cliques spanning more than 500 s, 17 are anomalous.

We deepen our understanding of large cliques by displaying in Fig. 5 the time span of all maximal balanced cliques of size 4 or more: We sort cliques according to their starting time, which gives their vertical position in the plot. We then represent each clique $(C_T, C_\perp, I)$ by a horizontal line from the beginning of I to its end. Colors indicate clique size. This plot confirms our previous observations; it shows the prevalence of smaller cliques among the longest ones, and it shows that we succeed in finding significant cliques during the whole time span of the dataset.

However, it also displays a sharp increase shortly after time 3000, corresponding to a large number (31, 201) of short balanced cliques that start then. Figure 5 right displays these cliques in more details and shows that they almost all involve anomalous nodes. This confirms that clique structures are related to anomalous activity.

Interestingly, it seems that these anomalies cannot easily be distinguished from other anomalies directly on simple plots like the number of distinct nodes or links over time; see Fig. 6. Although there is a peak in both plots at time 3060, it is not different from other peaks, yet it is the only one corresponding to such a sharp increase in Fig. 5. This indicates that cliques highlight specific features of this event that are worth investigating further.

We therefore display in Fig. 7 the graph induced by the balanced cliques we found at seconds 3059, 3060, or 3061. The nodes of this graph are the nodes involved in at least one of these cliques, and they are linked together if this link occurs in at least one of these cliques. The graph has a large connected component with two anomalous nodes linked to 228 distinct non-anomalous nodes, confirming that most of these cliques are the signature of a same event (they actually are parts of a much

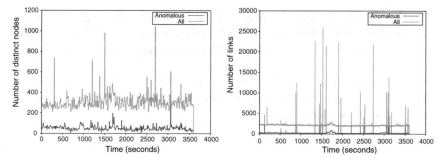

Fig. 6 Number of all and anomalous distinct nodes active in each second (left), and number of all and anomalous distinct links active in each second (right)

Fig. 7 The graph induced by all maximal balanced cliques starting between second 3059 and 3061, and involving at least 4 nodes. Anomalous nodes known from *MAWILab* are in red

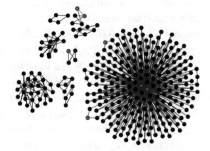

larger but unbalanced clique). This signature is typical of a coordinated scan in a class C network.

5 Related Work

IP traffic has been extensively studied for decades with powerful approaches relying on signal processing and machine learning [6], or graphs. In [7], the authors use graphs to represent temporal dependencies in traffic and characterize traffic behavior. In [8], the authors model the traffic as a bipartite graph and then used one-mode projections and clustering algorithms to discover behavioral clusters. Clustering communication behavior was also proposed in [9], where authors discuss the relevant features before analyzing significant nodes for long periods.

Graph-based approaches are however limited in their ability to capture temporal information, crucial for traffic analysis. Indeed, they generally rely on splitting data into time slices and then aggregate traffic occurring into each slice into a (possibly weighted, directed, and/or bipartite) graph. One obtains this way a sequence of graphs, and one may study the evolution of their properties; see for instance [10]. However, choosing small time slices leads to almost empty graphs and brings little information. Conversely, large slices lead to important loss of information as the dynamics within each slice is ignored. As a consequence, choosing appropriate sizes for time slices is extremely difficult and a research topic in itself [11]. There is currently an important interdisciplinary effort for solving these issues by defining formalisms able to deal with both the structure and dynamics of such data. The link stream approach is one of them [3], as well as temporal networks and time-varying graphs [12, 13]. Up to our knowledge, these other approaches have not yet been applied to network traffic analysis.

6 Discussion

We have shown that cliques in bipartite link streams modeling of IP traffic allow to detect anomalous activity in IP traffic: Long cliques of significant size involve anomalous nodes known from MAWILab, although they are rather small, and simultaneous apparition of many small cliques indicates coordinated activity like distributed scans. This work however only is a first step, and it raises many questions.

In particular, the computational cost of our method is prohibitive. Algorithmic work is therefore needed to design faster clique detection heuristics, and to search for quasi-cliques. One may also preprocess the stream by iteratively removing nodes of degree 1, which represent a large fraction of the whole and cannot be involved in non-trivial cliques. Going further, one may use link streams to define many other structures of interest regarding anomalous traffic. Exploring other modeling assumptions is also appealing, in particular the fact that we linked nodes together if they exchanged packets at least every second (other time limits may be interesting), the fact that we considered undirected links, or the use of port or protocol information present in the data.

Acknowledgements This work is funded in part by the European Commission H2020 FET-PROACT 2016-2017 program under grant 732942 (ODYCCEUS), by the ANR (French National Agency of Research) under grants ANR-15-CE38-0001 (AlgoDiv) and ANR-13-CORD-0017-01 (CODDDE), and by the Ile-de-France program FUI21 under grant 16010629 (iTRAC).

References

1. Latapy, M., Magnien, C., Vecchio, N.: Basic notions for the analysis of large two-mode networks. Soc. Netw. **30**(1), 31–48 (2008)
2. Latapy, M., Viard, T., Magnien, C.: Stream graphs and link streams for the modeling of interactions over time (2017). https://arxiv.org/abs/arXiv:1710.04073
3. Viard, T., Latapy, M., Magnien, C.: Computing maximal cliques in link streams. Theor. Comput. Sci. **609**, 245–252 (2016)
4. Himmel, A., Molter, H., Niedermeier, R., Sorge, M.: Enumerating maximal cliques in temporal graphs. In: IEEE/ACM International Conference on Advances in Social Networks Analysis and Mining. ASONAM (2016)

5. Fontugne, R., Borgnat, P., Abry, P., Fukuda, K.: Mawilab: Combining diverse anomaly detectors for automated anomaly labeling and performance benchmarking. In: ACM CoNext '10. (2010)
6. Himura, Y., Fukuda, K., Cho, K., Borgnat, P., Abry, P., Esaki, H.: Synoptic graphlet: bridg- ing the gap between supervised and unsupervised profiling of host-level network traffic. IEEE/ACM Trans. Netw. **21**(4), 1284–1297 (2013)
7. Asai, H., Fukuda, K., Abry, P., Borgnat, P., Esaki, H.: Network application profiling with traffic causality graphs. Int. J. Netw. Manag. **24**(4), 289–303 (2014)
8. Xu, K., Wang, F., Gu, L.: Behavior analysis of internet traffic via bipartite graphs and one-mode projections. IEEE/ACM Trans. Netw. **22**(3), 931–942 (2014)
9. Jakalan, A., Jian, G., Zhang, W., Qi, S.: Clustering and profiling ip hosts based on traffic behavior. J. Netw. **10**(2), 99–107 (2015)
10. Latapy, M., Hamzaoui, A., Magnien, C.: Detecting events in the dynamics of ego-centred measurements of the internet topology. J. Complex Netw. **2**(1), 38–59 (2014)
11. Leo, Y., Crespelle, C., Fleury, E.: Non-altering time scales for aggregation of dynamic networks into series of graphs. In: Proceedings of the ACM Conference on Emerging Networking Experiments and Technologies CoNEXT. (2015)
12. Wehmuth, K., Ziviani, A., Fleury, E.: A unifying model for representing time-varying graphs. In: 2015 IEEE International Conference on Data Science and Advanced Analytics, DSAA 2015, Campus des Cordeliers, pp. 1–10. Paris, France, 19–21 Oct 2015
13. Holme, P.: Modern temporal network theory: a colloquium. Eur. Phys. J. B **88**(9), 1–30 (2015)

Router-Level Topologies of Autonomous Systems

Muhammed Abdullah Canbaz, Khalid Bakhshaliyev
and Mehmet Hadi Gunes

Abstract In order to understand the Internet topology, it is important to analyze the underlying networks' characteristics. Internet is enabled by independently operating Autonomous Systems (ASes) that collaborate to provide end-to-end communication. In this paper, we investigate the network characteristics of backbone ASes that provide transit connectivity. We collect *router-level* probe data sets from all of the public Internet topology measurement platforms and obtain network topologies of the backbone ASes. We then analyze the network characteristics of each AS and perform an in-depth analysis of the high ranked ASes. Analyzing two snapshots, we observe disassortative network topologies in the majority of AS topologies independent of their network size. Also, most of the top-ranked ASes have a densely connected core and exhibit power-law degree distributions.

1 Introduction

A challenge for network practitioners is the lack of understanding topological characteristics of the Internet backbone as a plethora of new applications and systems is deployed over a growing network. Underlying infrastructure not only affects the scalability, robustness, and resiliency of the networks, but also impacts the dynamics of routing protocols and communication performance [2, 14, 18, 20].

As of March 2017, there are 45,382 Autonomous Systems (ASes) advertising their presence on the Internet [5]. These independently operated networks make up the Internet. ASes interconnect with others as customer–provider or as peers with the goal of maximizing their communication performance and profit. Each AS tries to optimize its own communication efficiency without oversight of a global entity.

M. A. Canbaz (✉) · K. Bakhshaliyev · M. H. Gunes
Computer Science and Engineering Department, University of Nevada, Reno, NV, USA
e-mail: mcanbaz@unr.edu

K. Bakhshaliyev
e-mail: kbakhshaliyev@unr.edu

M. H. Gunes
e-mail: mgunes@unr.edu

© Springer International Publishing AG 2018
S. Cornelius et al. (eds.), *Complex Networks IX*, Springer Proceedings
in Complexity, https://doi.org/10.1007/978-3-319-73198-8_21

AS relationships and their internal topologies are often proprietary, and the details are not public.

Even though each network component and protocol has been well engineered, there is a lack of understanding of the overall structure and operation of the Internet [7, 8, 10]. Hence, network researchers infer network topologies by probing IP addresses from diverse vantage points. Couple of research groups have built measurement platforms for the collection of the Internet topology data and share it with network practitioners [4]. Internet measurements can suffer from measurement artifacts such as incomplete observations, sampling bias, path accuracy, and unresponsive nodes due to the Internet dynamics and limited visibility of vantage points.

In this study, we investigate the router-level graphs of ASes by combining publicly available trace data sets provided by Internet topology measurement platforms. While earlier studies have analyzed a subset of ASes [22], to our knowledge, this is the first study to analyze AS topologies at Internet scale. We combined data from three measurement platforms CAIDA Archipelago (Ark) [23], Measurement Lab (M-Lab) [16], and RIPE NCC Atlas [21]. We analyzed the data for two samples, i.e., March 2017 and October 2016. Obtained router-level graphs along with the resolution results are provided at https://im.cse.unr.edu/data.

In our analysis of the basic network characteristics, we observe that there is a correlation between the number of nodes and the number of edges in an AS. While the maximum degree has a strong correlation with the number of nodes, the average degree is slightly correlated with the number of nodes. Similarly, the assortativity of topologies is independent of the network size. Majority of ASes have a disassortative topology indicating that their high-degree routers are connected to low-degree routers more than other high degree routers.

Moreover, we performed an in-depth analysis of the networks of the top-ranked ASes, as identified by the number of IPv4 prefixes in their customer cone size by CAIDA [3]. We observe that majority of the top-ranked ASes have a well-connected core. Many of them also exhibit a decaying clustering coefficient distribution indicating a hierarchical network topology. Similarly, we observe a rich club among the routers of these ASes.

The rest of this paper is organized as follows. In Sect. 2, we summarize the network measurement data collected from Internet measurement platforms and data processing to obtain router-level graphs. We present network characteristics of observed backbone ASes in Sect. 3 and then perform a detailed analysis of the top-ranked ASes in Sect. 4. Finally, we conclude with Sect. 5.

2 Methodology

In this section, we summarize utilized measurement data sets and data processing to obtain router-level topologies of backbone ASes.

Data Collection: Traceroute collects path traces from a source toward a destination IP address as a sequence of IP addresses belonging to routers. As load balancers could send packets over different paths, Paris traceroute enforces probes to follow the same path. We harvested trace data sets of three public platforms that provide continuous topology measurements, namely CAIDA Archipelago (Ark) [23], Measurement Lab (M-Lab) [16], and RIPE NCC Atlas [21]. These platforms perform different measurement campaigns using different sets of vantage points and destinations and hence have varying coverage [4]. By combining these data sets, we obtained *link level* connectivity information. We obtained trace data sets of a 5-day period, corresponding to an Ark probing cycle, for October 15–20, 2016 and March 1–5, 2017. In Oct-16 data, we obtained 189,343,288 path traces, generated from 5,859 vantage points toward over 148,605,967 destinations. Similarly, in Mar-17 data, we obtained 194,545,069 path traces, generated from 5,720 vantage points toward over 149,406,222 destinations. We then extracted edge pairs and performed IP to AS mapping to identify edges belonging to each AS. For IP to AS mapping, we utilized Caida's BGP Stream [17]. We also used the sister AS data reports to identify multiple ASes managed by the same organization. We then filtered *IP level* connectivity of each AS. Overall, we obtained IP level information of 38,566 ASes in Oct-16 and 39,101 ASes in Mar-17 data sets.

IP Alias Resolution: Each IP address captured in our path traces represents an interface of a router on a path from a vantage point toward a destination. As routers have multiple interfaces and the path traces may contain different IP addresses of a router, IP alias resolution is crucial to determine the router-level topology. After slicing path traces into AS regions, we performed IP alias resolution to obtain the router-level graph of each AS. We performed IP alias resolution using kapar [12], which performs analytical resolution [11], and midar [13], which combines address-based [9] and IP identification-based [22] probing methods. We then clustered alias IP addresses to obtain the underlying router-level topology of each AS.

3 Analysis of Observed as Topologies

In this section, we analyzed the router-level graphs of ASes observed in the trace data sets. While the measurements contained IP addresses belonging to 38,566 and 39,101 ASes, for Oct-16 and Mar-17 data, respectively, many had only couple of IP addresses. In Oct-16 data, 11,596 of ASes were transit ASes that provide connectivity to other ASes and 26,970 were stub ASes that only connect end users of the AS. Similarly, in Mar-17 data, 11,523 were transit ASes and 27,578 were stub ASes. We filtered ASes that did not have 10 or more routers after IP alias resolution and ended with 19,614 ASes in both Oct-16 and Mar-17 data sets. A reason for such low visibility of ASes is due to the fact that some of the ASes filter the traceroute probes at their border.

Figure 1 shows the **number of nodes and edges** along with **giant component sizes** for the router-level graphs of each of the ASes that had at least 10 nodes in the data sets. While Fig. 1a shows Mar-17 data ranked by the number of nodes from the highest to the lowest, Fig. 1b and c show the number of nodes and edges, respectively,

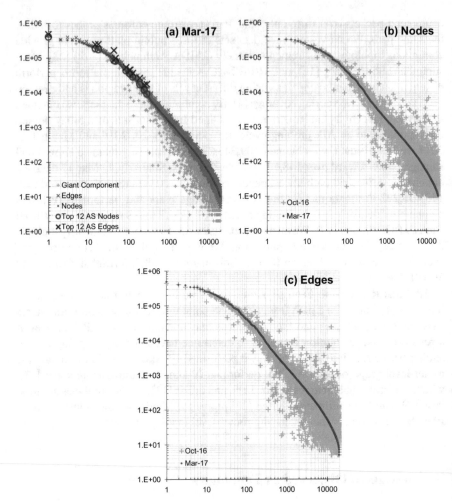

Fig. 1 Number of nodes, edges and giant components for all ASes—log–log scale. **a** *ranked by the number of nodes in Mar-17,* **b–c** *nodes and edges ranked by Mar-17 sample*

when they are independently ranked by the Mar-17 snapshot. Note that, the bigger marked blue crosses and red circles in the Mar-17 figure indicate the top-ranked ASes analyzed in depth in Sect. 4.

From the Mar-17 data set, we observe that there is a positive correlation between the number of routers and the number of links between them. Figure 1b and c indicates that captured graph size of some ASes varies between the two samples. We also observe a highly skewed distribution in the number of nodes and edges per AS. Moreover, 18.5 and 19.3% of the analyzed AS topologies in Mar-17 and Oct-16 data sets, respectively, seem to be tree-shaped, i.e., the number of nodes is exactly one more than the number of edges. This might be due to the tree-like discovery of

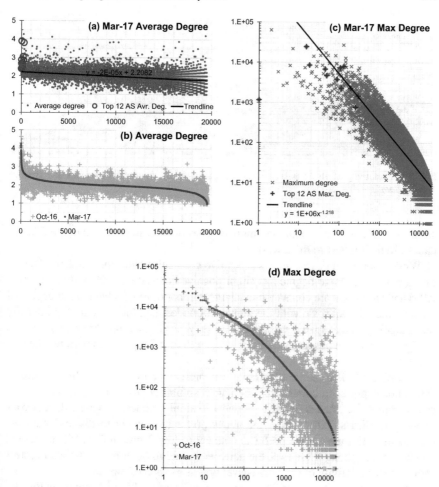

Fig. 2 Average and maximum degree for all ASes. **a** *Mar-17 average degrees ranked by the number of nodes—linear scale,* **b** *average degrees ranked by Mar-17 sample—linear scale,* **c** *Mar-17 maximum degrees ranked by the number of nodes—log–log scale,* **d** *maximum degrees ranked by Mar-17 sample—log–log scale*

traceroute paths or an actual tree topology of the AS. We observe that while the large ASes contain large giant components, the smaller ASes are fragmented indicating that the captured topologies were disconnected due to limited observations in those networks. This might also be due to lack of identification of some sister ASes. Finally, top-ranked ASes are not necessarily the largest topologies.

Figure 2 shows the **average and maximum degrees** of the topology of each of the ASes. Figure 2a and c is ranked by the number of nodes as shown in Fig. 1 and presents the trendline of the degrees and marks the top-ranked ASes. Figure 2b and d

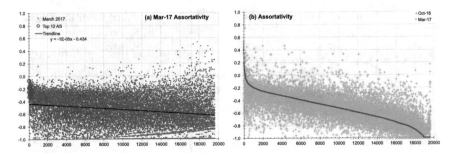

Fig. 3 Assortativity of all ASes. **a** *Mar-2017 data ranked by the number of nodes,* **b** *ranked by Mar-17 data*

show each of the average and maximum degrees of snapshots ranked by the Mar-17 data from the highest to the lowest.

While maximum degree shows a positive correlation with the number of nodes in the AS, the average degree has slight positive correlation. This indicates, some ASes have hubs that are connected to a large number of routers but overall degree of the networks do not scale with the number of nodes, i.e., there are no high-density mesh topologies that span the larger networks. When comparing different snapshots in Fig. 2b and d, we observe variations in degrees of AS topologies between the two data sets.

Assortativity is a metric illustrating the preference of a node to link to others. When nodes prefer to connect to others with similar degrees, the network is noted as assortative, i.e., nodes prefer to connect to similar nodes; however, the network is noted as disassortative when the nodes prefer to connect to dissimilar nodes, i.e., nodes with high degree prefer to connect to low-degree nodes. When there is no preference among these two, the network is said to be non-assortative. Figure 3 shows the assortativity of the router-level graphs of each of the AS.

When analyzing number of nodes-based ranking of the Mar-17 data set in Fig. 3a, we observe that assortativity of networks has a slight correlation with the number of nodes in the AS while there is considerable variance in the assortativity values. While larger networks are often disassortative, the smaller networks exhibit greater variation. Moreover, most of the top-ranked ASes are non-assortative with some being slightly disassortative. Autonomous System level networks shown to have different *local assortativity* profile then other networks (e.g., biological, social networks) [19]. Considering the top ASes presented in Sect. 4, the observed *disassortativity* indicates that the networks are *disassortative with assortative hubs*.

When analyzing Mar-17-based ranking Fig. 3b, we observe that ASes vary from being assortative to very highly disassortative with majority being within −0.1 to − 0.8 range, i.e., disassortative and non-assortative. This indicates ASes often have a hub-based star backbones and rarely contain core-based mesh backbones. Moreover, we observe assortativity values of networks vary between the two samples. This could be due to the change in the topological characteristics or due to the sampling of the underlying networks.

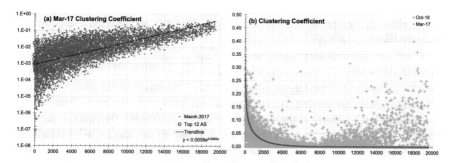

Fig. 4 Average of the clustering coefficient of routers in ASes. **a** *Mar-2017 data ranked by the number of nodes,* **b** *ranked by Mar-17 data*

Clustering Coefficient measures the degree to which nodes in a graph tend to cluster together. This measure, in social networks, reveals whether two friends of a node are also friends or not. High clustering indicates a higher ratio of triangles in the network. Figure 4 shows the average clustering coefficient of the networks of each of the AS.

In Fig. 4a ranked by the number of nodes, we observe a negative correlation between the clustering coefficient and the number of nodes in the AS topologies. This indicates smaller the AS network, the higher clustering coefficient they typically have. While clustering coefficient of the networks is low, especially for large graphs, they are often much higher than a random network of a similar size. In Fig. 4b ranked by Mar-17 data, we observe that only couple of AS networks have a clustering higher than 0.3. We observe greater variance of clustering for some ASes which could be due to sampling of the topologies.

In our analysis, we observed that about 62.7 and 60.8% of ASes have a clustering coefficient of 0, indicating no triangle in their topology. While this might be due to lack of comprehensive samples, it could be due to actual lack of triangles in their topologies. From Fig. 4b, we realize 8% of the ASes that had a nonzero clustering in one data set, have a zero clustering in the other. This indicates about than half of ASes have no triangle in their network in both samples. On the other hand, about 1.9% of ASes have a clustering of 0.1 or higher and 16.8% of ASes have a clustering of 0.01 or higher.

4 Analysis of the Top-Ranked ASes

In this section, we further analyze the characteristics of the top-ranked ASes. Caida provides ranking of ASes based on various topological characteristics [3]. We picked the top 12 ASes based on the ranking of *the number of IPv4 Addresses in their customer cone.* Table 1 presents the network characteristics of the chosen ASes ordered by the ranking for Mar-17 data. In addition to the earlier metrics, it shows the density, global network clustering, average path length, diameter, modularity, and number of communities for each AS.

When we analyzed the geographical coverage of these ASes, we observed that all of these ASes serve multiple countries and are intercontinental. While 9 out of 12 top-ranked ASes are based in the USA, the other 3 are based in the northern Europe. All of these ASes are serving as transit and access network providers in their regions.

Average Path Length is the average number of steps along the shortest paths for all possible pairs of nodes in the network. It is one of the most crucial metrics of network topology as it indicates how apart nodes are from each other. ASes try to minimize the average path length to reduce the latency between end systems they interconnect. Note that, in some of the topologies we have disconnected nodes. These are due to either observation of different parts of the network from multiple ingresses that do not overlap or due to unresponsive routers marked as unknown in the graph or lack of alias resolution between IP addresses of a router resulting in disconnected path segments. Nodes that are disconnected from the giant component, less than 1% on average, are ignored in the average path length calculations.

All of the top-ranked ASes have low average path lengths in the order of `log(n)` with only AS 1239 and AS 2828 having a value slightly larger than `log(nodes)`.

When considering **assortativity coefficient**, we observe that all top-ranked ASes have a non-assortative connectivity with values ranging between -0.068 and -0.32. This indicates their topologies contain both hubs surrounded with many small degree nodes and cores with high-degree clusters. Recall that significant majority of ASes, shown in Fig. 3b, had a disassortative network. Top-ranked ASes seem to be among the few that are non-assortative.

Community Detection tries to identify group of nodes that are highly interconnected and are close to be a clique. Detection of these communities can be achieved by various approaches such as minimum-cut method, hierarchical clustering, statistical inference, clique-based methods. We utilized [15] to detect communities in each AS. We observe the top-ranked ASes have 70–640 communities in their topologies.

Additionally, **modularity** reflects the density of the connections within the communities of the graphs. High modularity indicates highly dense connections between the nodes within communities with sparse connections between nodes of different communities. On the other hand, low modularity indicates lack of tightly knot communities. In our analysis, we observed high modularity in all ASes ranging between 0.59 and 0.88. This indicates the top-ranked ASes have regions of networks that are tightly connected as reflected in the communities.

Degree Distribution is one of the crucial metrics for a network. Many natural and man-made networks have shown to have a power-law degree distribution, which indicates there are very few hubs that have very high-degree nodes along with a very large number of nodes that have small degrees. Figure 5 presents the probability distribution function (PDF) of degrees of the top-ranked ASes for both Oct-16 and Mar-17 data. We observe a fat-tailed power-law behavior in all but one AS. To accurately classify degree distributions, we applied a goodness-of-fit test [1, 6]. Results confirm that except AS 701, ASes have a power-law degree distribution while ASes 3356, 1299, 174, and 6453 have a cutoff. AS 701 is the largest among top ASes,

Table 1 Network characteristics of the top-ranked ASes (Mar-17 data)

AS	Nodes	Edges	Giant component %	Degree		Density	Assortativity	Clustering		Avrg path length	Diameter	# of communities	Modularity
				Average	Maximum			Node average	Global				
3356	84,812	164,992	98.87	3.89	6,820	4.59e-05	−0.243	0.0602	0.0167	3.78	17	427	0.592
1299	186,588	248,112	99.85	2.66	24,698	1.43e-05	−0.068	0.0402	0.0010	3.64	12	139	0.800
2914	81,446	98,683	99.46	2.42	4,685	2.98e-05	−0.227	0.0168	0.0006	4.52	16	192	0.876
174	174,517	250,469	99.71	2.87	8,549	1.64e-05	−0.194	0.0548	0.0030	4.89	14	280	0.757
6453	34,206	45,501	99.40	2.66	2,303	7.78e-05	−0.181	0.0212	0.0012	4.45	15	81	0.802
6762	17,287	26,293	98.58	3.04	2,250	1.76e-04	−0.143	0.0290	0.0130	3.86	9	128	0.719
6939	36,219	44,670	99.67	2.47	7,329	6.81e-05	−0.323	0.0319	0.0004	4.07	15	82	0.813
2828	43,146	53,725	99.10	2.49	3,756	5.77e-05	−0.187	0.0220	0.0009	5.00	13	161	0.865
3491	9,234	17,551	99.26	3.80	1,462	4.12e-04	−0.257	0.0553	0.0252	3.56	12	70	0.613
701	410,701	489,928	99.89	2.39	1,175	5.81e-06	−0.303	0.0362	0.0016	4.37	16	640	0.876
1239	11,938	16,928	99.05	2.84	772	2.38e-04	−0.292	0.0638	0.0056	4.89	14	84	0.795
1273	28,365	36,524	98.99	2.58	7,939	9.08e-05	−0.253	0.0457	0.0010	3.87	17	102	0.796
mean	93,205	124,448	99.32	2.84	5,978	1.03e-04	−0.223	0.0398	0.0059	4.24	14	199	0.775
median	39,683	49,613	99.33	2.66	4,221	6.29e-05	−0.235	0.0382	0.0014	4.22	14	134	0.798

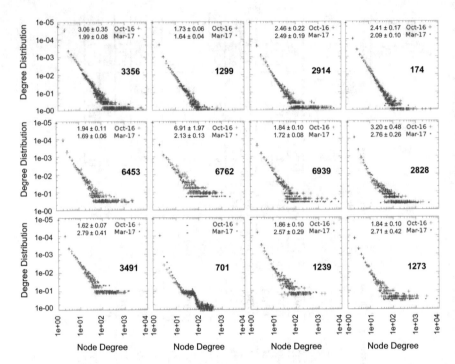

Fig. 5 Degree distributions of the top-ranked ASes (PDF in log–log scale. legend: $\alpha \pm$ error)

but it did not match any distributions of power-law, lognormal, exponential, stretched-exponential, or power-law with cutoff. This may be due to the traffic policies of this AS to handle measurement requests or could be due to limited visibility of the AS. Figure 5 legend also provides the power-law exponent and the standard error of each topology.

K-core is the subgraph of a network where every node in the subgraph has a degree of k or higher. Identification of k-core allows us to analyze cores of the topology and their connectivity. Figure 6 presents the k-cores, i.e., subgraphs where nodes with degree less than k are removed from the graph. The cutoff degree k is determined based on the median degree of the Mar-17 graph. The nodes are colored by node degree, with highest degree nodes in the center. The graphs indicate that most of the ASes have a well-connected core of high-degree nodes, as reflected in other metrics. Most of the topologies exhibit a core with a hierarchical or meshed periphery structure. We observed similar structures in Oct-16 snapshots (not shown) with majority getting denser in the Mar-17 snapshot. An interesting observation is that AS 1273 has two communities connected to each other with only a couple of routers in the k-core. It appears that this AS is serving in central Europe, North America, and Eastern Asia regions. Table 2 presents the subnetwork characteristics of k-cores with median and maximal, i.e., maximum k-value where the graph is non-empty, for both Oct-16 and Mar-17 data sets. In many ASes, the median degree is

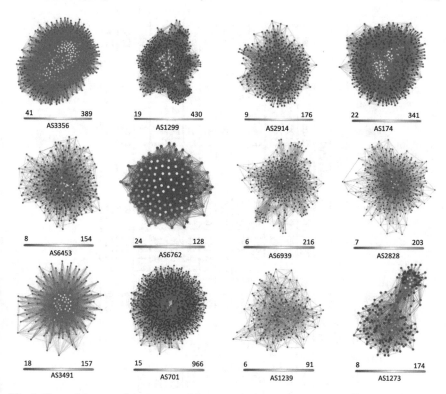

Fig. 6 Network layouts of the k-cores of the top-ranked ASes for Mar-17 data where k = median degree

close to the maximal k-value for k-core. Overall, the maximal k-cores indicate there are densely connected set of nodes that form a rich club.

Figure 7 displays the **clustering coefficient distribution** of the top-ranked ASes. We observe that most of the clustering distributions are inversely correlated with the node degree, indicating a hierarchical network topology. In some cases, such as AS 3356, the relation is not linear and in others, such as AS 174, the values are spread over a wider range. In most of the ASes, low-degree nodes tend to have higher clustering coefficient compared to the higher degree nodes. Average clustering values shown in Table 1 indicates that the top-ranked ASes are highly clustered. While a random network of similar size would have a clustering, in the order of 1E-05 the clustering values are in the range of 1E-02. Considering that they all have low average path length, the top-ranked ASes can be classified as small world networks.

Table 2 Median and maximal k-core of the top-ranked ASes for Oct-16 and Mar-17 data sets

AS	Oct-16								Mar-17							
	Median				Maximal				Median				Maximal			
	k	Nodes	Edges	Density	k	Nodes	Edges	Density	k	Nodes	Edges	Density	k	Nodes	Edges	Density
3356	41	408	23,058	0.278	82	165	8,452	0.625	41	420	23,456	0.267	77	190	10,077	0.561
1299	19	698	17,774	0.073	37	143	3,849	0.379	19	752	19,559	0.069	34	154	4,293	0.364
2914	9	78	541	0.180	9	78	541	0.180	9	501	5,516	0.044	15	94	1,078	0.247
174	22	158	3,231	0.261	24	123	2,447	0.326	22	583	17,499	0.103	41	152	4,597	0.401
6453	8	236	1,933	0.070	13	48	455	0.403	8	398	3,776	0.048	12	149	1,497	0.136
6762	24	108	2,777	0.481	38	82	1,949	0.587	24	142	4,157	0.415	44	99	2,694	0.555
6939	6	218	1,219	0.052	7	93	518	0.121	6	395	2,644	0.034	9	58	390	0.236
2828	7	40	199	0.255	7	40	199	0.255	7	386	3,376	0.045	11	151	1,447	0.128
3491	18	131	2,607	0.306	26	62	1,164	0.616	18	165	3,841	0.284	31	79	1,691	0.549
701	15	701	14,019	0.057	26	53	928	0.673	15	1,001	21,336	0.043	28	156	3,653	0.302
1239	6	196	1,136	0.059	7	108	627	0.109	6	255	1,612	0.050	8	99	622	0.128
1273	8	174	1,506	0.100	11	72	632	0.247	8	230	2,225	0.084	13	71	735	0.296

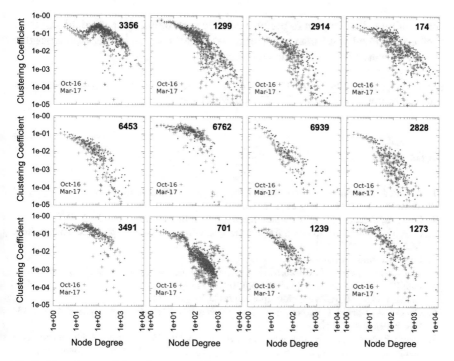

Fig. 7 Clustering coefficient distribution of the top-ranked ASes (log–log scale)

5 Conclusion and Future Work

Investigating topological infrastructure of the Internet at the macroscale can be achieved by understanding the network characteristics of the Autonomous Systems. In this paper, we utilized path trace data sets from multiple data sources to map the underlying topologies of ASes and obtained the router-level topologies. We then analyzed network metrics for each of the backbone AS topology.

We analyzed the graph characteristics of 19,614 backbone ASes that were mapped by measurement platforms in two snapshots. We observed that majority of ASes are disassortative and few are non-assortative. This indicates most ASes have star-like topologies where high-degree hubs connect low-degree nodes. Also, we observe that assortativity of graph is independent of its size. We also performed a detailed analysis of the top-ranked ASes. Majority of the top-ranked ASes have similar graph structures with a well-connected core and hierarchical or mesh-based peripheries. With only one exception, the top-ranked ASes have power-law degree distributions. Additionally, all of the top-ranked ASes are small worlds with high clustering and low average path length.

As some of our observations are affected by measurement artifacts, in the future, we plan to identify and adjust for the measurement artifacts in measurement data sets. Additionally, we plan to perform detailed measurements that would minimize such errors. Finally, we plan to perform these analyses at link level topologies that include subnetworks along with the routers as a 2-mode graph.

Acknowledgements This material is based upon work supported by the National Science Foundation under grant number CNS-1321164 and EPS-IIA-1301726.

References

1. Alstott, J., Bullmore, E., Plenz, D.: Powerlaw: a python package for analysis of heavy-tailed distributions. PloS one **9**(1), e85777 (2014)
2. Basu, A., Riecke, J.: Stability issues in ospf routing. In: Proceedings of the 2001 Conference on Applications, Technologies, Architectures, and Protocols for Computer Communications. ACM SIGCOMM (2001)
3. Caida's as rank. http://as-rank.caida.org/
4. Canbaz, M.A., Thom, J., Gunes, M.H.: Comparative analysis of internet topology data sets. In: 20th IEEE Global Internet Symposium (GI) (2017)
5. Cidrreports.org. http://www.cidr-report.org/as2.0/
6. Clauset, A., Shalizi, C.R., Newman, M.E.J.: Power-law distributions in empirical data. SIAM Rev. **51**(4), 661–703 (2009)
7. Faggiani, A., Gregori, E., Improta, A., Lenzini, L., Luconi, V., Sani, L.: A study on traceroute potentiality in revealing the internet as-level topology. In: IEEE Networking Conference, IFIP (2014)
8. Giotsas, V., Luckie, M., Huffaker, B., claffy, k.: Inferring complex as relationships. In: Proceedings of the 2014 Conference on Internet Measurement Conference. ACM IMC (2014)
9. Govindan, R., Tangmunarunkit, H.: Heuristics for internet map discovery. In: Proceedings IEEE INFOCOM 2000
10. Gregori, E., Improta, A., Lenzini, L., Rossi, L., Sani, L.: A novel methodology to address the internet as-level data incompleteness. IEEE/ACM Trans. Netw. (TON) **23**(4), 1314–1327 (2015)
11. Gunes, M.H., Sarac, K.: Resolving ip aliases in building traceroute-based internet maps. IEEE/ACM Trans. Netw. (ToN) **17**(6), 1738–1751 (2009)
12. Keys, K.: Internet-scale ip alias resolution techniques. ACM SIGCOMM Comput. Commun. Rev. **40**(1), 50–55 (2010)
13. Keys, K., Hyun, Y., Luckie, M., Claffy, K.: Internet-scale ipv4 alias resolution with midar. IEEE/ACM Trans. Netw. **21**(2), 383–399 (2013)
14. Labovitz, C., Ahuja, A., Bose, A., Jahanian, F.: Delayed internet routing convergence. In: Proceedings of the Conference on Applications, Technologies, Architectures, and Protocols for Computer Communication. ACM SIGCOMM (2000)
15. Lambiotte, R., Delvenne, J.C., Barahona, M.: Random walks, markov processes and the multiscale modular organization of complex networks. IEEE Trans. Netw. Sci. Eng. **1**(2), 76–90 (2014)
16. Measurement lab. https://www.measurementlab.net/about/
17. Orsini, C., King, A., Giordano, D., Giotsas, V., Dainotti, A.: Bgpstream: a software framework for live and historical bgp data analysis. In: Proceedings of the 2016 ACM on Internet Measurement Conference. ACM IMC (2016)
18. Park, K., Lee, H.: On the effectiveness of route-based packet filtering for distributed dos attack prevention in power-law internets. ACM SIGCOMM Comput. Commun. Rev. **31**, 15–25 (2001)
19. Piraveenan, M., Prokopenko, M., Zomaya, A.Y.: Local assortativity and growth of internet. Euro. Phys. J. B.-Condens. Matter Complex Syst. **70**(2), 275–285 (2009)

20. Radoslavov, P., Tangmunarunkit, H., Yu, H., Govindan, R., Shenker, S., Estrin, D.: On characterizing network topologies and analyzing their impact on protocol design (2000)
21. Ripe atlas probes. https://atlas.ripe.net/about/probes/
22. Spring, N., Mahajan, R., Wetherall, D., Anderson, T.: Measuring isp topologies with rocketfuel. IEEE/ACM Trans. Netw. **12**(1), 2–16 (2004)
23. University of california san diego, the ark measurement platform. https://ucsd.edu/ark/measurement/

Part V
Human Behavior and Social Networks

Social Influence (Deep) Learning for Human Behavior Prediction

Luca Luceri, Torsten Braun and Silvia Giordano

Abstract Influence propagation in social networks has recently received large interest. In fact, the understanding of how influence propagates among subjects in a social network opens the way to a growing number of applications. Many efforts have been made to quantitatively measure the influence probability between pairs of subjects. Existing approaches have two main drawbacks: (*i*) they assume that the influence probabilities are independent of each other, and (*ii*) they do not consider the actions not performed by the subject (but performed by her/his friends) to learn these probabilities. In this paper, we propose to address these limitations by employing a deep learning approach. We introduce a Deep Neural Network (DNN) framework that has the capability for both modeling social influence and for predicting human behavior. To empirically validate the proposed framework, we conduct experiments on a real-life (offline) dataset of an Event-Based Social Network (EBSN). Results indicate that our approach outperforms existing solutions, by efficiently resolving the limitations previously described.

1 Introduction

Influence propagation in social networks has recently received large interest, both in academia and industry. In fact, the understanding of how influence propagates in a social network opens the door to a wide range of applications, as targeted advertising, viral marketing, and recommendation. In this context, social networks play an important role as a medium for spreading processes [1, 2]. As an example,

L. Luceri (✉) · S. Giordano
University of Applied Sciences and Arts of Southern Switzerland (SUPSI),
Manno, Switzerland
e-mail: luca.luceri@supsi.ch

S. Giordano
e-mail: silvia.giordano@supsi.ch

L. Luceri · T. Braun
University of Bern, Bern, Switzerland
e-mail: braun@inf.unibe.ch

© Springer International Publishing AG 2018
S. Cornelius et al. (eds.), *Complex Networks IX*, Springer Proceedings
in Complexity, https://doi.org/10.1007/978-3-319-73198-8_22

261

a new idea can spread through a social network in the form of "word-of-mouth" communication [3]. In the last decade, particular attention has been devoted to the comprehension and modeling of the social influence phenomenon. Social influence is recognized as a key factor that governs human behavior. It indicates the attitude of certain individuals to be affected by other subjects' actions and decisions. The idea is that the interaction with other individuals (or a group) may result in a change of subject's thoughts, feelings, or behavior. In other words, a subject may take a decision, e.g., to buy a new product or to watch a TV show, when she/he sees her/his friends taking that decision.

A considerable amount of work has been conducted to investigate social influence and analyze its effect. In [4, 5], the authors propose how to qualitatively measure the existence of social influence, whereas in [6], the correlation between social similarity and influence is examined. In [7], we introduce a novel interpretation of physical, homophily, and social community, as sources of social influence. Other relevant works focused on the problem of influence maximization [8–11]. This problem aims to find the most influential individuals in a social network in order to maximize the number of influenced subjects. Viral marketing is a strategy that exploits this idea to promote new products. Kempe et al. [10] focus on two fundamental propagation models, referred to as Independent Cascade (IC) model and Linear Threshold (LT) model. In the IC model, each subject independently influences her/his friends with given influence probabilities. In the LT model, a subject is influenced by her/his friends if the combination of their total influence probabilities exceeds a threshold. Both models assume to have as input a social network whose edges are weighted by a measure of influence probability. However, these values are not known in practice and, thus, should be estimated. Many efforts have been made to quantitatively measure the influence strength between pairs of friends [12–17]. In particular, Goyal et al. [15] and Saito et al. [13] investigate how to learn the influence probabilities using only the history of subjects' actions. Such approaches have two main drawbacks: (i) they assume that the probability of friends influencing a subject is independent of each other, and (ii) they do not consider the actions not performed by the subject (but performed by her/his friends) to learn the influence probabilities.

In this paper, we propose to address the aforementioned drawbacks by employing a deep learning approach. Our objective is to learn subjects interplay for modeling social influence and predicting their behavior. We summarize our contributions as follows:

- We introduce a Deep Neural Network (DNN) framework that has the capability for both learning social influence and predicting human behavior. To the best of our knowledge, our solution is the first architecture that accomplishes these two tasks in one shot.
- We model social influence among subjects overcoming the assumptions introduced by previous works. We design a DNN taking into account both (i) the relationship between the subject and her/his friends and (ii) the interactions among them. Further, we learn social influence considering also the actions not performed by the subject (but performed by her/his friends) to understand who really affects subject's decisions.

- We evaluate the performance of our approach using data from an Event-Based Social Network (EBSN). This allows us to investigate social influence considering together *online* (through the social network) and *offline* (real-life) social interactions. Previous works conducted their experiments analyzing social influence only in Online Social Networks (OSNs). We compare our approach with existing solutions, achieving a remarkable improvement.

2 Problem Definition

In this paper, we aim to learn social influence in a social network in order to predict human behavior, in terms of decision and actions performed by individuals. Let $G = (V, E)$ be a directed graph, which represents the social network, where $V = \{u_1, u_2, \ldots, u_N\}$ is the set of subjects and E is the set of edges connecting them. Subject u_j is considered a *friend* of subject u_i if $(u_j, u_i) \in E$. To model social influence we measure the strength of friends' influence on subject's actions. We define A as the whole set of actions. For each action $a \in A$, each subject is either *active*, if she/he has performed the action, or *inactive*, otherwise. It should be noticed that inactive subjects may become active, but not the opposite. We define $S_{u_i,a}$ as the set of active friends of u_i for the action a. The objective is to predict whether a subject becomes active based on her/his active friends. To achieve this purpose, previous works determine the influence probability $p_{u_i}(S_{u_i,a})$, i.e., the influence exerted on subject u_i by the active friends $S_{u_i,a}$, by exploiting the history of u_i actions. The main assumption in these works is that the probability of various friends influencing u_i is independent of each other. Thereby, the probability $p_{u_i}(S_{u_i,a})$ is computed as $p_{u_i}(S_{u_i,a}) = 1 - \prod_{u_j \in S_{u_i,a}} (1 - p_{u_j,u_i})$, where p_{u_j,u_i} is the influence probability of u_j on u_i.

As an example, Fig. 1a represents the social network of subject u_5. To simplify the reading, only the incoming edges of node u_5 are represented. Each edge is weighted by the influence probability p_{u_j,u_i}. A red node represents an inactive subject. The decision of u_5 to perform an action a is a function (1) of the active friends (u_1, u_2, u_4) and related influence probabilities.

Existing approaches learn the probability p_{u_j,u_i}, $\forall(u_j, u_i) \in E$, from the actions performed by both u_j and u_i. In particular, they consider u_i as influenced by u_j if

Fig. 1 Example of influence probabilities in a social network

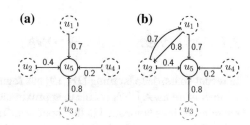

the latter performed the action before the former. Such approaches have two main drawbacks. The probability of friends influencing a subject is considered independent of each other. This assumption may not be always true, especially when two friends of a subject are in turn friends, as for the nodes u_1 and u_2 in the example of Fig. 1b. The fact that subject u_1 and u_2 are both active can differently affect subject u_5 decision. In this instance, the joint probability of influencing u_5 should be higher if compared to the combination of the independent probabilities (1). Further, previous works in the literature learn the influence probability by considering only the actions performed by the subject (*positive samples*). However, it may be relevant to take into account the actions not performed by the subject (*negative samples*), but performed by her/his friends, so as to understand who really affects subject's decisions. As an example, we consider the scenario where subject u_5 does not buy a certain product, while some of her/his friends do. In this instance, considering also negative samples can improve the influence modeling, as u_5 may be affected by the friends that share the same *negative* decision.

Previous works differ from each other for the way the probabilities p_{u_j,u_i} are estimated. In this paper, we study the LT models proposed by Goyal et al. [15] and the IC model of Saito et al. [13]. Other works in the literature model social influence at topic level, i.e., considering influence among subjects with respect to a set of OSN topics. We are not only interested in online scenarios, thus, we aim to model social influence among subjects independently of the topics. In the LT models of Goyal et al., a node becomes active if $p_{u_i}(S_{u_i,a}) \geq \theta$, where θ is the activation threshold. They propose different probabilistic models to capture the influence probability p_{u_j,u_i}, referred to as Bernoulli Distribution (BD), Jaccard Index (JI), Partial Credits-Bernoulli (PC-B), and Partial Credits-Jaccard (PC-J). We do not describe them in details for the lack of space. In the IC model of Saito et al., each active subject independently influences her/his inactive friends with influence probabilities estimated by maximizing a likelihood function with the Expectation Maximization (EM) algorithm.

3 Proposed Solution

This work addresses the aforementioned drawbacks by formalizing a deep learning approach for modeling social influence and predicting subject's behavior. In this section, we present the proposed approach based on a DNN architecture.

3.1 Deep Neural Network (DNN)

In recent years, deep learning [18, 19] has found successful application in a growing number of areas. A DNN is able to approximate any continuous function by learning the relationships embedded in the input data. Thereby, it replaces the manual feature extraction procedure by building up a complex hierarchy of concepts through the multiple layers of the network to automatically extract discriminative and abstractive features of data [20]. A DNN is defined by a combination of three layers: input layer (**x**), hidden layers ($\mathbf{h}_1, \mathbf{h}_2, \ldots, \mathbf{h}_L$), and output layer (**y**). These layers are fully connected in a weighted way as follows

$$\mathbf{h}_j = \begin{cases} \phi_j(\mathbf{x}\mathbf{W}_{xh_j}) & \text{if } j = 1 \\ \phi_j(\mathbf{h}_{j-1}\mathbf{W}_{h_{j-1}h_j}) & \text{if } 1 < j \le L \end{cases}$$

$$\mathbf{y} = \phi_o(\mathbf{h}_L\mathbf{W}_{h_Ly}) ,$$

where $\mathbf{W}_{kl}$ indicates the weights of the connections between layer k and l, while ϕ_j is a nonlinear activation function (e.g., sigmoid, ReLU, tanh, softmax) of each hidden node at layer j, and ϕ_o is a nonlinear activation function of each output node. The predictive model of a DNN can be formulated as $\hat{\mathbf{y}} = f(\mathbf{x}|\Theta)$, where $\hat{\mathbf{y}}$ denotes the predicted output, Θ represents the model parameters (i.e., the inter-layers weights), and f indicates the function that maps the input **x** to the output $\hat{\mathbf{y}}$ based on the DNN architecture, i.e., $f(\mathbf{x}) = \phi_o(\phi_L(\ldots \phi_2(\phi_1(\mathbf{x}))\ldots))$.

3.2 Social Influence Deep Learning

In this work, we address the limitations of existing approaches by learning the interplay among subjects using a DNN. The proposed approach has the capability for both modeling social influence and predicting human behavior in one shot. It should be noticed that the DNN does not explicitly produce a mathematical model, but it learns abstractive feature to implicitly model and learn the interaction of the data in input. Our task can be formulated as the problem of predicting whether subject u_i performed action a as a function of the active friends $S_{u_i,a}$. We address this task as a binary classification problem. Thereby, the output $y_{u_i,a}$ of the DNN is a Boolean variable that is equal to 1 if u_i performed a, and is 0 otherwise. The input layer consists of two vectors $\mathbf{v}_{u_i}^U$ and $\mathbf{v}_{u_i}^{F_a}$ that indicate subject u_i and her/his active friends for the action a, respectively. Both of them have length $N = |V|$. The former is a one-hot vector that uniquely identifies each subject $u_i \in V$. The vector consists of all *zeros* with the exception of a single *one* that identifies one element of the set. In this instance, subject u_i is represented by the vector $\mathbf{v}_{u_i}^U$, which has only the ith element equal to one. The latter represents the active friends of subject u_i for the action a. The j-th element of $\mathbf{v}_{u_i}^{F_a}$ corresponds to subject u_j and it equals one only if u_j is active and $(u_j, u_i) \in E$, otherwise is equal to 0. These two vectors are first concatenated and then fed into a multi-layer architecture, as depicted in Fig. 2. For the sake of simplicity, a DNN with only one hidden layer ($L = 1$) is depicted. In our experiments, we design a network with a tower structure, where the bottom layer is the largest and the number of nodes of each successive layer is half of its precedent. In such a way, higher layers with few nodes can learn more abstractive features from the

Fig. 2 DNN framework

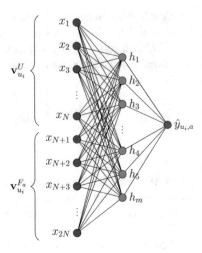

input data [20, 21]. Details about the implementation will be given in Sect. 4.2. The training is performed by minimizing the binary cross-entropy loss between $\hat{y}_{u_i,a}$ and $y_{u_i,a}$, where $\hat{y}_{u_i,a} = f(\mathbf{v}_{u_i}^U, \mathbf{v}_{u_i}^{F_a}|\Theta)$ is the predicted output of our DNN framework.

The rationale of this approach is based on the attempt of overcoming the drawbacks of previous works described in Sect. 2. We model social influence by considering the inter-dependencies among friends. In fact, according to the DNN architecture presented above, we take into account both (i) the relationship between the subject and her/his friends and (ii) the interactions among them. We accomplish this task by placing the social network in a neural network, letting the DNN learn the influence strengths and the interplay among the subjects in the social network. We learn social influence including in the training phase also actions not performed by the subject. For each subject, we train our DNN with an equal number of positive ($y_{u_i,a} = 1$) and negative samples ($y_{u_i,a} = 0$). In such a way, the DNN framework has the capability for both modeling social influence and predicting human behavior in one shot.

4 Experimental Evaluation

To empirically evaluate our framework, we conduct experiments using data of an EBSN. This dataset allows us to investigate social influence considering both *online* (through the social network) and *offline* (real-life) social interactions.

4.1 Dataset Description

An EBSN is a Web platform where users can create events, promote them, and invite friends to participate. Events range from small get-together activities, e.g., Sunday

brunch or movie night, to bigger events, e.g., concerts or conferences [22]. The rationale behind the choice of utilizing an EBSN is based on the intrinsic agglomerative
power of the events. In fact, participating in an event represent a direct and explicit
form of social interaction, other than a personal interest. An EBSN provides a social network service so as to connect friends and users with common interests. In
the event main page, a user can see the information related to the event, e.g., date,
location, and description, along with the confirmed participants. This information
may activate processes of social influence, which can drive user participation in the
events [23].

In this study, we use a dataset from *Plancast*, an EBSN for sharing upcoming
plans with friends. Plancast allows users to subscribe each other providing direct
connections among them. Subscription is similar to the concept of *following* in OSNs,
e.g., Twitter. We utilize a dataset [22] that includes 93041 users and 401634 events,
combined in 1702058 user subscriptions and 869200 user–event participations. We
restrict our analysis to the USA, as most of the events have been organized there. We
filter out users without any subscription and that attended less than 20 events. We
set this threshold in order to build, per each user, a reasonable training and test set
to predict her/his behavior.

4.2 DNN Implementation

In this section, we describe how we implement and design our DNN framework. The
actions set A is defined by the user–events participation in the EBSN dataset, while
$A_{u_i} \subseteq A$ is the set of events attended by subject $u_i \in V$. A subject is considered active
for the event a if she/he decided to participate in the event $a \in A$. For each subject
u_i, we randomly select n_{u_i} events not attended by u_i so as to consider also negative
samples, where $n_{u_i} = |A_{u_i}|$. In order to limit overfitting and to reduce variability,
we utilize a tenfold cross-validation to split the dataset into training and test set. We
build the folds so as to preserve the percentage of positive and negative samples for
each subject in the dataset.

We implement our DNN framework in Keras [24], following a tower pattern composed of $L = 3$ layers with $\{128, 64, 32\}$ nodes, respectively. We train the network
for 25 epochs using RMSProp as optimization function, employing the ReLu as activation function at the hidden layers and the sigmoid as activation function at the
output layer. Moreover, we apply a dropout technique, with a dropout equal to 0.1,
to avoid overfitting. We tune these hyper-parameter performing a grid search on a
validation set (10% of the data).

4.3 Performance Comparison

To validate the performance of our approach, we compare our proposed method
(DNN) with the following baseline algorithms: the LT models (BD, JI, PC-B, and

Table 1 Prediction performances comparison: DNN versus LT models versus IC model

	DNN(%)	BD(%)	JI(%)	PC-B(%)	PC-J(%)	IC(%)
Accuracy	85	78	77	78	77	77
TPR	75	74	75	66	61	60
FPR	5	14	15	6	5	5

PC-J) proposed by Goyal et al. [15], and the IC model of Saito et al. [13]. To find the best threshold θ in the LT model, we measured two metrics: the Youden's index and the closest point to (0,1) in the Receiver Operating Characteristic (ROC) curve. We show only the performance related to the Youden's index as it achieves better results. To examine the performance of these models, we employ widely used metrics in the evaluation of classification problem: Accuracy, True-Positive Rate (TPR), and False-Positive Rate (FPR).

Table 1 depicts the performance of the different solutions. Results indicate that the DNN framework achieves the best Accuracy, TPR, and FPR. We empirically show that the proposed approach outperforms the baseline algorithms, by efficiently resolving the limitations related to the existing works. This result highlights the importance of (i) the interplay among subject's friends, in terms of dependent influence probabilities, and of (ii) the negative samples to detect influential friends and learn influence strengths. Our DNN framework has the capability for both modeling social influence taking into account these aspects and for predicting human behavior, achieving remarkable results.

5 Conclusions

In this paper, we investigated social influence and how it impacts human behavior. We propose to address the limitations of existing approaches by employing a deep learning approach. We introduced a DNN framework that has the capability for both modeling social influence and predicting human behavior. We implemented an architecture that allows the DNN to learn the interplay among friends and to consider both positive and negative samples. To empirically validate the proposed framework, we evaluated our approach using real-life data of an EBSN. Performances exhibit a significant improvement with respect to the state of the art, showing that the proposed approach efficiently resolves the limitations related to existing works.

References

1. Newman, M.E.: The structure and function of complex networks. SIAM (2003)
2. Albert, R., Barabási, A.L.: Statistical mechanics of complex networks. Rev. Mod. Phys. **74**(1), 47 (2002)

3. Goldenberg, J., Libai, B., Muller, E.: Talk of the network: a complex systems look at the underlying process of word-of-mouth. Mark. Lett. **12**(3), 211–223 (2001)
4. Singla, P., Richardson, M.: Yes, there is a correlation:-from social networks to personal behavior on the web. In: International Conference on World Wide Web (2008)
5. Anagnostopoulos, A., Kumar, R., Mahdian, M.: Influence and correlation in social networks. In: International Conference on Knowledge Discovery and Data Mining (2006)
6. Crandall, D., Cosley, D., Huttenlocher, D., Kleinberg, J., Suri, S.: Feedback effects between similarity and social influence in online communities. In: 14th ACM SIGKDD International Conference on Knowledge Discovery and Data Mining, ACM (2008)
7. Luceri, L., Vancheri, A., Braun, T., Giordano, S.: On the social influence in human behavior: physical, homophily, and social communities. In: International Conference on Complex Networks and Their Applications, Springer (2017)
8. Domingos, P., Richardson, M.: Mining the network value of customers. In: ACM International Conference on Knowledge Discovery and Data Mining, ACM (2001)
9. Richardson, M., Domingos, P.: Mining knowledge-sharing sites for viral marketing. In: ACM International Conference on Knowledge Discovery and Data Mining, ACM (2002)
10. Kempe, D., Kleinberg, J., Tardos, É.: Maximizing the spread of influence through a social network. In: Proceedings of the International Conference on Knowledge Discovery and Data Mining, ACM (2003)
11. Kimura, M., Saito, K., Nakano, R.: Extracting influential nodes for information diffusion on a social network. AAAI **7**, 1371–1376 (2007)
12. Gruhl, D., Guha, R., Liben-Nowell, D., Tomkins, A.: Information diffusion through blogspace. In: International Conference on World Wide Web, ACM (2004)
13. Saito, K., Nakano, R., Kimura, M.: Prediction of information diffusion probabilities for independent cascade model. Knowledge-Based Intelligent Information and Engineering Systems (2008)
14. Tang, J., Sun, J., Wang, C., Yang, Z.: Social influence analysis in large-scale networks. In: International Conference on Knowledge Discovery and Data Mining, ACM (2009)
15. Goyal, A., Bonchi, F., Lakshmanan, L.V.: Learning influence probabilities in social networks. In: ACM International Conference on Web Search and Data Mining, ACM (2010)
16. Liu, L., Tang, J., Han, J., Yang, S.: Learning influence from heterogeneous social networks. Data Min. Knowl. Discov. **25**(3), 511–544 (2012)
17. Fang, X., Hu, P.J.H., Li, Z., Tsai, W.: Predicting adoption probabilities in social networks. Inf. Syst. Res. (2013)
18. LeCun, Y., Bengio, Y., Hinton, G.: Deep learning. Nature **521**, 436–444 (2015)
19. Schmidhuber, J.: Deep learning in neural networks. Neural Netw. (2015)
20. He, X., Liao, L., Zhang, H., Nie, L., Hu, X., Chua, T.S.: Neural collaborative filtering. In: the 26th International Conference on World Wide Web (2017)
21. He, K., Zhang, X., Ren, S., Sun, J.: Deep residual learning for image recognition. In: IEEE Conference on Computer Vision and Pattern Recognition, IEEE (2016)
22. Liu, X., He, Q., Tian, Y., Lee, W.C., McPherson, J., Han, J.: Event-based social networks: linking the online and offline social worlds. In: ACM International Conference on Knowledge Discovery and Data Mining, ACM (2012)
23. Georgiev, P., Noulas, A., Mascolo, C.: The call of the crowd: event participation in location-based social services. arXiv:1403.7657 (2014)
24. Chollet, F.: Keras (2015). http://keras.io (2017)

Inspiration, Captivation, and Misdirection: Emergent Properties in Networks of Online Navigation

Patrick Gildersleve and Taha Yasseri

Abstract The World Wide Web (WWW) has fundamentally changed the ways billions of people are able to access information. Thus, understanding how people seek information online is an important issue of study. Wikipedia is a hugely important part of information provision on the Web, with hundreds of millions of users browsing and contributing to its network of knowledge. The study of navigational behavior on Wikipedia, due to the site's popularity and breadth of content, can reveal more general information seeking patterns that may be applied beyond Wikipedia and the Web. Our work addresses the relative shortcomings of existing literature in relating how information structure influences patterns of navigation online. We study aggregated clickstream data for articles on the English Wikipedia in the form of a weighted, directed navigational network. We introduce two parameters that describe how articles act to source and spread traffic through the network, based on their in/out strength and entropy. From these, we construct a navigational phase space where different article types occupy different, distinct regions, indicating how the structure of information online has differential effects on patterns of navigation. Finally, we go on to suggest applications for this analysis in identifying and correcting deficiencies in the Wikipedia page network that may also be adapted to more general information networks.

1 Introduction

The Internet and particularly the World Wide Web (WWW) have brought a vast sea of information to the fingertips of billions of people, fundamentally changing the ways that we seek and gain information. Given the scale and importance of WWW, it is important that people are able to navigate through it effectively, thus understanding how people seek information online is vital for the design of such information systems. Affordances of the platform (including Web page content, Web

P. Gildersleve · T. Yasseri (✉)
Oxford Internet Institute, University of Oxford, 1 St Giles, Oxford OX1 3JS, UK
e-mail: taha.yaseri@gmail.com

T. Yasseri
Alan Turing Institute, The British Library, 96 Euston Road, London NW1 2DB, UK

© Springer International Publishing AG 2018
S. Cornelius et al. (eds.), *Complex Networks IX*, Springer Proceedings
in Complexity, https://doi.org/10.1007/978-3-319-73198-8_23

page design, and Web site hyperlink structure) and user desires determine whether users are assisted, misdirected, or manipulated in navigation. We may record traces of how users navigate online in clickstream logs—the sequence of clicks a user makes within and between Web pages. These navigational data are an important representation for the quality and utility of a Web site and may be harnessed by Webmasters for improvements in Web site design [1, 2].

Wikipedia, the free, online, collaborative encyclopedia has become a hugely important part of information provision on WWW. As the Web's fifth most popular Web site, it is the largest and most popular general reference work on the Internet [3]. While it is not immune to criticism on its accuracy [4], biases [5], and coverage [6], it has been meaningfully compared to the Encyclopedia Britannica [7]. Analyzing clickstream data for Wikipedia is of particular interest, since we may study navigational behavior for its hundreds of millions of users across the huge collaboratively generated network of knowledge. These data on navigation between the vast number of articles can reveal general patterns of information seeking behavior as well as the influence of the article network's structure. These insights can be used by both the Wikimedia Foundation and editors to improve the Web site such as in regular error correction, more fundamental Web site design changes as well as in addressing important issues such as how regular users are affected by imbalance and systemic bias [8–10] of content across Wikipedia. More generally, this analysis can help us to understand human information seeking patterns beyond Wikipedia and even WWW.

Analysis of clickstream data, in the form of Web usage mining, has been practiced with the aim of improving Web site design and targeting users more effectively to increase use of the service on offer. For example, online shopping Web sites track users to understand the patterns of behavior that may lead to a purchase [11], online social networks analyze navigation patterns within their Web site to provide customized experiences to increase user interaction and retention [12, 13], and clickstream data may be used to detect fake or automated accounts on online services [13].

Past work that covers more general patterns of navigation on WWW has tried to identify how people use the Web (both within and between Web sites and individual pages), as compared to its design and structure. Weinreich et al. find that 'Link following has remained the most common navigation action, accounting for about 45% of all page transitions' [14]. However, the structure of the Web alone does not give a full picture of how it acts as a medium for the discovery of information for billions of people. Wu and Ackland identify a 'mismatch between hyperlinks and clickstreams' for navigation between the 1000 most popular Web sites (as ranked by Alexa.com), finding a marked difference between the network of hyperlinks and the navigational network of clickstreams [2]. The authors comment that as we move through Web 2.0 and Web 3.0 that this mismatch between hyperlinks and clickstreams will be alleviated, as users and algorithms ostensibly designed to serve users' interest (rather than individual Webmasters) provide more relevant hyperlinks across the Web.

Wikipedia is one instance where users' engagement in the form of editors' writing, correcting, and warring over articles shapes the form and structure of the Web site itself. The editorial and traffic statistics of Wikipedia pages have been used to predict movies box office revenue [15], elections turnout and outcome [16, 17], and disease outbreaks [18]. Clickstream analysis can provide important insights that both the Wikimedia Foundation and editors may draw on when improving the Web site. For example, while many link prediction techniques rely solely on hyperlink structure and semantic features [19, 20], West et al. use navigation paths from the Wikispeedia game to create a link suggestion model [21]. Unfortunately, this approach is limited to navigation where the eventual target page be explicitly known to both the algorithm and the user. A more general approach presented in [22] utilizes implicit signals from server logs to maximize objective functions under 3 Web browsing models in suggesting the most useful links for the future, validated on the desktop English Wikipedia.

Lamprecht et al. investigate how different naïve link selection models compare to data from the Wikipedia clickstream, finding that a model based on article structure best explains user navigation choices [23]. This work is built on by Dimitrov et al. by using a more complex model utilizing Bayesian inference, supplemented by mixed effects hurdle models using network, semantic, and visual features to predict transition counts in clickstream data [24]. Finally, 'Why We Read Wikipedia' [25] provides a comprehensive overview of navigation on Wikipedia to create a taxonomy of users, their behaviors, and their motivations by matching survey responses with data including user clickstreams. A wide range of navigational patterns are observed, including fast-paced random exploration, current events driven navigation, and long sessions of work and research.

The existing literature well covers the analysis of user behavior from clickstream data including its applications in improving Web site design and user experience across the WWW and specific to Wikipedia. However, relatively little research covers the properties and utility of the navigational graph itself. As a complement to user-focussed clickstream analysis, traces of user navigation from clickstream data may be used as part of a page-focused analysis. Put simply, instead of asking 'how do users behave?', we shift the lens of focus to the Web pages in order to ask 'how are articles used?'. This change in reference frame allows us to directly investigate how the structure of information on Wikipedia (and the Web at large) supports and hinders different kinds of navigational behavior. Formally, our main contributions are to introduce two metrics to describe the sourcing and spreading of user traffic through Web pages and to use these metrics to construct a phase space that is used to analyze patterns of user behavior. Different page types introduce different, distinct navigational structures into the page network. We finish by recommending use cases for this navigational phase space in identifying errors and deficiencies in the Wikipedia page network.

2 Results and Discussion

The Wikipedia clickstream dataset contains monthly aggregates for the number of clicks on hyperlinks between articles on Wikipedia. From a network perspective, these data act as an edge list and may be used to construct a weighted, directed

Fig. 1 An illustration of the different sourcing/spreading configurations. The intensity of shading schematically codes the volume of traffic across links to/from the range of an article's neighbors

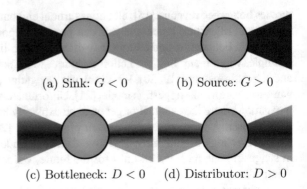

(a) Sink: $G < 0$ (b) Source: $G > 0$

(c) Bottleneck: $D < 0$ (d) Distributor: $D > 0$

graph. We use measures of how much a page acts as a relative traffic source or sink (sourcing coefficient G) and how much a page acts to spread or focus traffic (spreading coefficient D) as part of a 'Navigational Phase Space' to describe and visualize how users navigate through pages on Wikipedia. Illustrations of different sourcing and spreading configurations are provided in Fig. 1.

The sourcing coefficient of a given node (article) is defined as

$$G = \frac{S_{\text{out}} - S_{\text{in}}}{S_{\text{out}} + S_{\text{in}}}, \tag{1}$$

where $S_{\text{in/out}}$ is the in/out strength of the node, i.e., the total number of clicks into/out from the article. Flow is not conserved in this network, since users may arrive at a node from an external source and may also click on multiple links within one page, so S_{out} may be larger than S_{in} for a given node.

The distribution of G for all articles in the month of September 2016 is shown in Fig. 2a. A wide range of values is observed for G. However, on the whole, more pages act as traffic sinks ($G < 0$, Fig. 1a) rather than sources ($G > 0$, Fig. 1b) since users eventually stop browsing.

How the shape of traffic changes as it passes through nodes is another important emergent feature for each article. An article may receive disperse traffic from a range of neighbors or focussed traffic from a narrow subset of neighbors and then go on to send disperse or focussed traffic.

The spreading coefficient is defined as

$$D = \bar{\sigma}_{\text{out}} - \bar{\sigma}_{\text{in}}, \tag{2}$$

where $\bar{\sigma}_{\text{in/out}}$ is the normalized in/out entropy of the node. This describes the spread of traffic, whether it is focussed over a relatively narrow subset of neighbors, or dispersed across many neighbors; for further details on node entropy, see Sect. 4.

The distribution of D for articles in September 2016 is shown in Fig. 2b. We note that most articles act as 'channels' ($D \sim 0$), with no noticeable effect on the shape

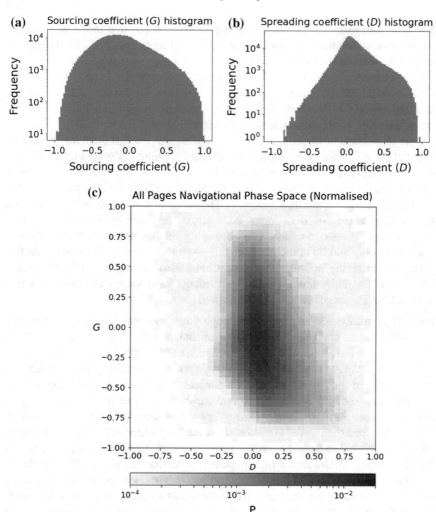

Fig. 2 Distributions of all articles' sourcing coefficient (**a**) and spreading coefficient (**b**) are combined to form a navigational phase space (**c**). P is the probability of an article existing at that point in phase space

of traffic and that more articles act as distributors ($D > 0$), Fig. 1d, than bottlenecks ($D < 0$), Fig. 1c.

We use these measures to construct a 2D navigational phase space for articles, as shown in Fig. 2c. The point that an article exists in within this space describes the nature of the traffic through it.

This navigational phase space may also be used to identify errors and deficiencies in the Wikipedia page network, and in principle the network of pages of any Web site.

Extreme Behavior: An article occupying an extreme position in the navigational phase space is usually indicative of some error relating to hyperlinks on the page or hyperlinks leading to the page. Sourcing coefficient $G \sim -1$ may be a result of an article being erroneously linked to by another, far more popular, page that distorts the data for traffic into the article. Spreading coefficient $D \sim -1$ would indicate that the vast majority of users seek the information of another article following or instead of the information on the current article. This information could be used to inform the merger of articles, or identify erroneous hyperlinks to the first page that should be targeted at the second. An example is the article 'Passenger' (*a person who travels in a vehicle but bears little or no responsibility...*)—$D = -0.79$. A brief investigation reveals that the article is linked to by a template[1] for 'Nettwerk', a popular music label that manages 'Passenger (singer)'. The result of this is that many articles on acts relating to the Nettwerk music label erroneously link to 'Passenger' instead of 'Passenger (singer)' and that users browsing acts relating to Nettwerk traverse this link to only end up at the page relating to people traveling in vehicles. At this point, after receiving incorrect information, most users' strategy is to click through to 'Passenger (disambiguation)' (resulting in strong article bottleneck behavior) which does link to 'Passenger (singer)'—the desired page.

Page-Type Analysis: We observed through page title text that many of the articles with extreme navigational phase space behavior were of particular page types. This motivated us to study the navigational phase spaces for different page types— list pages, -ography pages, disambiguation pages, and trending pages. List articles provide a list of links to all articles of a particular class (e.g., *List of common misconceptions*). -ography articles act to summarize the body of work of creative professionals (e.g., *Paul McCartney Discography, Leonardo DiCaprio Filmography*). Disambiguation articles exist to resolve any ambiguity when articles on several topics might be expected to have the same 'natural' page title (e.g., *Wooden (disambiguation))*. Trending articles are the most popular articles that receive a peak in popularity over the month of study (for more details on article types, see Sect. 4). By examining the difference between the navigational phase spaces for these particular article types against that for all Wikipedia articles, we observe distinct navigational patterns. The results are shown in Fig. 3.

List pages act as both relatively strong sources and distributors of traffic, as indicated in Fig. 3a. Users may arrive at a list page from an external Web site or from a relatively narrow range of Wikipedia articles and be inspired to open a wide range of articles from the list.

-ography pages (Fig. 3b) act as strong sources and distributors of traffic, more so than regular list pages. Traffic toward -ography pages is predominantly focussed from the respective actor/author/band, etc., and, as with list pages, users are inspired to open a wide range of articles from the -ography page.

One would expect disambiguation pages to spread traffic to a range of articles, since they are designed to act as navigation aids when a user may be searching for any of a range of similarly titled articles. We indeed observe this with for mode of articles

---

[1]A mini-page that can be automatically copied and updated across many articles.

with $D > 0$ in Fig. 3c. However, there is a second mode with $D < 0$ that act to focus traffic. This is atypical and likely undesirable behavior, evidence of misdirection that is later elaborated on.

Trending articles (Fig. 3d) and related/linked articles are popular, by definition, receiving a large amount of traffic, predominantly from external sources. It is of no surprise that these articles then act as strong sources of traffic to the rest of the article network; however, they do not act to focus or distribute traffic; captivated users come from and navigate toward a similar spread of articles.

Since there are distinct patterns of navigation for different types of article, we go on to identify two further examples of errors and deficiencies in the article network through page navigational behavior.

Atypical Behavior: As previously observed, certain page types typically occupy distinct areas in the navigational phase space. Aside from more general extreme behavior, some articles of particular types may exhibit behavior that is atypical and perhaps undesirable for their type.

Consider the aforementioned 2-mode distribution for disambiguation pages in Fig. 3c. Articles in the mode with $D < 0$ act as bottlenecks, focussing traffic toward one page—suggesting that on the whole, users find no ambiguity and mostly require one page in particular. An example of this is for the page 'Tinder (disambiguation)' ($D = -0.70$) from which traffic is heavily focussed toward 'Tinder (app)', rather than the combustible material. This behavior would suggest that a disambiguation page might not be necessary and that 'Tinder (app)' should be the default.

Mimetic Behavior: Occasionally, a general article will behave very similarly to articles of a particular type without explicitly being named or set out as such. We have observed this most clearly with articles behaving like list pages. Examples of these include 'Saturday Night Live cast members' ($G = 0.61$, $D = 0.71$), 'Allied leaders of World War II' ($G = 0.16$, $D = 0.76$), and 'Bollywood horror films' ($G = 0.96$, $D = 0.41$). This could be the basis for simple name changes or even splitting these kind of articles into separate dedicated list and descriptive pages.

Clearly, the way that information is structured on Wikipedia has differential, non-trivial effects on patterns of navigation on the Web site. Future research on the production and structure of information on Wikipedia must clearly detail the impact on regular users' navigational behavior and ability to access information.

3 Conclusion

This work emphasizes that information structure is an important factor for navigation online and that hyperlinks are not created equal. Moreover, how information is organized between and within pages shapes both the volume and shape of traffic that flows across hyperlinks between them. Structure in the information network translates non-trivially to patterns of user navigational behavior so it is important that this information structure acts to appropriately direct users, rather than to manipulate or

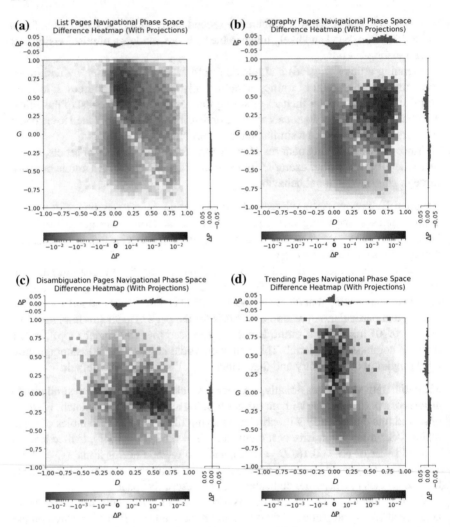

Fig. 3 Navigational phase space difference heatmaps for **a** list pages, **b** -ography pages, **c** disambiguation pages, and **d** trending pages. ΔP is the difference between the values of the probability distributions at that point for a given page type and the distribution for all articles (see Fig. 5). Separate probability difference plots for G and D are also provided for each subfigure. Distinct patterns of navigation are observed for different article types

misdirect them. Analyzing pages' sourcing and spreading behavior as part of a navigational phase space can act to further inform approaches to incorporating insights from data on user desires into the production and design of content online.

We have provided several practical suggestions for using this analysis to improve the content and structure of Wikipedia. There is certainly further work that may utilize these methods for studying patterns of navigation on and improving Wikipedia and other Web sites. Similar analysis could be applied to how patterns of navigation vary among different categories of content (politics, religion, sport, etc.), as well as deeper analysis on how navigation behavior is related to properties of pages such as their popularity, rankings of article quality, and controversiality [26]. We hope that this work inspires further research and application of clickstream data to issues of information structure and navigation online, emphasizing that work on Web structure should not be conducted in isolation from assessing its impact on regular users' search for information.

4 Data and Methods

Data: The Wikipedia clickstream dataset [27] contains several month-long counts of the total number of clicks between pairs of pages (referrer, resource), including traffic from external sites such as Google and Facebook, as well as the type of click (internal link, external link, redlink for missing pages, and other for searches and referrer spoofing). While these data do not offer a complete picture of individual user navigation paths, the Markovian aggregated representation offers informative in studying navigation through individual pages. Measures to filter out activity from bots are enacted, and any (referrer, resource) pair with fewer than 11 clicks is cut from the dataset. An example of the records from one month is shown in Fig. 4. From a network perspective, these data function as an edge list, detailing edges between source and target nodes with their respective weights. From this edge list, we construct a weighted, directed navigation network of pages for 1 month of user navigation on Wikipedia. For the purpose of this project, we only consider navigation and information seeking behavior across links within Wikipedia for the month of September 2016. This leaves us with a network that has 2,227,070 nodes and 13,951,247 edges. The total number of clicks (sum of all weights) in this dataset is 1,187,607,386.

Node-Level Statistics: For each article (node) in the network, we consider 3 basic properties for traffic both in and out: degree k, strength S, and entropy σ. In/out degree describes the number of neighboring articles that an article receives traffic from/sends traffic to. In/out strength describes the total volume of traffic into/out of an article. Finally, a node's in/out entropy, given by

$$\sigma_{in/out} = - \sum_{i \in in/out \text{ edges}} \frac{w_i}{S_{in/out}} \ln\left(\frac{w_i}{S_{in/out}}\right) , \qquad (3)$$

Fig. 4 A sample of
Wikipedia clickstream data

```
                         prev       curr   type   n
0        Economics_(Aristotle)   Aristotle   link   12
1        Greek_words_for_love   Aristotle   link   14
2                     Pythias   Aristotle   link   12
3                  Psychology   Aristotle   link   37
4       Aristotelian_theology   Aristotle   link   16
```

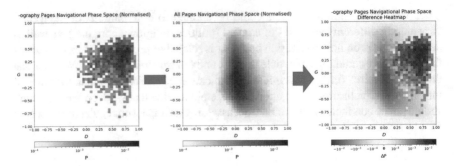

Fig. 5 By subtracting the navigational phase space for all articles from that for articles of a particular type, we highlight what patterns of navigational behavior are more (or less) likely to be associated with articles of that particular type

where w_i is the weight of a given edge in/out, describes the spread of traffic into/out from a page. That is, whether traffic is focussed from/toward a narrow set of an article's neighbors or whether traffic from/toward an article is relatively evenly distributed across a wide set of its neighbors. This measure is normalized according to the maximum possible entropy for a page of given degree k;

$$\bar{\sigma} = \frac{\sigma}{\ln(k)} . \qquad (4)$$

We note that the dataset filter for minimum edge weight (number of clicks) between articles introduces boundary effects for articles with low degree and low strength. For articles in the dataset with low degree and/or strength, any links they might have with slightly fewer than 11 clicks that are filtered out have a larger relative effect on that page's overall recorded properties. To mitigate these effects, while all edges in the dataset from internal links in a month-long Wikipedia article network are present in our analysis, we do not study articles with in/out degree or strength below certain defined thresholds. Firstly, an article must have traffic both into and out from it. Secondly, we take the peak from the navigational network's in/out degree and strength distributions, and only consider articles with degree/strength above this value. Minimum values are typically $k \sim 2$ and $S \sim 150$. The remaining set of articles (13,76230 for September 2016) is where we focus our analysis.

Type Selection and Analysis: By analyzing the strings of text of the page title, we detect the article types: List pages (n = 17997), disambiguation pages (n = 2379), -ography pages (n = 1337). We also introduce a group of trending articles (n = 651) based on the volume of external traffic toward an article—so as not to directly influence S_{in} which is based on internal traffic. A trending article must be one of the 5000 most popular articles in a particular month and must also receive its peak volume of traffic within said month compared to the other months in the dataset. This removes consistently popular articles, preserving those which receive a spike in popularity.

For each of these article types, we find the distributions of sourcing parameter and spreading parameter and form a normalized navigational phase space—equivalent to a 2D probability distribution. We then subtract the equivalent 2D probability distribution for all articles. This generates a heatmap highlighting where particular article types are more or less likely to exist as compared to the distribution of all Wikipedia articles and is our main tool in the analysis of different types of Wikipedia article. An example of this process is shown in Fig. 5.

References

1. Meiss, M.R., Gonçalves, B., Ramasco, J.J., Flammini, A. Menczer, F.: Agents, bookmarks and clicks: a topical model of web navigation. In: Proceedings of the 21st ACM conference on Hypertext and hypermedia, pp. 229–234 (2010)
2. Wu, L., Ackland, R.: How Web 1.0 fails: the mismatch between hyperlinks and clickstreams. Soc. Netw. Anal. Min. **4**, 1–17 (2014). ISSN: 18695469
3. Alexa Top 500 Global Sites Retrieved: 14:05, October 07, 2017 (GMT). https://www.alexa.com/topsites
4. Waters, N.L.: Why you can't cite Wikipedia in my class. Commun. ACM **50**, 15–17 (2007)
5. Wagner, C., Graells-Garrido, E., Garcia, D., Menczer, F.: Women through the glass ceiling: gender asymmetries in Wikipedia. EPJ Data Sci. **5**, 5 (2016)
6. Samoilenko, A., Yasseri, T.: The distorted mirror of Wikipedia: a quantitative analysis of Wikipedia coverage of academics. EPJ Data Sci. **3**, 1 (2014)
7. Giles, J.: Internet encyclopaedias go head to head. Nature **438**, 900–901 (2005). ISSN: 0028-0836
8. Reagle, J., Rhue, L.: Gender Bias in Wikipedia and Britannica. Int. J. Commun. **5**, 00 (2011). http://ijoc.org/index.php/ijoc/article/view/777. ISSN: 1932-8036
9. Callahan, E.S., Herring, S.C.: Cultural bias in Wikipedia content on famous persons. J. Assoc. Inf. Sci. Technol. **62**, 1899–1915 (2011)
10. Lam, S. T. K., et al.: WP: clubhouse?: an exploration of Wikipedia's gender imbalance. In: Proceedings of the 7th international symposium on Wikis and open collaboration, pp. 1–10 (2011)
11. Bucklin, R.E., et al.: Choice and the Internet: from clickstream to research stream. Mark. Lett. **13**, 245–258 (2002). ISSN: 09230645
12. Benevenuto, F., Rodrigues, T., Cha, M., Almeida, V.: Characterizing user behavior in online social networks. In: Proceedings of the 9th ACM SIGCOMM conference on Internet measurement conference - IMC '09, vol. 49 (2009). https://doi.org/10.1145/1644893.1644900. ISBN: 9781605587714
13. Wang, G., Zhang, X., Tang, S., Zheng, H., Zhao, B.Y.: Unsupervised clickstream clustering for user behavior analysis. In: Proceedings of the 2016 CHI Conference on Human Factors in Computing Systems - CHI '16, pp. 225–236 (2016). ISSN: 10495258
14. Weinreich, H., Obendorf, H., Herder, E., Mayer, M.: Off the beaten tracks: exploring three aspects of web navigation, In: Proceedings of the 15th International Conference on World Wide Web, pp. 133–142, ACM, USA, 2006. https://doi.org/10.1145/1135777.1135802. ISBN: 1-59593-323-9
15. Mestyán, M., Yasseri, T., Kertész, J.: Early prediction of movie box office success based on Wikipedia activity big data. PloS one **8**, e71226 (2013)
16. Yasseri, T., Bright, J.: Can electoral popularity be predicted using socially generated big data? it-Inf. Technol. **56**, 246–253 (2014)
17. Yasseri, T., Bright, J.: Wikipedia traffic data and electoral prediction: towards theoretically informed models. EPJ Data Sci. **5**, 1–15 (2016)

18. Generous, N., Fairchild, G., Deshpande, A., Del Valle, S.Y., Priedhorsky, R.: Global disease monitoring and forecasting with Wikipedia. PLoS Comput. Biol. **10**, e1003892 (2014)
19. Milne, D., Witten, I.H.: Learning to link with wikipedia, In: Proceedings of the 17th ACM conference on Information and knowledge management, pp. 509–518 (2008)
20. Noraset, T., Bhagavatula, C., Downey, D.: Adding high-precision links to Wikipedia. In: EMNLP, pp. 651–656 (2014)
21. West, R., Paranjape, A., Leskovec, J.: Mining missing hyperlinks from human navigation traces: a case study of Wikipedia, In: Proceedings of the 24th international conference on World Wide Web, pp. 1242–1252 (2015)
22. Paranjape, A., West, R., Zia, L., Leskovec, J.: Improving website hyperlink structure using server logs, In: Proceedings of the Ninth ACM International Conference on Web Search and Data Mining, pp. 615–624 (2016)
23. Lamprecht, D., Lerman, K., Helic, D., Strohmaier, M.: How the structure of Wikipedia articles inuences user navigation. New Rev. Hypermedia Multimed. **4568**, 1–22 (2016). ISSN: 1361–4568
24. Dimitrov, D., Singer, P., Lemmerich, F., Strohmaier, M.: What Makes a Link Successful on Wikipedia?, In: Proceedings of the 26th International Conference on World Wide Web, pp. 917–926 (2017)
25. Singer, P., et al.: Why We Read Wikipedia. In: Proceedings of the 26th International Conference on World Wide Web - WWW '17, pp. 1591–1600 (2017)
26. Yasseri, T., Sumi, R., Rung, A., Kornai, A., Kertész, J.: Dynamics of conicts in Wikipedia. PloS one **7**, e38869 (2012)
27. Ellery, W., Taraborelli, D.: Wikipedia Clickstream. figshare. Retrieved: 21 22, May 07, 2017 (GMT). https://doi.org/10.6084/m9.figshare.1305770.v22

Are Crisis Platforms Supporting Citizen Participation?

Gonzalo Bacigalupe and Javier Velasco-Martin

Abstract Information systems are central to disaster management, and getting the right information to everyone is fundamental. Besides research, digital systems for disaster management have gained a central role in public and private disaster management organizations. The blogging and social media platforms popularized a decade ago were built around a user-generated content model in which users are not only readers but also producers of information, and their use is now pervasive. During a natural disaster crisis, massive amounts of information are generated via social media, including messages of caution and advice, information about affected individuals, infrastructure damage, volunteering and donations, among others. Based on a review of the literature and a systematic analysis of crisis platforms, we assess the ways in which participation is defined, propose a participation categorization, and evaluate the role that digital platforms may play in supporting community resilience for crisis and extreme events. The present study reviews what kinds of participation crisis computing projects are offered to the citizens of the regions they are scoping, and will evaluate how crowdsourcing is framed and how it is made available to citizens.

1 Crisis Platforms for Participation

Information systems are central to disaster management, and getting the right information to everyone is fundamental. The field of crisis computing/informatics grows strongly [1], with contributions from multiple areas of knowledge, including social computing [2], artificial intelligence [3, 4], geographical information systems [5], and social sciences [6]. Besides research, digital systems for disaster management have

G. Bacigalupe (✉) · J. Velasco-Martin
CIGIDEN, National Research Center for Disaster Risk Management, Edificio Hernan Briones,
Campus San Joaquin, Universidad Catolica de Chile, Santiago, Chile
e-mail: gonzalo.bacigalupe@cigiden.cl

J. Velasco-Martin
e-mail: javier.velasco@cigiden.cl

G. Bacigalupe
CSP, CEHD, University of Massachusetts Boston, Boston, MA, USA

© Springer International Publishing AG 2018
S. Cornelius et al. (eds.), *Complex Networks IX*, Springer Proceedings
in Complexity, https://doi.org/10.1007/978-3-319-73198-8_24

283

Table 1 Research Process

Prodecure		Stage One—descriptive analysis	Stage two—participation levels analysis	Stage three—participation modes analysis
Project collection		Continuous, adjusted to filtering criteria.		
Filter	Inclusion criteria	Web or Mobile interface	Citizen participation	Currently active
		Dynamic information		Public-access web interface to show results
		Focus on natural hazards		
	Exclusion criteria	Simulation interoperability	N/A	N/A
Category development		Type,	Participation type	N/A
		Sponsor		
		Audience		
		Scope		
		Risk type		
		Citizen participation		
Analysis factors		Type	Participation type	Participation
		Sponsor		Functions/features
		Audience		
		Scope		
		Risk type		
		Citizen participation		

gained a central role in public and private disaster management organizations. The blogging and social media platforms popularized a decade ago were built around a user-generated content model in which users are not only readers but also producers of information, and their use is now pervasive. During a natural disaster crisis, massive amounts of information are generated via social media, including messages of caution and advice, information about affected individuals, infrastructure damage, volunteering and donations, among others [7]. The Sendai Framework endorses the departure from the Emergency Management paradigm toward disaster risk reduction—DRR [8]. Citizens have an active role in the DRR framework, and they are aware of the risks in their community and engage in the development of solutions to increase their resilience and reduce their vulnerability: Participation and resilience are central tenets of the DRR model. Community resilience is strengthened by connectedness [8–10]. The exchange of information that takes place in human communication is always modulating relationships between the participants [11], and the same dynamics apply to computer-mediated communication: The Internet

Table 2 Categories for general analysis

Dimension	Category	Description
Type	Interoperability	Enables different telecommunication systems to work together
	Simulation	Uses data analysis to estimate hypothetical or historic scenarios
	Frameworks	Offer a set of tools to build tailored disaster management solutions
	Dashboards	Provide visual representations of a scenario or emergency in present time, often driven with real-time data and combining multiple information sources
	Other	Present more specific functions, these include memorial archives, refuge setup app, and a phone call handling system
Sponsor	Public	Government institutions, including EMAs and armed forces
	Private	Business, usually software/technology companies
	Academic	University researchers
	NGO	Formal and informal volunteer organizations, including software and emergency oriented teams
Audience	Emergency management	Central organization coordinators and first responders, people officially assigned to manage the crisis, with access to sensitive information through private interfaces
	Public access	The affected community and anyone of interest, either nationally or globally
	Clients	Only paying customers can see information
	NGO	Formal and informal disaster-related civil and volunteer organizations focused on response and volunteering
	Variable	The audience will depend on how the system is implemented on particular cases
	Government	System explicitly developed for national-wide decision making at central government
Risk	General	Covers four or more types of emergency, frequently all, regardless of the risk it was created from
	Multi-risk	Is focused on two or three types of risk
	Specific	Is invested on a particular type of risk (e.g. earthquakes)
Scope	Global	Has worldwide reach, works with global data
	Multi-country	Reaches some countries, usually adjacent, based on specific data from such countries
	Country	Is specifically designed for certain country, adjusts to local reality
	Local	Covers a specific subset of a country, either a province or a city
Citizen Input	Yes	Has some type of input from citizens and volunteers who are not part of the emergency management institutions
	No	Has no citizen input

Table 3 Levels of participation in DRR crowdsourcing

Participation level	Definition
Harvesting	Citizens share information on social media, not intending the collection for a particular system. This information is mined by the system [47–50]. Information needs to be validated before processing. Citizens contribute passively to the project
Sensors	Citizens actively share information user-generated content to a project; frequently through a standalone app. Information has to be validated before processing [48, 49, 51]
Trusted sensors	Verified citizens send reports to a specific project, information does not require validation before processing [40]
Nodes	Verified citizens collaborate in some form of manually processing information. This may include relaying, translation [40], and tagging
Agents	Trusted citizens with identified skills or training perform specific tasks under supervision of civil protection agencies [40]
Community	Citizens share information with the system as part of communities, revealing relationships between members, including relationships of closeness and authority. Participation is not only individual but also explicitly collective [15, 22, 24]

is as much about information exchange as it is about maintaining and building new relationships [12–15]. Human connections are what determines the flow of information on social media [16–21]. Taken together, these ideas suggest social media and social software represents an important opportunity for nurturing community resilience in the context of disaster risk reduction [22–25].

2 Crowdsourcing in Crisis Computing

The importance of citizen-generated information has been promoted by international disaster-related organizations [8, 26]. A growing series of projects for disaster management capture this citizen-generated information and analyze it for disaster management under a crowdsourcing model [27]. There are important technological challenges in these projects, and they frequently include: The real-time processing of massive information streams [1, 28–30], information ranking: How to find relevant information [1, 4, 16, 31], information validation: How to verify information authenticity [16], rumor control: How to stop the spread of unverified information [32–34], geolocalization: Determining geographic locations discussed on messages in order to place the information in the context of a map [35, 36], information design: How to present information in the most understandable way [37], citizen participation and community resilience: How digital tools can empower citizens to be active agents of disaster risk management [10, 13, 38–41]. The crowdsourcing model is rooted in the concept of coproduction [40]: The incorporation of a group of individuals outside an organization to collaborate in the production of goods. Coproduction implies the deliberate participation of external members, who collaborate based on their skills and are compensated for their contribution. The degree of citizen participation in disaster crowdsourcing varies. Substantive work addressing the technical issues of crisis computing exists, but fewer studies examine how this participation manifests and how crowdsourcing platforms enable significant agency in disaster risk management. To assess how participation emerges in these platforms, this paper characterizes crisis crowdsourcing platforms in relation to a key dimension of disaster risk reduction strategies: citizen participation. In order to measure citizen participation, a variety of models have been developed in face-to-face settings [41–43], computer-mediated settings [35, 44, 45], as well as specifically in crisis technologies in crisis computing as well [27, 46]. The broadest and most practical model of participation for the purpose of this study was a taxonomy of participation roles for a disaster-oriented crowdsourcing by Diaz, Carroll [40]. Text mining techniques are often used to collect citizen-generated information to assist disaster management agencies in decision making and planning; this method does not require the will of citizens to participate, and it is passive participation [47–49] or harvesting [50]. Crowdsourcing implies participants' collaboration toward a shared goal [48, 49, 51]. Aside from reporting events, other projects allow citizens to assist in information classification by way of volunteer tagging, manually categorizing data, as is the case of creating training data for artificial intelligence systems [3] or in geocoding [52]. Environments that promote interaction between citizens allow spontaneous communities to emerge [53]. In online communities, citizens organize, debate, and make decisions, communities may generate collective intelligence, solving crisis-related problems such as victim identification [54]. Communities formed online by citizens around disasters can be considered communities of practice. For communities of practice, the relationships between participants, including factors of trust, closeness, and authority, are vital for the community structure and participation [24, 54–56]. Citizen online communities for crisis can be spontaneous, improvised, self-organized and are able to support collective behavior [53, 54]. The present study reviews what kinds of participation crisis-computing-projects offer to the citizens of the regions they are scoping, and will evaluate how crowdsourcing is framed, and how it is made available to citizens. Table 1 highlights the methodological choices for the review of platforms. Table 2 offers a taxonomy of the dimensions studied while Table 3 addresses the question of what participation means in the types of platforms.

References

1. Castillo, C.: Big Crisis Data: Social Media in Disasters and Time-Critical Situations. Cambridge University Press, New York (2016)
2. Schafer, W.A., Ganoe, C.H., Carroll, J.M.: Supporting community emergency management planning through a geocollaboration software architecture. Comput. Support. Coop. Work (CSCW) 16(4), 501–537 (2007)

3. Imran, M., Castillo, C., Lucas, J., Meier, P., Vieweg, S.: Aidr: Artificial intelligence for disaster response. In: 23rd International Conference on World Wide Web, pp. 159–162. ACM, 07 –11 April 2014
4. Ofli, F., Meier, P., Imran, M., Castillo, C., Tuia, D., Rey, N., Briant, J., Millet, P., Reinhard, F., Parkan, M., Joost, S.: Combining human computing and machine learning to make sense of big (aerial) data for disaster response. Big Data 4(1), 47–59 (2016)
5. de Albuquerque, J.P., Eckle, M., Herfort, B., Zipf, A.: Crowdsourcing geographic information for disaster management and improving urban resilience: an overview of recent developments and lessons learned. European Handbook of Crowdsourced Geographic Information, pp. 309–321. Ubiquity Press, London (2016)
6. Palen, L., Anderson, K.M.: Crisis informatics-new data for extraordinary times. Science 353(6296), 224–225 (2016)
7. Olteanu, A., Vieweg, S., Castillo, C.: What to expect when the unexpected happens. In: Computer Supported Collaborative Work (CSCW), pp. 994–1009. ACM (2015)
8. United Nations Office for Disaster Risk Reduction.: Sendai framework for disaster risk reduction 2015–2030. Report (2015)
9. Cutter, S.L., Ash, K.D., Emrich, C.T.: The geographies of community disaster resilience. Glob. Environ. Change 29, 65–77 (2014)
10. Whittaker, J., McLennan, B., Handmer, J.: A review of informal volunteerism in emergencies and disasters: definition, opportunities and challenges. Int. J. Disaster Risk Reduct. 13, 358–368 (2015)
11. Watzlawick, P., Bavelas, J.B., Jackson, D.D.: Pragmatics of Human Communication: A study of Interactional Patterns. Pathologies and Paradoxes. Faber and Faber, London (1967)
12. Council, N.R.: Applications of Social Network Analysis for Building Community Disaster Resilience: Workshop Summary. The National Academies Press, Washington (2009)
13. Gimenez, R., Hernantes, J., Labaka, L., Hiltz, S.R., Turoff, M.: Improving the resilience of disaster management organizations through virtual communities of practice: a delphi study. J. Conting. Crisis Manag. 25(3), 160–170 (2017)
14. Licklider, J.C., Taylor, R.W.: The computer as a communication device. Sci. Technol. 76(2), 21–41 (1968)
15. Taylor, M., Wells, G., Howell, G., Raphael, B.: The role of social media as psychological first aid as a support to community resilience building. Aust. J. Emerg. Manag. 27(1), 20–26 (2012)
16. Castillo, C., Mendoza, M., Poblete, B.: Predicting information credibility in time-sensitive social media. Internet Res. 23(5), 560–588 (2013)
17. Fenton, N.: Social media is. Soc. Media + Soc. 1(1) (2015)
18. Haynes, N.: Social Media in Northern Chile. UCL Press, London (2016)
19. Papacharissi, Z.: We have always been social. Soc. Media + Soc. 1(1) (2015)
20. Wellman, B., Gulia, M.: Net Surfers Don't Ride Alone: Virtual Communities As Communities, pp. 331–366. Westview Press, Boulder (1999)
21. Wellman, B., Salaff, J., Dimitrova, D., Garton, L., Gulia, M., Haythornthwaite, C.: Computer networks as social networks: collaborative work, telework, and virtual community. Annu. Rev. Sociol. 22, 213–238 (1996)
22. Dufty, N.: Using social media to build community disaster resilience. Aust. J. Emerg. Manag. 27(1), 40 (2012)
23. Ghose, R.: Use of information technology for community empowerment: transforming geographic information systems into community information systems. Trans. GIS 5(2), 141–163 (2001)
24. Shklovski, I., Palen, L., Sutton, J.: Finding community through information and communication technology in disaster response. In: 2008 Conference on Computer Supported Cooperative Work (CSCW), pp. 127–136. ACM (2008)
25. Simon, T., Goldberg, A., Adini, B.: Socializing in emergencies-a review of the use of social media in emergency situations. Int. J. Inf. Manag. 35(5), 609–619 (2015)
26. International Federation of Red Cross And Red Crescent Societies.: World disasters report: focus on information in disasters. IFRCRCS (2005)

27. Poblet, M., García-Cuesta, E., Casanovas, P.: Crowdsourcing tools for disaster management: a review of platforms and methods. AI Approaches to the Complexity of Legal Systems, pp. 261–274. Springer, Berlin (2014)
28. Abu-Elkheir, M., Hassanein, H.S., Oteafy, M.A.: Enhancing emergency response systems through leveraging crowdsensing and heterogeneous data. In: 2016 International Wireless Communications and Mobile Computing Conference (IWCMC), pp. 188–193. IEEE (2016)
29. Besaleva, L.I., Weaver, A.C.: Crowdhelp: a crowdsourcing application for improving disaster management. In: 2013 IEEE Global Humanitarian Technology Conference (GHTC), pp. 185–190. IEEE (2013)
30. Gruhl, D., Nagarajan, M., Pieper, J., Robson, C., Sheth, A.: Multimodal social intelligence in a real-time dashboard system. VLDB J. 19(6), 825–848 (2010)
31. Cobo, A., Parra, D., Navón, J.: Identifying relevant messages in a twitter-based citizen channel for natural disaster situations. In: 24th International Conference on World Wide Web, pp. 1189–1194. ACM (2015)
32. Liang, G., He, W., Xu, C., Chen, L., Zeng, J.: Rumor identification in microblogging systems based on users' behavior. IEEE Trans. Comput. Soc. Syst. 2(3), 99–108 (2015)
33. Starbird, K., Maddock, J., Orand, M., Achterman, P., Mason, R.M.: Rumors, false flags, and digital vigilantes: misinformation on twitter after the 2013 boston marathon bombing. In: iConference 2014 Proceedings, pp. 654 – 662. iSchools, 4–7 March 2014
34. Tripathy, R.M., Bagchi, A., Mehta, S.: Towards combating rumors in social networks: models and metrics. Intell. Data Anal. 17(1), 149–175 (2013)
35. Ardichvili, A., Page, V., Wentling, T.: Motivation and barriers to participation in virtual knowledge-sharing communities of practice. J. Knowl. Manag. 7(1), 64–77 (2003)
36. Power, R., Robinson, B., Ratcliffe, D.: Finding fires with twitter. In: Australasian Language Technology Workshop (ALTA 2013), pp. 80–89. Association of Computational Linguistics (ACL), 4–6 December 2013
37. Bica, M., Palen, L., Bopp, C.: Visual representations of disaster. In: Proceedings of the 2017 ACM Conference on Computer Supported Cooperative Work and Social Computing, CSCW '17, pp. 1262–1276. ACM, New York (2017)
38. David, L., Soon Ae, C., Jaideep, V., Basit, S., Vijay, A., Nabil, R.A.: Peer: a framework for public engagement in emergency response. Int. J. E-Plan. Res. (IJEPR) 4(3), 29–46 (2015)
39. Díaz, P., Aedo, I., Herranz, S.: Citizen participation and social technologies: Exploring the perspective of emergency organizations. In: Hanachi, C.B., Charoy, F., and François (eds) ISCRAM-med 2014: Information Systems for Crisis Response and Management in Mediterranean Countries: First International Conference, pp. 85–97. Springer, 15-17 October 2014
40. Díaz, P., Carroll, J.M., Aedo, I.: Coproduction as an approach to technology-mediated citizen participation in emergency management. Future Internet 8(3), 41 (2016)
41. Pearce, L.: Disaster management and community planning, and public participation: How to achieve sustainable hazard mitigation. Nat. Hazards 28(2), 211–228 (2003)
42. Arnstein, S.R.: A ladder of citizen participation. J. Am. Inst. Plan. 35(4), 216–224 (1969)
43. Chilvers, J.: Environmental risk, uncertainty, and participation: mapping an emergent epistemic community. Environ. Plan. A 40(12), 2990–3008 (2008)
44. Haklay, M.: Citizen science and volunteered geographic Information: overview and typology of participation. Crowdsourcing Geographic Knowledge, pp. 105–122. Springer, Dordrecht (2013)
45. Stewart, O., Lubensky, D., Huerta, J.M.: Crowdsourcing participation inequality: a scout model for the enterprise domain. In: Proceedings of the ACM SIGKDD Workshop on Human Computation, HCOMP '10, pp. 30–33. ACM, New York (2010)
46. Park, C.H., Johnston, E.W.: A framework for analyzing digital volunteer contributions in emergent crisis response efforts. N. Media Soc. 19(8), 1308–1327 (2017)
47. Briscoe, E., Appling, S., Schlosser, J.: Technology futures from passive crowdsourcing. IEEE Trans. Comput. Soc. Syst. 3(1), 23–31 (2016)
48. Loukis, E., Charalabidis, Y.: Active and passive crowdsourcing in government. Policy Practice and Digital Science, pp. 261–289. Springer, Cham (2015)

49. See, L., Mooney, P., Foody, G., Bastin, L., Comber, A., Estima, J., Fritz, S., Kerle, N., Jiang, B., Laakso, M., Liu, H.-Y., Milčinski, G., Nikšič, M., Painho, M., Pődör, A., Olteanu-Raimond, A.-M., Rutzinger, M.: Crowdsourcing, citizen science or volunteered geographic information? the current state of crowdsourced geographic information. ISPRS Int. J. Geo-Inf. 5(5), 55 (2016)
50. Stefanidis, A., Crooks, A., Radzikowski, J.: Harvesting ambient geospatial information from social media feeds. GeoJournal **78**(2), 319–338 (2013)
51. Shirky, C.: Here Comes Everybody: The Power of Organizing Without Organizations. Penguin, New York (2008)
52. Sierra, J., Garrido, R.: Volunteering assistance to online geocoding services through a distributed knowledge ssolution. In: The RICH-VGI Workshop at 18th AGILE Conference on Geographic Information Science (2015)
53. Starbird, K., Palen, L.: "Voluntweeters": Self-organizing by digital volunteers in times of crisis, pp. 1071–1080 (2011)
54. Vieweg, S., Palen, L., Liu, S.B., Hughes, A.L., Sutton, J.: Collective intelligence in disaster: An examination of the phenomenon in the aftermath of the 2007 virginia tech shootings. In: Fifth Information Systems for Crisis Response and Management Conference (ISCRAM) (2008)
55. Purohit, H., Hampton, A., Bhatt, S., Shalin, V.L., Sheth, A.P., Flach, J.M.: Identifying seekers and suppliers in social media communities to support crisis coordination. Comput. Support. Coop. Work (CSCW) **23**(4), 513–545 (2014)
56. Wenger, E., McDermott, R., Snyder, W.M.: Seven principles for cultivating communities of practice. HBS Working Knowledge, vol. 4. Harvard Business School Press, Boston (2002)

Dynamic Visualization of Citation Networks and Detection of Influential Node Addition

Takayasu Fushimi, Tetsuji Satoh and Noriko Kando

Abstract In this paper, to effectively visualize the browsing order of scientific articles, we propose a visualization method for citation networks focusing on the directed acyclic graph (DAG) structure. In our method, all article nodes are embedded into polar coordinate plane, where angular and radial coordinates express the citation relations and order relations among articles, respectively. Furthermore, the proposed method is equipped with a dynamic property to update coordinates of all nodes at low cost when a new article node and citation links are added to the citation network. From experimental evaluations using real citation networks, we confirm that our method explicitly reflects citation relations and browsing order compared with existing methods. Furthermore, focusing on changes in visualization results when new nodes and links are added to the citation network, our method can detect influential node and links addition by angular displacement of each node.

1 Introduction

In recent years, documents such as news articles, blog articles, and scientific articles are posted on the Web every moment. Some newly posted documents are related to past documents, and users can find relevant documents by following hyperlinks,

T. Fushimi (✉)
School of Computer Science, Tokyo University of Technology,
1404-1 Katakuramachi, Hachioji-city, Tokyo 192–0982, Japan
e-mail: takayasu.fushimi@gmail.com

T. Satoh
Faculty of Library, Information and Media Science, University of Tsukuba,
1-2 Kasuga, Tsukuba-city, Ibaraki 305–8550, Japan
e-mail: satoh@ce.slis.tsukuba.ac.jp

N. Kando
Information and Society Research Division, National Institute of Informatics,
2-1-2 Hitotsubashi, Chiyoda-ku, Tokyo 101–8430, Japan
e-mail: kando@nii.ac.jp

© Springer International Publishing AG 2018
S. Cornelius et al. (eds.), *Complex Networks IX*, Springer Proceedings
in Complexity, https://doi.org/10.1007/978-3-319-73198-8_25

trackbacks, and citations. This allows many users to acquire not only single documents but also peripheral knowledge. However, there are multiple related documents, some of which have many related documents, and then users including researchers face the problem of which documents to view from.

As effective visualization methods of the graph structure, Cross-entropy method [9], spring-force model [6], multi-dimensional scaling [8], and spectral embedding [2] have been proposed. In the former two methods, since the layout coordinates of each node are calculated by solving the optimization problem of the nonlinear objective function, it is possible to obtain a very rich output result, but a large computation time is required. On the other hand, the latter two methods can obtain output results at a very high speed because of solving the linear eigenvalue problem when computing the layout coordinates of all nodes, but some nodes and links are overlapped in visualization results and then the quality of the output result is very poor. In general, these existing visualization methods are difficult to output results reflecting the order relation among articles. Therefore, in this paper, we focus on the directed acyclic graph (DAG) structure of the citation network and propose a visualization method considering the browsing order of the articles. Specifically, the layer of each node is determined based on the order obtained by topological sorting with respect to the DAG structure, and the embedding coordinates are determined by the citation relation with the node group of the adjacent layers. By doing this, we express citation relations by angle and order relations by radius, and realize visualization to polar coordinate plane. Furthermore, the proposed method is equipped with a dynamic property to update the visualization result at low cost when a new article node and citation links are added to the citation network. Utilizing the advantage of low-cost updating of coordinates by the proposed method, we quantify the influence degree of the node that caused structural changes by coordinate displacement of each node.

This paper is organized as follows. After revisiting the details of our existing method in Sect. 2, we explain our proposed method in Sect. 3. Next, we mention basic statistics of our experimental dataset in Sect. 4 and show evaluation results in Sect. 5 and Sect. 6. Finally, we describe related work in Sect. 7 and conclude in Sect. 8.

2 Revisit of TF+PCE Method

The proposed method of this paper is an extension of our previous method [4] which targets a hierarchical tree structure so as to consider connection between non-contiguous layers like DAG. Therefore, in this section, we revisit our previous method.

Our previous method, TF+PCE method, consists of the following two steps:

1. Construct a Topic Forest (TF) based on the similarity and the posting order of the documents;
2. Embed a TF into the polar coordinates.

For general time-series documents, the relevance among documents do not explicitly represented unlink hyperlinks among Web pages, trackbacks among blog articles. Therefore, our previous method constructs tree structures, which we call Topic Forest, where relatively similar documents are connected keeping posted order of documents. Then the TF is embedded into polar coordinates by PCE method, which is based on a method [5] that embeds two node sets of a bipartite graph in concentric circles. In polar coordinates, the angular and radial coordinates, respectively, express the similarity of the documents and their posted order.

Now, for a given node set $\mathcal{V}$ and link set $\mathcal{E}$ of a Topic Forest $G = (\mathcal{V}, \mathcal{E})$, let root node $root$ in each tree be the zeroth layer, and assign the layers to all the nodes based on graph distance $g(\cdot, \cdot)$ from $root$, $\mathcal{V}_s = \{u; g(u, root(u)) = s\}$. We consider the sth and $(s + 1)$th layer node groups, $\mathcal{V}_s$ and $\mathcal{V}_{s+1}$, in a tree as a bipartite graph and calculate coordinate vectors $\mathbf{X}_s$ and $\mathbf{X}_{s+1}$. The nodes of $\mathcal{V}_s$ and $\mathcal{V}_{s+1}$ are respectively embedded into the concentric circles of radii r_s and r_{s+1}. For adjacency matrix $\mathbf{A} = [a_{u,v}]_{u \in \mathcal{V}_s, v \in \mathcal{V}_{s+1}}$, we apply a double-centering operation on the adjacency matrix as well as MDS method [8] and obtain a centered adjacency matrix, $\tilde{\mathbf{A}} = [\tilde{a}_{u,v}]$. Then after initializing coordinate vectors $\mathbf{X}_s$ and $\mathbf{X}_{s+1}$, the method iteratively calculates the coordinate vectors to maximize the following objective function:

$$
J(\mathbf{X}_s, \mathbf{X}_{s+1}) = \sum_{u \in \mathcal{V}_s} \sum_{v \in \mathcal{V}_{s+1}} \tilde{a}_{u,v} \frac{\mathbf{x}_u^T \mathbf{x}_v}{r_s \ r_{s+1}}
$$
$$
+ \frac{1}{2} \sum_{u \in \mathcal{V}_s} \lambda_u (r_s^2 - \mathbf{x}_u^T \mathbf{x}_u) + \frac{1}{2} \sum_{v \in \mathcal{V}_{s+1}} \mu_v (r_{s+1}^2 - \mathbf{x}_v^T \mathbf{x}_v), \quad (1)
$$

where λ_u and μ_v are Lagrangian multipliers that represent the placement constraints on each circumference. In (1), $\frac{\mathbf{x}_u^T \mathbf{x}_v}{r_s \ r_{s+1}} = \cos \theta_{u,v}$ holds, and $J(\mathbf{X}_s, \mathbf{X}_{s+1})$ is maximized by placing adjacent nodes in the same direction from the origin.

If vectors $\mathbf{X}_{s+1}$ are fixed, optimal coordinate vector $\mathbf{x}_u$ of node $u \in \mathcal{V}_s$ is also obtained:

$$
\mathbf{x}_u = \frac{r_s}{\|\tilde{\mathbf{x}}_u\|} \tilde{\mathbf{x}}_u, \quad \tilde{\mathbf{x}}_u = \sum_{v \in \mathcal{V}_{s+1}} \tilde{a}_{u,v} \mathbf{x}_v. \quad (2)
$$

From (2), it can be seen that coordinate vector $\tilde{\mathbf{x}}_u$ is calculated as the resultant vector of the coordinate vectors of *all nodes* in $\mathcal{V}_{s+1}$. However, we can calculate the resultant vector *only from the adjacent nodes* of node u if the resultant vector is calculated before double-centering of the adjacency matrix. That is because the value of $a_{u,v}$ can have 1 only if u and v are connected. After calculating the resultant vector, we execute double-centering and normalizing operations to it.

Likewise, if vectors $\mathbf{X}_s$ are fixed, optimal coordinate vector $\mathbf{x}_v$ of node $v \in \mathcal{V}_{s+1}$ is obtained as $\mathbf{x}_v = \frac{r_{s+1}}{\|\tilde{\mathbf{x}}_v\|} \tilde{\mathbf{x}}_v$, $\tilde{\mathbf{x}}_v = \sum_{u \in \mathcal{V}_s} \tilde{a}_{u,v} \mathbf{x}_u$.

According to the above procedure, we determined optimal coordinate vectors $\mathbf{X}_s$ and $\mathbf{X}_{s+1}$. Next, analogous with the above procedure, we can calculate the optimal coordinate vectors of $\mathbf{X}_{s+1}$ and $\mathbf{X}_{s+2}$ from the relationship between $\mathcal{V}_{s+1}$ and $\mathcal{V}_{s+2}$. As a basic framework, this algorithm employs a power iteration, just like the HITS algorithm [7]. Clearly, the main computational complexity of one-iteration comes from multiplication by matrix $\mathbf{A}$, which is the most intensive part and is proportional to the number of links in the bipartite graph. Thus, the PCE method is expected to work much faster.

3 Proposed Method

In our newly proposed method, we first assign the layer to each article node in a citation network. Concretely, given a node (article) set $\mathcal{V}$ which consists of $N = |\mathcal{V}|$ nodes and a link (citation) set $\mathcal{E}$, we define a citation network structure $G = (\mathcal{V}, \mathcal{E})$ which has the directed acyclic graph (DAG) structure. For each node $v \in \mathcal{V}$, we calculate the maximum distance not minimum distance from the root node $root \in \mathcal{V}$. For instance, we show an example of assigning layers in Fig. 1. In Fig. 1, a node whose name is "Jan." is $root$ which is assigned the layer 0, and nodes which cite only $root$ are assigned the layer 1, $\mathcal{V}_1$, like node "Feb." and node "Mar." Although node "Apr." and node "Jul." cite $root$, they also cite nodes of $\mathcal{V}_1$. Therefore, they are assigned the layer 2, $\mathcal{V}_2$, based on maximum distance from $root$. Similarly to the above, we assign a layer to a node $v \in \mathcal{V}$ as follows:

$$L(v) = \max_{(v,u) \in \mathcal{E}} L(u) + 1,$$

where (v, u) means the node v refers to the node u. For all the nodes, we can calculate the layers in order from the $root$ with $O(|\mathcal{V}| + |\mathcal{E}|)$ based on the topological

Fig. 1 Assigning layers for DAG structure. The month name in the node and the color of the node indicate the month when the corresponding article was posted. The older one is bluish, and the newer one is reddish

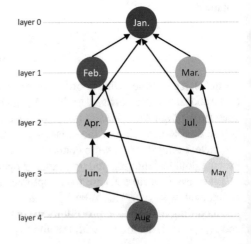

sorting. This assignment method of layers is intuitive because researchers tend to survey reference articles from newer ones than older ones considering dependence relationships among them.

At the second step of our newly proposed method, we embed all nodes into the polar coordinates considering the layers and connectivity of nodes based on our previous study, PCE method. Our original PCE method only considers the node connection between adjacent layers like V_s and V_{s+1}. As can be seen from the Fig. 1, there also exist some links between non-contiguous layers. Thus, we extend our previous PCE method to handle such a DAG structure by introducing the following objective function:

$$J(\mathbf{X}) = \sum_{u \in V} \sum_{v \in V \setminus \{u\}} \tilde{a}_{u,v} \frac{\mathbf{x}_u^T}{L(u)} \frac{\mathbf{x}_v}{L(v)} + \frac{1}{2} \sum_{u \in V} v_u (L(u)^2 - \mathbf{x}_u^T \mathbf{x}_u), \qquad (3)$$

where v_u stands for the Lagrangian multipliers that represent the radius constraints on each circumference.

Furthermore, we explain the dynamic property of our method. Now let node w be the newly added node, and we consider the situation that the node w refers to some nodes $\Gamma(w) \subset V$. Then our method decides the coordinate vector $\mathbf{x}_w$ of node w *only from those of connected nodes* $\Gamma(w)$, because $\mathbf{x}_w$ is obtained by calculating the resultant vector of coordinate vectors of connected nodes, double-centering, and normalizing like (2):

$$\mathbf{x}_w = \frac{L(w)}{\|\tilde{\mathbf{x}}_w\|} \tilde{\mathbf{x}}_w, \quad \tilde{\mathbf{x}}_w = \sum_{v \in V} \tilde{a}_{w,v} \mathbf{x}_v.$$

Although we have also to update coordinates of the other nodes like $\{u \in V | (v, u) \in \mathcal{E} \wedge v \in \Gamma(w)\}$, we need only to a few iteration to update them by using already calculated coordinate vectors as initial vectors.

4 Dataset

In order to evaluate our proposed method, we utilize the real citations of scientific articles obtained by CiteSeerX[1] and construct citation networks according to the bibliography in each article. The total numbers of nodes (articles) and directed links (citations) are 281,977 and 1,187,204, respectively. In our experiments, we randomly select a seed node[2] and extract nodes which the seed node can reach through the citation paths, and then we set the oldest article node to *root*. As a result, the numbers of nodes and directed links are 3,734 and 6,595, respectively.

---

[1]http://citeseerx.ist.psu.edu/.
[2]The article title is "The Graham Scan Triangulates Simple Polygons."

5 Static Visualization Results

5.1 Existing Visualization Methods Used for Comparison

In our experiments, we compare the visualization results of our method to those of four well-known existing methods. Now, let $K = 2$ be the dimensionality of embedding space and $\mathbf{z}_k = (x_{1,k}, \ldots, x_{N,k})^T$ be a vector whose element stands for the kth coordinate of each node.

We first explain the spectral embedding (SE) [2], which calculates the coordinate vectors of all nodes by directly minimizing the following objective function $\mathcal{S}(\mathbf{X}) = \sum_{k=1}^{K} \mathbf{z}_k^T (\mathbf{D} - \mathbf{A}) \mathbf{z}_k$, where $\mathbf{D}$ is a diagonal matrix, each element of which is the degree of node, and $(\mathbf{D} - \mathbf{A})$ is referred to as a Laplacian matrix. Actually, we compute the coordinate vectors as eigenvectors corresponding to the two minimum eigenvalues excluding the trivial eigenvector.

Second one is multi-dimensional scaling (MDS) [8], which first calculates the distance matrix $\mathbf{G} = [g(u, v)]_{u,v \in \mathcal{V}}$ and then performs the double-centering operation by multiplying the matrix $\mathbf{H}_N = \mathbf{I}_N - \frac{1}{N} \mathbf{1}_N \mathbf{1}_N^T$ to the distance matrix. Mathematically it is formulated as minimizing the following objective function, $\mathcal{M}(\mathbf{X}) = \frac{1}{2} \sum_{k=1}^{K} \mathbf{z}_k^T (\mathbf{H}_N \mathbf{G} \mathbf{H}_N) \mathbf{z}_k$.

Third model (KK) [6], which minimizes the following objective function, $\mathcal{K}(\mathbf{X}) = \sum_{(u,v) \in \mathcal{V} \times \mathcal{V}} \alpha_{u,v} (g(u, v) - \|\mathbf{x}_u - \mathbf{x}_v\|)^2$, where $\alpha_{u,v}$ is a spring constant which is normally set to $1/2g(u, v)^2$.

Last one is cross-entropy method (CE) [9], which first defines a similarity $\rho(\mathbf{x}_u, \mathbf{x}_v)$ between the embedding positions $\mathbf{x}_u$ and $\mathbf{x}_v$, uses the corresponding element $a_{u,v}$ of the adjacency matrix as a measure of distance between the node pair, and tries to minimize the total cross-entropy between these two represented as the following objective function, $\mathcal{C}(\mathbf{X}) = - \sum_{(u,v) \in \mathcal{V} \times \mathcal{V}} \left(a_{u,v} \log \rho(\mathbf{x}_u, \mathbf{x}_v) + (1 - a_{u,v}) \log(1 - \rho(\mathbf{x}_u, \mathbf{x}_v)) \right)$. In this study, we used the function $\rho(\mathbf{x}_u, \mathbf{x}_v) = \exp(-\frac{1}{2} \|\mathbf{x}_u - \mathbf{x}_v\|^2)$.

Here the former two perform a power iteration with respect to either a graph Laplacian matrix or a double-centered distance matrix which is calculated from a given graph while the latter two repeatedly move each position vector by using the Newton method in a framework of nonlinear optimization.

5.2 Results and Discussion

Figures 2 and 3 show the static visualization results of our proposed and other existing methods. In each visualization result, node color is assigned according to the posted order of corresponding articles and belonging CNM [3] communities. From Fig. 2, we can observe that in the results of SE and MDS, many nodes overlap, and the relationship between articles is unclear. Since these methods are linear methods, the

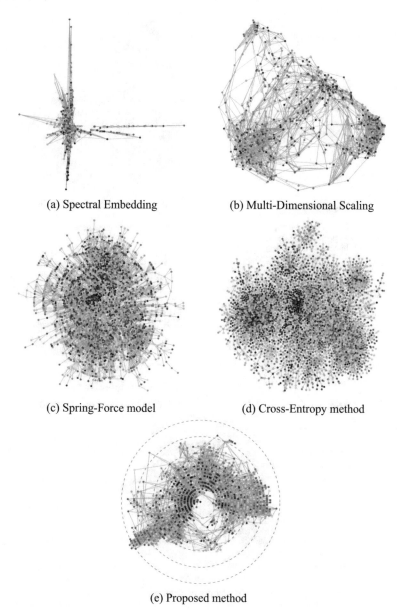

(a) Spectral Embedding   (b) Multi-Dimensional Scaling

(c) Spring-Force model   (d) Cross-Entropy method

(e) Proposed method

Fig. 2 Static visualization results. Node is gradationally colored according to the posted order of corresponding articles. The older one is bluish, and the newer one is reddish

coordinate vectors can be obtained quickly, but the quality of the solution is somewhat poor. We can also see that in the results of KK and CE, nodes are efficiently distributed on the plane since the citation network can be embedded in the plane with few link

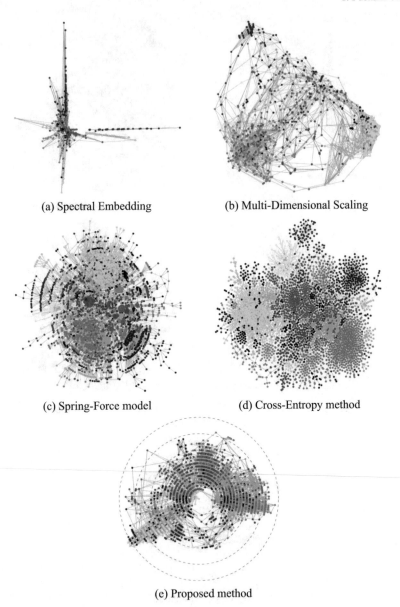

(a) Spectral Embedding   (b) Multi-Dimensional Scaling

(c) Spring-Force model   (d) Cross-Entropy method

(e) Proposed method

Fig. 3 Static visualization results. Node color stands for the CNM community. The number of communities is set to 20

overlapping. Compared to the result of our proposed method, where almost all nodes are embedded according to the posted order of corresponding articles, the order of the posting times is not explicitly reflected in the visualization results of these existing methods.

From Fig. 3, in all the visualization results, we can confirm that densely connected nodes assigned the same color are placed in the vicinity of each other. Especially for the result of our proposed method, same color nodes are embedded into the almost same directions. From these results, our method can produce decipherable visualization results where connected nodes are located in the same direction and posted order of corresponding articles are almost preserved. Our visualization results allow researchers to efficiently access to the desired articles, because users can visually confirm whether an article refers to multi-disciplinary articles or not, authoritative articles or not, articles of new development fields or not, and articles of the state-of-the-art techniques or not from our visualization results.

6 Dynamic Visualization Results

6.1 Average Moving Angle

In this section, we evaluate dynamic property of our proposed method. In evaluating, we consider the following factors: a newly added node refers to the nodes (1) belonging to the lower layers (like layer 25) or (2) the middle layers (like layer 12); and (i) embedded into a certain direction or (ii) multi-directional.

Then we evaluate our method from the viewpoint of moving angle of each node when adding a node. Concretely, we calculate the coordinate vector of a newly added node w and re-calculate all the nodes of the original network by our proposed method. Let $\mathbf{y}_u$ be the coordinate vector of node $u \in V \cup \{w\}$ in the augmented network. We calculate the average of moving angles between coordinate vectors $\mathbf{x}_u$ and $\mathbf{y}_u$ of each node $u \in V \setminus \{root\}$ in the original and the augmented networks:

$$\bar{\Theta} = \frac{1}{N-1} \sum_{u \in V \setminus \{root\}} \arccos \left(\frac{\mathbf{x}_u^T \, \mathbf{y}_u}{|\mathbf{x}_u| \, |\mathbf{y}_u|} \right).$$

6.2 Results and Discussion

Figure 4 shows the dynamic visualization results, where a big black node with ten links was added to the original network described in previous section, and Fig. 5 shows the moving angle averaged for each layer. From these results, we can observe that (1) the average of moving angles represents relatively small value when links to nodes located in a certain direction are added like case1 and case3, (2) the variance of moving angles represents relatively large value when links to nodes located in various direction are added like case2 and case4, and (3) the average moving angles represent the larger values not only at target layers but also at related layers especially in

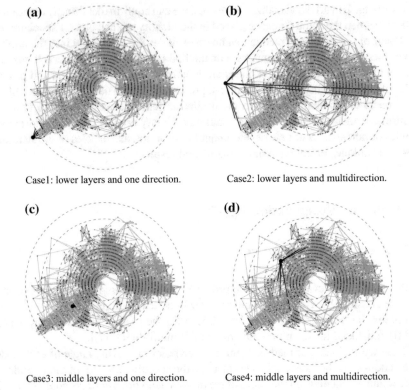

(a)

Case1: lower layers and one direction.

(b)

Case2: lower layers and multidirection.

(c)

Case3: middle layers and one direction.

(d)

Case4: middle layers and multidirection.

Fig. 4 Visualization results of augmented network. A big black circle and black solid lines are newly added node and links

Fig. 5 Average moving angles in each layer

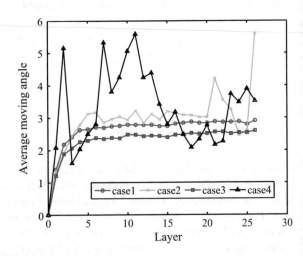

case4. These results indicate that when an influential article which refers to articles of various fields or articles cited/citing a lot is added, the average and standard deviation of moving angle represent somewhat larger and very larger values, respectively. Actually the average and standard deviation of moving angles are 2.7° and 0.15 in case1, 3.0° and 2.68 in case2, 2.4° and 0.16 in case3, and 3.6° and 66.06 in case4, respectively. These promising property of our method is derived from its algorithm similar to the HITS algorithm, where a good hub represented a node that linked to a lot of other nodes, and a good authority represented a node that was linked by a lot of hubs. Thus, we can conclude that our method detects influential node and link addition by average and standard deviation of moving angles of coordinate vectors. Furthermore, we emphasize that each visualization result in Fig. 4 was obtained at a few second computation with a few iteration times.

7 Related Work

In this paper, we propose a method that effectively visualizes citation relationships among time-series documents like scientific articles. A related study of time-series document visualization is Alsakran et al.'s STREAMIT [1]. This method treats documents in a certain snapshot as a particle and calculates an optimal location in two-dimensional space by a dynamics model based on the similarity between document particles. Similar document particles are located in the vicinity of each other, and non-similar document particles are located far away. STREAMIT dynamically computes optimal locations for newly added documents and constructs a graph by the Delaunay triangle based on the optimal location of each document particle. Then for the constructed graph, the method cuts a link that is longer than the threshold parameter and forms document clusters. Finally, it displays each graph of each snapshot by animation. Unlike our method that represents the time axis (layer) by the polar coordinate plane's radius, STREAMIT greatly differs because it animates the change of the corresponding document cluster, which is a disadvantage from the viewpoint of the manifestness of time-series data.

8 Conclusion

In this paper, focusing on the DAG structure of citation networks, we proposed a visualization method considering the browsing order of the articles. In our method, article nodes are embedded into the polar coordinate where citation relations and posted order relations are represented by angle and radius, respectively. Furthermore, our method is equipped with a dynamic function to update coordinates of all nodes at low cost when a new node and links are added to the citation network. From experimental evaluations using real citation networks, we confirmed that our method explicitly reflects citation relations and browsing order compared with existing methods. Furthermore, we verify that our method can detect influential node and link addition by average and standard deviation of moving angles of coordinate vectors. In future, we plan to verify usefulness of our method using more various and larger-scale networks.

Acknowledgements This work was supported by JSPS KAKENHI Grant No.16K16154 and by NII's strategic open-type collaborative research.

References

1. Alsakran, J., Chen, Y., Luo, D., Zhao, Y., Yang, J., Dou, W., Liu, S.: Real-time visualization of streaming text with a force-based dynamic system. IEEE Comput. Graph. Appl. **32**(1), 34–45 (2012). Jan
2. Chung, F.R.K.: Spectral Graph Theory. American Mathematical Society, Providence (1997)
3. Clauset, A., Newman, M.E.J., Moore, C.: Finding community structure in very large networks. Phys. Rev. E **70**(6), 066111 (2004)
4. Fushimi, T., Satoh, T.: Constructing and visualizing topic forests for text streams. In: Proceedings of the 2017 IEEE/WIC/ACM International Conference on Web Intelligence (WI'17), pp. 10–17 (2017)
5. Fushimi, T., Kubota, Y., Saito, K., Kimura, M., Ohara, K., Motoda, H.: Speeding up bipartite graph visualization method. In: AI 2011: Advances in Artificial Intelligence: 24th Australasian Joint Conference, Perth, Australia, 5–8 December 2011. Proceedings, pp. 697–706. Springer, Berlin (2011)
6. Kamada, T., Kawai, S.: An algorithm for drawing general undirected graphs. Inf. Process. Lett. **31**, 7–15 (1989)
7. Kleinberg, J.M.: Authoritative sources in a hyperlinked environment. J. ACM **46**, 604–632 (1999)
8. Torgerson, W.: Multidimensional scaling: I theory and method. Psychometrika **17**, 401–419 (1952)
9. Yamada, T., Saito, K., Ueda, N.: Cross-entropy directed embedding of network data. In: Proceedings of the 20th International Conference on Machine Learning (ICML03), pp. 832–839 (2003)

A Trust-Based News Spreading Model

V. Carchiolo, A. Longheu, M. Malgeri, G. Mangioni and M. Previti

Abstract Online social networks are overwhelmed with information spreading, phenomena whose underlying mechanisms require deeper analysis. In this paper, we introduce the direct credibility among nodes, a parameter that takes into account news trustworthiness. We exploit this amount into a well-known epidemic model to study the news diffusion process and to discover which elements affect the decision of individuals to propagate or not the news. In addition, we also consider how credibility evolves over time, in particular how news spreading process and nodes credibility assessment mutually influence themselves. Simulations on synthesized social networks show that the proposed approach seems a good starting point to define a realistic news spreading model.

1 Introduction

Disseminating information is a pervasive activity in nowadays society, indeed people interact each other continuously to exchange information in the real world as well as in virtual spaces, e.g., online social networks. Understanding information spreading mechanisms have several advantages, e.g., epidemic control [1], political topics analysis [2] or marketing optimization [3] among others.

V. Carchiolo · A. Longheu · M. Malgeri · G. Mangioni · M. Previti (✉)
Dip. di Ingegneria Elettrica Elettronica Informatica (DIEEI),
Università degli Studi di Catania, Catania, Italy
e-mail: marialaura.previti@unict.it

V. Carchiolo
e-mail: vincenza.carchiolo@dieei.unict.it

A. Longheu
e-mail: alessandro.longheu@dieei.unict.it

M. Malgeri
e-mail: michele.malgeri@dieei.unict.it

G. Mangioni
e-mail: giuseppe.mangioni@dieei.unict.it

© Springer International Publishing AG 2018   303
S. Cornelius et al. (eds.), *Complex Networks IX*, Springer Proceedings
in Complexity, https://doi.org/10.1007/978-3-319-73198-8_26

Data traversing (online) social networks can be classified as news, rumors, opinion, or even simple information (for instance about market items, real estates, and so on); sometimes it needs a sort of users approval to be propagated along the network; therefore, it is possible to label such information as either false or true. The assessment of information truthfulness can be actually performed on a per-user basis, i.e., each individual that receives a news can perceive it as false or true, and consequently, he can decide to forward it or not. Moreover, the quality of an information generally also affects the source it comes from; indeed if a person usually sends fake news, people he contacted probably will tend to discard such information, considering him untrustworthy.

Modeling news spreading in social networks is not a really new question [4, 5]; one of the most effective approach leverages epidemic models [6, 7], where replacing diseases with news is adopted [8–10], leading to a social contagion metaphor where ideas *infect* people, whose judgment though still allows to propagate or stop the infection process (in contrast with real-world diseases contagion).

In this paper, we want to model news spreading using an epidemic approach, also taking into account individuals trust relationships described previously, i.e., how much *credibility* users assign each other as inferred from the judgment about news they spread.

The concept of trust (here we will use the term *trust* or *credibility* interchangeably) is widely adopted in modeling social networks [11–14]; in particular, we want to investigate on how trust and the information contagion process mutually affect each other, as in real world where if an unreliable person tries to propagate an information, it will probably stopped by receivers since they suppose this is a (yet another) fake news; or conversely, the news spreading process influences trust since a set of real (respectively, false) news posted by a person over time will reasonably rise (respectively, lower) the credibility he receives from others.

The model we leverage is a modified version of the susceptible-infective-removed (SIR) epidemic proposed by Newman [15], where we introduce nodes direct credibility. We exploit such credibility to build a multiplex network where one level is the contact network and the latter we named the *credibility network*; the multiplex network allows us to model both the news spreading and the trust assessment processes, as well as their mutual interaction.

To validate our approach, we ran a 1000-nodes duplex network simulations where trust-based news spreading process is examined in several scenarios; results showed that our proposal models social media users real behavior.

Our final goal is to define a realistic news spreading model to improve these real-world phenomena comprehension, also achieving their predictability and control, e.g., leveraging the model in large-scale simulations or for "bad spreaders" detection/removal.

The paper is organized as follows. In the next section, we introduce the credibility parameter and its role and dynamics. In Sect. 3 we present and discuss simulations results, while in Sect. 4 we finally present our conclusions and future works.

2 Credibility in News Propagation

In [15], the contagion propagation occurs in a *contact network*, where each node is an individual and each link represents a contact between individuals through which the infection is propagated. To describe the spread of the infection, a SIR model is used, where a population of N individuals is divided into three states (susceptible, infective, and removed) and the state transitions are guided by the transmissibility formula T_{ij} where two main parameters are used for each pair of nodes i and j: the contact rate and the time interval in which node i is infectious (see [15] Sect. 2 for more details).

In our scenario instead, we want to describe how news is propagated in an online social network, how their propagation is influenced by the proponents credibilities, and how in turn news spreading affects credibilities. The interaction between these processes can be easily thought as a multiplex network with two layers, i.e., the contact network and the *credibility* network, having the latter two directed edges for each node pair to model their different mutual trust.

In addition, in real social networks, individuals behavior can change over time; so, for instance, people that often reposted false news could decide to verify received contents and stop fake information forwarding for a particular source, or conversely, a person could start to trust another bad one he receives news from, if the *bad guy* becomes reliable by sending a significant number of true news over time. To model this, in our proposal, an individual updates the trustworthiness he assigned to each of the nodes he received news from, according to the truth of news being sent. In addition, people usually tend to weight more the most recent news, disregarding older ones; in other words, as time passes people increasingly neglect news (a sort of *aging* process).

To take into account all these issues, every time that an ignorant node j has to decide if repost or not a new news spread by its spreader neighbor i, we use a modified version of abovementioned Newman's transmissibility formula T_{ij}. In our version, we add the direct credibility; the new equation hence is:

$$T_{ij} = 1 - e^{-\beta_{ij}\tau_i Cr_{ji}} \tag{1}$$

where β_{ij} is the contact rate between i and j, τ_i is the time interval in which node i is infectious; hence, its news is visible (and sharable) for its neighborhood, and Cr_{ji} is the credibility that susceptible node j assigned to i before the current transmission attempt of news from i itself.

As argued in [16], the direct credibility should also be updated every time a node receives a news; therefore, we also introduce the equation:

$$Cr_{ji} = \frac{\sum_{x=0}^{n}(x+1)C_{news_x}}{\sum_{x=0}^{n}(x+1)} \tag{2}$$

where C_{news_x} is the credibility j assigns to the news x according to its credibility, in particular:

- if the news x is false $C_{news_x} = 0$,
- if x is true $C_{news_x} = 1$,
- if x is false, but it is perceived as true (e.g., it cites false authoritative sources, and/or it seems coming from logical reasoning or it starts from a real event), then $C_{news_x} \in [0, 0.5[$,
- if x is true, but it is perceived as false (e.g., it does not cites any sources, it is poorly described or its content is strange enough to sound fake), then $C_{news_x} \in]0.5, 1]$.

x is used to number news according to their temporal order (the more recent is a news, the higher is x); moreover, it is also used to weight C_{news_x}, thus easily modeling the aging process cited above. Finally, Cr_{ji} is normalized so that it falls within the range $[0, 1]$.

3 Simulations

To validate the proposal described in previous section, we implemented a simulator to generate a 1000-nodes duplex network (the same we used in [16]) composed by:

- the undirected contact network, modeled as a scale-free (typically used for social networks),
- the directed credibility network, in which for each news propagated through a contact network edge, if receiver reposts the news (according to the transmissibility formula 1), the credibility of the trust network edge in the opposite direction is updated using (2).

As previously explained in [16], after various attempts that we made with nodes of different degrees, we identified a well-connected node (adjacent to three hubs) with high degree (32 in our simulation), so we used it as unique seed of all the 150 news propagated in each simulation in order to test the influence of credibility parameter in news spreading over this network.

Our simulator works as follows. First, all credibilities are set to 0.5 that represents a *neutral* credibility, i.e., the confidence that an individual gives to a new contact. When the simulation starts, the seed node injects a news in social network and each of his neighbors use (1) to decide whether repost the news. If the receiving node does not overcome the transmissibility threshold value, he reposts the news and a credibility update is performed using (2) on the corresponding credibility network edge. At next step, infected nodes attempt their neighborhood contagion using the same mechanism. If they succeed to spread the news, the simulator updates directed credibilities and another diffusion step is performed.

The spreading process goes on until the contagion cycle ends; i.e., there are no new infected nodes. After that, the simulator counts the number of infected nodes and resets the network to allow further news to be spread.

The only parameters retained by the simulator are the directed credibilities Cr_{ji} that are used in (1) to evaluate the transmissibility on a specific edge for next news spreading, and they represent the memory of past experiences of a node's neighborhood.

These local changes on credibility network cause a global change in news propagation on the contacts network. Indeed, if an individual's local credibilities decrease due to the increment of false posted news, as well as the credibility of his neighbors, this implies that "bad spreader" inserted news progressively reach a lesser number of nodes, since individuals decide to not propagate false news coming from unreliable neighbors. Conversely, if posted news is true, individuals' local credibilities increase because individuals decide to repost such news.

To test these behaviors, we performed five groups of 10 different simulations:

- all news are false, i.e., $C_{news_x} \in [0, 0.5[$ (case A);
- all news are true, i.e., $C_{news_x} \in]0.5, 1]$ (case B);
- 50 news are false, other 50 are true and the last 50 are false (case C);
- 50 news are true, other 50 are false and the last 50 are true (case D); cases C and D allow to better investigate on the dynamics of credibility;
- all news are propagated with Newman transmissibility formula (to use it as a reference case where all credibilities are set to 1).

For each simulation group, we calculated the average number of infected nodes for each propagated news in order to discover the temporal evolution of the network. Such values are presented in Fig. 1.

Since all simulations where credibility matters (cases A, B, C, and D) leverage on initial neutral credibility, the number of infected nodes for the first news is similar in all these simulation groups. In further steps, i.e., when other news are inserted in the network, the parameter Cr_{ji} plays a fundamental role in the network evolution, indeed:

- In case A, the number of infected nodes for each news progressively decrease as the number of spread news increments (from an average of 188 infected nodes at starting time to a final value of 22).
- In case B, the number of infected nodes for each news progressively increases as the number of spread news increments (moving from an average of 184 infected nodes to a final value of 342).
- In case C, the number of infected nodes for each news decreases with the same trend as case A for the first 50 news (an average of 28 infected nodes in case A versus 27 in case C). Afterward, with individual behavior changes from spreading false news to true ones, the curve changes consequently, reaching a peak of 201 infected nodes at the 101st news. Similarly, when the behavior switches from good to bad again, the curve decreases.
- In case D, the number of infected nodes for each news increase with the same trend as case B for the first 50 propagated news (an average of 320 infected nodes in case B versus 312 in case D). Then, when the seed becomes a *bad guy*, the curve trend

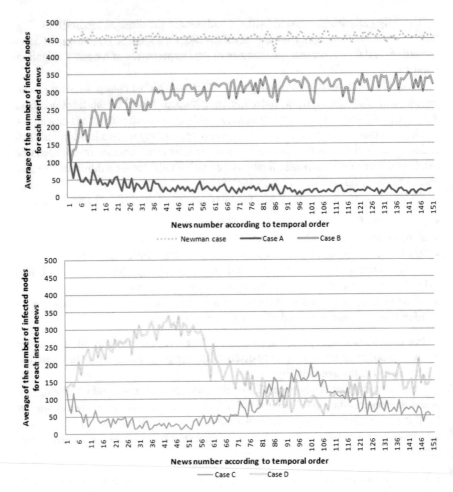

Fig. 1 Temporal evolution average of the network for each group of simulations

decreases, reaching a minimum of 54 infected nodes at the 106th news, whereas it increases as soon as the seed behavior becomes good again (true news).

- In the Newman case, the credibility does not affect the curve, so it is used only as reference to show the improvements introduced to the model. Indeed, one would expect that this curve is (at least asymptotical) reached by the case B curve; in fact, it could be compared to the case where credibility is always equal to 1. Instead, at a steady state, the case B simulations perform an average of about 130 infected nodes less than the Newman ones. This is because not all the true news is perceived as such and some received a low credibility value although true, for instance, because they are without sources or relevant details. Furthermore, our credibility update formula keeps track of past inserted news, albeit in a decreasing way due to aging factor; hence, it will never reach $Cr_{ji} = 1$.

Cases C and D show another real social media users behavior; i.e., losing credibility is easier than achieving it again. In the case C indeed, the curve increases slowly after the first behavior change at the 50th news (from an average of 27 infected nodes at 50th news to 52 at the 60th news), whereas in the case D, at the same point the curve decreases very rapidly (from an average of 312 infected nodes at 50th news to 154 at the 60th news). The same curve variations can be observed in the other behavior changing points at the 100th news.

4 Conclusions and Future Works

In this work, we introduced a model to describe how news is propagated on an online social network, how this is influenced by the proponents credibilities, and how in turn the proponents credibilities are affected by the news propagation mechanism over the time.

Starting from the Newman epidemic model, we modified the transmissibility formula adding the direct credibility, also providing a related updating formula triggered for each news shared or reposted by users; such formula takes into account the credibility of each news posted by users.

To validate our approach, we simulated a 1000-nodes duplex network (a contact and a credibility network) where a hub node spread 150 news through its contact in several scenarios. Results showed that our proposal models social media users real behavior for what concern credibility dynamics and news propagation.

In our further works, we will refine the model distinguishing also the case in which some nodes are disconnected in one or more temporal steps during which the news is visible on social network from the case in which nodes read that news but decide to no repost it; we also aim at validating the model on larger networks to apply it on real social media data.

References

1. Funk, S., Gilad, E., Watkins, C., Jansen, V.A.A.: The spread of awareness and its impact on epidemic outbreaks. Proc. Natl. Acad. Sci. **106**(16), 6872–6877 (2009)
2. Romero, D.M., Meeder, B., Kleinberg, J.: Differences in the mechanics of information diffusion across topics: Idioms, political hashtags, and complex contagion on twitter. In: Proceedings of the 20th International Conference on World Wide Web. WWW '11, pp. 695–704. ACM, New York, NY, USA (2011)
3. Watts, D.J., Dodds, P.S., served as editor, J.D., served as associate editor for this article., T.E.: Influentials, networks, and public opinion formation. J. Consum. Res. **34**(4):441–458 (2007)
4. Moreno, Y., Nekovee, M., Pacheco, A.F.: Dynamics of rumor spreading in complex networks. Phys. Rev. E **69** (2004)
5. Lind, P.G., da Silva, L.R., Jr., J.S.A., Herrmann, H.J.: Spreading gossip in social networks. Phys. Rev. E **76**, 036117 (2007)

6. Pastor-Satorras, R., Vespignani, A.: Epidemic spreading in scale-free networks. Phys. Rev. Lett. **86**, 3200–3203 (2001)
7. May, R.M., Lloyd, A.L.: Infection dynamics on scale-free networks. Phys. Rev. E **64**, 066112 (2001). Nov
8. Guille, A., Hacid, H., Favre, C., Zighed, D.A.: Information diffusion in online social networks: a survey. SIGMOD Rec. **42**(2), 17–28 (2013). July
9. Pastor-Satorras, R., Castellano, C., Van Mieghem, P., Vespignani, A.: Epidemic processes in complex networks. Rev. Mod. Phys. **87**(3), 925 (2015)
10. Vega-Oliveros, D.A., da F Costa, L., Rodrigues, F.A.: Rumor propagation with heterogeneous transmission in social networks. J. Stat. Mech.: Theory Exp. **2017**(2), 023401 (2017)
11. Carchiolo, V., Longheu, A., Malgeri, M., Mangioni, G.: Trust assessment: a personalized, distributed, and secure approach. Concurr. Comput.: Pract. Exp. **24**(6), 605–617 (2012)
12. Buzzanca, M., Carchiolo, V., Longheu, A., Malgeri, M., Mangioni, G.: Direct trust assignment using social reputation and aging. J. Ambient Intell. Humaniz. Comput. **8**(2), 167–175 (2017)
13. Carchiolo, V., Longheu, A., Malgeri, M., Mangioni, G.: Users' attachment in trust networks: Reputation vs. effort. Int. J. Bio-Inspir. Comput. **5**(4), 199–209 (2013)
14. Carchiolo, V., Longheu, A., Malgeri, M., Mangioni, G.: The cost of trust in the dynamics of best attachment. Comput. Inf. **34**(1), 167–184 (2015)
15. Newman, M.E.: Spread of epidemic disease on networks. Phys. rev. E **66**(1), 016128 (2002)
16. Carchiolo, V., Longheu, A., Malgeri, M., Mangioni, G., Previti, M.: Introducing credibility to model news spreading. Accepted for publication in Complex Networks 2017. In: The 6th International Conference on Complex Networks and Their Applications (2017)

Discovering Mobility Functional Areas: A Mobility Data Analysis Approach

Lorenzo Gabrielli, Daniele Fadda, Giulio Rossetti, Mirco Nanni,
Leonardo Piccinini, Dino Pedreschi, Fosca Giannotti
and Patrizia Lattarulo

Abstract How do we measure the borders of urban areas and therefore decide
which are the functional units of the territory? Nowadays, we typically do that just
looking at census data, while in this work we aim to identify functional areas for
mobility in a completely data-driven way. Our solution makes use of human mobility
data (vehicle trajectories) and consists in an agglomerative process which gradually
groups together those municipalities that maximize internal vehicular traffic while
minimizing external one. The approach is tested against a dataset of trips involving
individuals of an Italian Region, obtaining a new territorial division which allows us
to identify mobility attractors. Leveraging such partitioning and external knowledge,
we show that our method outperforms the state-of-the-art algorithms. Indeed, the
outcome of our approach is of great value to public administrations for creating
synergies within the aggregations of the territories obtained.

L. Gabrielli (✉) · D. Fadda · G. Rossetti · M. Nanni · F. Giannotti
KDD Lab. ISTI-CNR, via G. Moruzzi, 1 Pisa, Italy
e-mail: Lorenzo.Gabrielli@isti.cnr.it

D. Fadda
e-mail: Daniele.Fadda@isti.cnr.it

G. Rossetti
e-mail: Giulio.Rossetti@di.unipi.it; Giulio.Rossetti@isti.cnr.it

M. Nanni
e-mail: Mirco.Nanni@isti.cnr.it

F. Giannotti
e-mail: Fosca.Giannotti@isti.cnr.it

G. Rossetti · D. Pedreschi
University of Pisa, Largo Bruno Pontecorvo, 2 Pisa, Italy
e-mail: Dino.Pedreschi@di.unipi.it

L. Piccinini · P. Lattarulo
IRPET, Regione Toscana Via Pietro Dazzi, 1, Firenze, Italy
e-mail: Leonardo.Piccinini@irpet.it

P. Lattarulo
e-mail: Patrizia.Lattarulo@irpet.it

© Springer International Publishing AG 2018
S. Cornelius et al. (eds.), *Complex Networks IX*, Springer Proceedings
in Complexity, https://doi.org/10.1007/978-3-319-73198-8_27

311

1 Introduction

The traditional interpretation of the urban hierarchy refers merely to the size of the city, with its population and boundaries. From the theoretical point of view, a slightly different perspective is given by the concept of polycentrism [1]: Urban areas are often evolving from mono-centric agglomerations to more complex systems made of integrated urban centers (cores) and subcenters. In other territories, some cities and towns are increasingly linking up, forming polycentric integrated areas.

The understanding of the spatial organization of similar regions and of how places link among them can improve analytical approaches when facing governance challenges such as the economic development of complex nationwide systems. Indeed, policymakers are paying increasing attention to the role of homogeneous economic agglomeration and to the capacity of local areas to contribute to social growth [2].

Moreover, the contraction of public expenditure has driven a process of service concentration toward denser urban areas.

Our work aims at contributing to this debate by providing a tool to researchers and policymakers to build a novel definition of regions seen as functional areas of similar behaviors [3].

The questions that drive our research are thus the following:

Q1: Can we identify mobility functional units just looking at human vehicular movement data?

Q2: Are such units mono-centric agglomerations or more complex polycentric integrated areas?

Q3: Which are the most relevant characteristics of such areas?

To answer such questions, we developed a methodology which identifies a reasonable number of well-knit subregions that are significantly self-contained regarding mobility fluxes and therefore represent candidate functional areas w.r.t. mobility. The approach also tries to be not influenced by marginal municipalities that are substantially disconnected from the others and/or less appealing from the decision-maker point of view, e.g., because of low traffic flows or small population. This latter characteristic is often neglected by traditional group discovery algorithms, whose final goal is to partition a generic set of linked elements disregarding any semantics attached to it.

As in [4], we model movements between municipalities as a network, and we compare our approach with competitors taken from network analysis studies (i.e., *community detection*) as well as from data mining ones (i.e., *clustering*).

2 Background

In this section, we discuss some works in the literature that are adopted—or might be adapted— to identify functional areas.

		Local Function	Inner traffic		
a d		localQ(a,b)	50%	a d	
		localQ(a,d)	0%		
		localQ(a,c)	0%	b+c	
b c		**localQ(b,c)**	**52%**		
		localQ(d,c)	48%		
(a) Iteration n		(b) $Evaluation(n)$		(c) Iteration $n+1$	

Fig. 1 An iteration of *MFAD*: Considering a generic iteration n (**a**), we evaluate all possible combination of *localQ* (**b**). Once selected the best pair, i.e., (b, c), we proceed to the union of nodes and updating the data structures (**c**)

From a statistical and economical point of view, in [3] are illustrated different methodologies used to solve the problem of redefining urban areas. Among these, Dynamic Metropolitan Areas (DMAs) are specifically designed to deal with the characteristics of polycentricity. The first stage of the DMA algorithm has a top-down approach: It identifies first-order centers (seeds) which have at least 50,000 inhabitants and merges the surrounding municipalities that commute at least 15% of their inhabitants.

While (to the best of our knowledge) there are no works on data science tackling our specific problem, several group discovery methods might be applied to it, following a clustering of network-based perspective. Here, we briefly mention some basic approaches in the field, while Fig. 1b will provide a detailed description of those we compared to.

Clustering methods generally aim to group a set of objects putting together those that are similar to each other under some specific notion of similarity. The three classical and most frequently adopted examples are: *k-means*, representing a family of partitioning methods that create compact clusters, trying to minimize the diversity within a bunch and to maximize it across different groups; *hierarchical clustering*, producing several different partitioning at various levels of aggregation; *density-based clustering*, which puts together groups of objects that form locally dense areas, not enforcing any constraint on the size and shape of clusters. The solution we proposed belongs to the hierarchical methods, yet basing the aggregation of groups on complex self-containment considerations rather than on the standard maximization of mutual similarities within the cluster.

Network-based methods search for *communities*, i.e., groups of linked nodes that share common properties, defining them w.r.t. several objective functions [5]. In the context of territorial partitioning, community discovery has become an essential tool for decision-makers that need to study social complex systems, e.g., in grouping together municipalities showing similar characteristics [2]. Indeed, by adopting a community discovery approach, we can obtain, in a bottom-up way, an unsupervised classification of territories. In this work, we realize a dedicated method, which we called Mobility Functional Areas Discovery (MFAD), based on a context-specific combination of objective functions. As we will show in the experiment section, results prove the superiority of our solution.

Algorithm 1 *MFAD(OD)*

1: Inputs: OD represents the mobility flows among municipality.
2: Output: The territorial tessellation $\tilde{T}$.
3: $G = CreateMobilityGraph(OD)$ ▷ loading the graph
4: $T = \emptyset$
5: **while** $|G.V| > 1$ **do**
6:     $C = ComputeConfigurations(G)$ ▷ computing all the possible fusions
7:     $bestPair = \arg\max_{(a,b)\in C} localQ(a, b, G)$ ▷ selecting the best fusion
8:     $G = update(G, bestPair)$ ▷ updating G
9:     $T = T \cup \{G\}$ ▷ Saving configuration
10: $\tilde{T} = \arg\max_{G\in T} globalQ(G)$ ▷ Evaluating configurations
     return $\tilde{T}$

---

3 Mobility Functional Areas Discovery (*MFAD*)

Our final goal is to partition the territory considering mobility habits. We model the mobility between municipalities as a network graph $G = (V, E, F)$ and approach the task of defining a meaningful tessellation as the problem of identifying a community coverage of G. The municipalities define the set of the nodes V of G, while edges E represent the flows between nodes, their weights being denoted with $F(e)$ for each $e \in E$.

The general criterion we follow to obtain an optimal solution is self-containment of traffic flows: The traffic within a group of nodes should be much higher than that across different groups. The literature on community detection over networks provides a measure, called *modularity*, that seems to approximate this notion. However, directly maximizing modularity in our context leads to results that violate some basics expectations of the domain expert; e.g., it causes the appearance of geographically discontinuous groups (which is counterintuitive) and the fact that densely populated areas tend to dominate the whole process (undesirable).

To overcome modularity limitations, adopt an agglomerative process, summarized in Algorithm 1, driven by a *local* measure that at each step of the process evaluates the benefits—regarding self-containment of flows—of aggregating a pair of groups into a single larger one. Modularity is then used as a global measure to decide when the aggregation process should stop. The process starts from a situation where each input node in the network G is kept separated from the others, meaning that each node forms a cluster by itself. Iteratively, two clusters are selected and merged together, thus reducing the number of clusters by one unit, and stop only when G contains just one cluster. In order to choose which clusters to merge, all possible pairs are taken into consideration, and for each of them our local measure *localQ* is computed, selecting the best one. Such measure, in particular, is computed as the fraction of the *local* traffic (i.e., the flows involving either node of the pair) that would be converted into internal traffic thanks to the merger:

$$localQ(a, b, G) = \frac{F(a, b) + F(b, a)}{\sum_{(x,y)\in E \wedge \{a,b\}\cap\{x,y\}\neq\emptyset} F(x, y)} \tag{1}$$

Considering the example in Fig. 1, $localQ(b, c, G)$ would compute the ratio between the total traffic between b and c, i.e., $F(b, c) + F(c, b)$ (the same traffic that, if b and c are merged, will move from inter-group traffic to intra-group) and the total traffic from/to any of them, i.e., $F(a, b) + F(b, a) + F(b, c) + F(c, b) + F(c, d) + F(d, c)$.

When the best pair of nodes a, b is found, they are merged thus replacing them in G by a single node $a\&b$. The edges to/from the new node are the union of those to/from either of the original nodes, and the flows associated with them are computed as the sum of the original flows, e.g., $F(a\&b, c) = F(a, c) + F(b, c)$.

When the iterative process comes to an end, T will contain the collection of all graphs obtained at each step, including the original graph G and the last one where only two (big) nodes are left. In order to identify the most promising aggregation level, we adopt the *modularity* measure as global evaluation criterion and find the graph $G \in T$ that maximizes it. That is computed by function $globalQ$:

$$globalQ(G) = \sum_{(i,j)\in E} F(i, j) - F(i \to) * \frac{F(\to j)}{K} \tag{2}$$

where $F(\to i)$ represents the total sum of outgoing flows from node i and $F(j \to)$ is the total of incoming flows to node j. Finally, $K = \sum_{e \in E} F(e)$ represents the total flows in the network. Overall, the rightmost part of the formula provides the expected flow from i to j.

Computational costs: From a computational point of view, the algorithm costs $o(n^3)$ where n is the cardinality of V. While high, the cost is not a real issue in our application, since n is typically a low number; in our case study, covering the municipalities of a region, we have around 300 nodes, and running the whole process on a standard computer takes about 20 min. It is worth to notice that the expensive part of the algorithm is easily parallelizable.

4 Experiments

We apply our methodology on a dataset of trajectories capturing the mobility of individuals in a region. The dataset consists of 5 million trips produced by around 70 thousand cars within Tuscany (Italy) in a period of observation of 5 weeks.[1] Tuscany has about 4.8 million residents and 287 municipalities with a population density of 163 residents/Km^2.

MFAD is applied to the origin and destination matrix (OD Matrix) at the municipality level. An OD Matrix is a network that describes the number of trajectories that start in a municipality and end in another one (not necessarily adjacent): The Tuscany dataset is composed by 287 nodes and 30 thousands of arcs. The average

---

[1] The analyzed trajectories are generated from raw GPS data using a tool called M-Atlas [6].

degree of the network is 119.74, and the average clustering coefficient is 0.74, while the average shortest path is 1.64. The final output of *MFAD* is a set of 25 contiguous areas that, as will be discussed in broader detail later, highlight some impressive structures in the region.

4.1 Competitors

Here, we introduce some of the main state-of-the-art methods for partitioning sets of elements into groups, which are then used as competitors against our proposed solution. Following the literature, we refer to partitioning methods based on networks (CD) or clustering. Finally, we introduce a simple random model as baseline solution.

Louvain, described in [7], is a fast and scalable algorithm based on a greedy modularity approach. It has been shown that modularity-based approaches suffer a resolution limit, and therefore, Louvain is unable to detect medium-sized communities [8]. This produces communities with high average density, due to the identification of a predominant set of very small communities and a few huge communities. **Demon**, introduced in [9], is an incremental and limited time complexity algorithm for community discovery. It extracts ego networks, i.e., the set of nodes connected to an ego node u, and identifies the real communities by adopting a democratic, bottom-up merging approach of such structures. **Infohiermap** is one of the most accurate and best performing hierarchical non-overlapping clustering algorithms for community discovery [10] studied to optimize community conductance. The graph structure is explored with some random walks of a given length and with a given probability of jumping into a random node.

K-medoids and **DBSCAN** are two of the existing methods for obtaining a homogeneous grouping of elements. We choose these algorithms because they can easily accommodate any distance function, which is a crucial feature for our problem, since standard measures (Euclidean, etc.) would not model it in a meaningful way. In particular, since the goal of our analysis is to identify groups that maximize the local score introduced in Sect. 3, we feed the algorithms with a distance which is the complement of *localQ*, i.e., $distance(a, b) = 1 - localQ(a, b, Q)$, which has the same range of values [0, 1]. **K-medoids** is a partitional clustering algorithm that clusters the dataset into k clusters, where k is known a priori. It minimizes the overall distance (more specifically, the sum of squared distances) between the points of a cluster and its center. In contrast to the K-means algorithm, K-medoids chooses a real point as center and works with an arbitrary metrics of distances between data-points. We determine k using the standard *silhouette* score [11]. **DBSCAN** is a density-based clustering algorithm. It identifies each input point as *core point*, *border point*, or *outliers*. A point p is a core point if at least $minPts$ other points are within distance ϵ from it. Border points are those that are not core but have a core point within distance ϵ. Finally, all remaining points labeled as outliers. The $minPts$ parameter is known to be not critical and thus was set to the standard default value of 3. The (more critical) parameter ϵ was instead chosen through a grid search, selecting the one that optimizes the *globalQ* function introduced in 2.

We compute a baseline method called **NM1** to test if there is a random configuration that generates a better territorial partitioning than *MFAD*. $NM1$ fixes a priori the number of territorial partitions (k) and then randomly chooses k elements that represent the seed of clusters. The remaining municipalities are assigned sequentially, according to three criteria: The candidate municipality is assigned to an adjacent seed; if not possible, the municipality is assigned to an existing group that contains adjacent municipalities; if all fails, the municipality is assigned randomly to a seed (even if not adjacent).

4.2 Results

Here, we show the territorial partitions obtained with network-based methods and how they provide a good partitioning, yet not satisfying some requirements needed to answer our research questions. We will see the territorial partitions obtained with clustering methods and how DBSCAN proves to be the best competitor for our approach. Finally, we evaluate the territorial partitions obtained with a null model w.r.t. the final configuration of *MFAD*, proving that its random process fails to find better partitioning.

Since **Louvain** optimizes the partition modularity, its modularity score is higher than *MFAD*, yet the latter provides communities with higher average densities and higher conductance. The main drawback of Louvain is the reduced number of communities it produces, which comes from the tendency of modularity-based approaches to build up few huge communities along with small-sized ones (Fig. 2a). **Demon** is a good method because it manages to handle noise and overlapping communities. Yet, *MFAD* improves both density and conductance (Fig. 2b). In our context, moreover, offering crisp and non-overlapping partitions is a plus, since it simplifies the interpretation of results. Demon's overall very good results are therefore not very appealing to our goal. **Infohiermap** creates communities that are on average less dense than *MFAD*. Also, it produces a comparable conductance (*MFAD*= 0.88, Info-hiermap = 0.95), while not optimizing this measure explicitly. Finally, Infohiermap groups are consistently non-contiguous, which is a counterintuitive result from the application viewpoint (Fig. 2c). Overall, *MFAD* produces results that outperform CD approaches, since the latter tend to find either too few and big communities, or non-contiguous ones.

Now, we show the territorial partitions obtained with K-medoids and DBSCAN cluster methods. For **K-medoids**, we selected the k value that optimizes the *silhouette* coefficient, which results to be $k = 6$. As shown in Fig. 3, the algorithm produces non-contiguous areas and a fragmented spatial partitioning, which also happens for any other value of k. **DBSCAN** was performed for several values of ϵ in the interval [0, 1], computing the *globalQ* score for each result, as reported in Fig. 3c. The best score is obtained for $\epsilon = 0.79$. Figure 3a depicts the corresponding

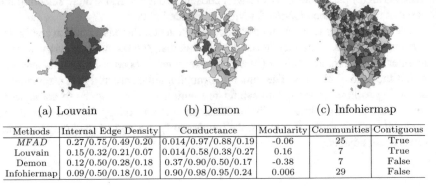

(a) Louvain   (b) Demon   (c) Infohiermap

Methods	Internal Edge Density	Conductance	Modularity	Communities	Contiguous
MFAD	0.27/0.75/0.49/0.20	0.014/0.97/0.88/0.19	-0.06	25	True
Louvain	0.15/0.32/0.21/0.07	0.014/0.58/0.38/0.27	0.16	7	True
Demon	0.12/0.50/0.28/0.18	0.37/0.90/0.50/0.17	-0.38	7	False
Infohiermap	0.09/0.50/0.18/0.10	0.90/0.98/0.95/0.24	0.006	29	False

Fig. 2 Communities identified. **a** Louvain produces very few, and large communities; **b** Demon communities are slightly dispersed and show a significant overlapping (not visible from the figure); **c** Infohiermap produces several non-contiguous areas. We report the min., max., avg., and std. deviation of the measures. *MFAD* communities are denser on average and have a good value of conductance even though it was not explicitly among its optimization criteria

Fig. 3 Territorial partitioning. K-medoids (**a**) generates a fragmented partitioning useless for our purpose. DBSCAN (**b**) provides 21 contiguous communities with $\epsilon = 0.79$ (the optimal value). Visually, the result is good and comparable to *MFAD* (**c**). As shown in **d**, however, the optimal value of *globalQ* for DBSCAN is much lower than *MFAD*

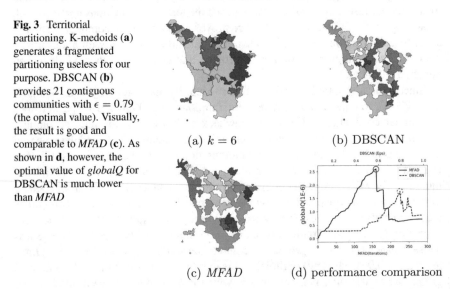

(a) $k = 6$ (b) DBSCAN

(c) *MFAD*   (d) performance comparison

territorial partitioning, showing that DBSCAN basically satisfies the requirements we mentioned before, producing a reasonable number of contiguous communities and isolating/removing uninteresting (noisy) municipalities. In terms of *globalQ* score, however, we can see that the best DBSCAN can reach is still largely inferior to *MFAD* (around 35% smaller). Also in this case, although DBSCAN is so far the best competitors, *MFAD* remains, however, the best option, since it reaches much higher values of our *globalQ* quality function.

Finally, the random heuristics called $NM1$ has been applied with all possible values of k between 10 and 30, running the method 100 times for each k. Varying the number of areas produced by the model $NM1$, which assigns each municipality randomly to one of the k groups, we obtain lower values of $globalQ$ w.r.t. $MFAD$. We can see that the random approach consistently behaves much worse than our solution, regardless of the number of groups it seeks.

5 Evaluation

In this section, we evaluate the functional areas obtained by $MFAD$ with the aid of domain experts (co-authors of this paper) working for a public agency on topics related to territorial policies. For this kind of problems, the expert needs a complete tessellation of the territory; therefore, in Sect. 5.1, we show an assignment criterion for municipalities not grouped by $MFAD$. In Sect. 5.2, we show the internal structure of the main areas identified and, finally, we report some domain expert's comments regarding how the obtained results can be used (Sect. 5.3).

5.1 Saturation

$MFAD$ produces clusters which do not include the totality of the municipalities. For some applications, the domain expert requires the assignment of all the municipalities in a cluster. This may be the case, for example, if we use the partition to redefine the perimeters of universal public services (health care, education, transport). In this scenario, we must assign every municipality to a cluster, since we cannot have a territory where the service is not provided. For this reason, we applied a *saturation* process that iteratively (i) selects the unassigned municipality m and the area a such that they are adjacent and their merger maximizes the $globalQ$ function; (ii) assigns m to a; (iii) reiterates the process until all areas have been assigned. Geographically isolated municipalities, if any, form singleton areas (Fig. 4).

5.2 The Polycentric Structure of the Urban Areas

As requested by the domain expert, we analyzed the structure of the communities identified, with particular reference to the highly populated areas. In Fig. 5a, we note that rural areas (mainly situated in the South) are defined by larger aggregations, while the central areas are comprised of smaller ones. The Northern border, which is mainly mountainous, shows more fragmentation than the population density would suggest. This could be due to a combination of two factors: insulated communities and a border effect (since our data are trimmed at the administrative regional border). After

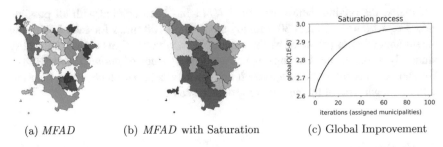

(a) *MFAD*   (b) *MFAD* with Saturation   (c) Global Improvement

Fig. 4 Figures **a** and **b** show the result provided by *MFAD*, before and after the saturation process applied to include also the unclustered municipalities in the detected areas. In **c**, we show the growth of *globalQ* value for each reallocated municipality

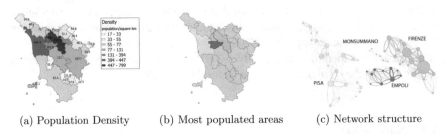

(a) Population Density   (b) Most populated areas   (c) Network structure

Fig. 5 Here, we evaluate the characteristics of the areas obtained by *MFAD*. **a** The size and density of communities depend on the socioeconomic characteristics of the territories, observing a very low density in rural areas (mainly in the South of Tuscany and mountainous locations) and very high one in the most urbanized zones. **b** Among the 25 areas, we select those belonging the more urbanized part of the region. **c** Observing the internal structure of the network generated by vehicular flows, we can observe several polycentric subareas

selecting some interesting areas for the domain expert (Fig. 5b), we can observe the internal structure of the communities with the highest population density. Figure 5c shows the structure of four communities, depicting the flow between municipalities and the in/out flows of every single municipality as the size of lines and points. The densest area has a main hub in Firenze, which is accompanied by a second, slightly smaller one at North-West, and together they keep all the area tightly connected to them. The area on the West is centered on Pisa and has a different and more diffuse structure. Indeed, there are several poles of comparable size (Pisa being slightly larger), each capturing a part of the area, and in most cases, they are only weakly connected to other poles. The area around Empoli is quite similar to Pisa, at a smaller scale. Finally, the small area around Monsummano is very homogeneous and made of municipalities of approximately the same weight.

5.3   Exploitation Potential

The proposed bottom-up approach defines a partition of the selected region that can be interpreted in different ways. The first application could be an analytical approach: Mobility patterns tell us a story about territorial integration that goes beyond the administrative borders. If we want to analyze the socioeconomic dynamics and the determinants of local development, the algorithm can suggest us which might be the boundaries of our analysis.

From the public administration perspective, this method of clustering territories could be helpful in the policy design phase. Since we are looking at highly integrated areas, we might want to tailor the intervention on the characteristics of the aggregated partition, since we expect that the outcome at the municipality level can have a spillover effect on the surrounding territories and, vice versa, that the socioeconomic conditions of surrounding territories affect the potential outcome of the single municipality. Therefore, an integrated and coordinated policy implementation approach can maximize the desired outcome and prevent potential drawbacks.

Moreover, as we mentioned in the introduction, public service provision can be more cost-effective when implemented at an aggregated level. This is especially true in the case of Italian municipalities, where excessive fragmentation of administrative units has been recognized as one of the sources of inefficient public expenditure.

6   Conclusion

The evolution of the economy and society affects the way metropolitan areas change over space and time. It is, therefore, necessary an accurate boundary delimitation of services to increase the efficiency of public administrations without marginalizing the surrounding territories. We propose to use Big Data, to be precise GPS data produced by vehicles, to overcome the limitations of traditional sources in the measurement of the real boundaries of the city. The position of our work is to contribute to provide a tool to policymakers for building a novel definition of regions considered as mobility functional areas [3].

The results highlighted in the paper show on a real dataset that *MFAD* outperforms state-of-the-art methods optimizing an objective function defined by domain experts. We have shown that 25 communities emerge from data, observing only the private vehicle mobility (ref. question Q1 in the introduction). The identified communities show a polycentric structure, with centers apparently corresponding to highest population density and presence of transport infrastructures that facilitate connections to/from other municipalities (ref. Q2). Finally, the areas found have a very diverse population density and size, the tighter connections corresponding to highest populated areas (ref. Q3, and see Fig. 5a).

Planned developments of the work include the exploration further aspects with domain experts, in particular how communities change by applying our method only to the systematic versus occasional traffic and by including public transportation. Also, it would be very helpful including social and productive aspects of the global objective function. Comparison with the Local Labour Communities defined by the Italian Statistical Institution (ISTAT) shows that the border effect (i.e., the influence of neighboring municipalities outside our dataset) is relevant in those areas. This suggests that further developments should include cross-border data. Finally, we plan to modify the initialization phase of our method—now consisting in putting each municipality in a separated group—by following the approach in [3], which might provide better initial seeds for the computation, injected through local domain[2] knowledge.

Acknowledgements This work is partially supported by the European Community's H2020 Program under the scheme 'INFRAIA-1-2014-2015: Research Infrastructures,' grant agreement #654024 'SoBigData: Social Mining and Big Data Ecosystem'.

References

1. Brezzi, M.: Redefining "Urban": A New Way to Measure Metropolitan Areas. OECD (2012)
2. ISTAT: Local labour system
3. Boix, R., Veneri, P., Almenar, V.: Polycentric metropolitan areas in europe: towards a unified proposal of delimitation. Defining the Spatial Scale in Modern Regional Analysis, pp. 45–70. Springer, Berlin (2012)
4. Rinzivillo, S., Mainardi, S., Pezzoni, F., Coscia, M., Pedreschi, D., Giannotti, F.: Discovering the geographical borders of human mobility. KI - Künstliche Intell. **26**(3), 253–260 (2012)
5. Fortunato, S.: Community detection in graphs. Phys. Rep. **486**(3–5), 75–174 (2010)
6. Trasarti, R., Rinzivillo, S., Pinelli, F., Nanni, M., Monreale, A., Renso, C., Pedreschi, D., Giannotti, F.: Exploring real mobility data with m-atlas. In: Joint European Conference on Machine Learning and Knowledge Discovery in Databases, pp. 624–627. Springer, Berlin, Heidelberg (2010)
7. Blondel, V.D., Guillaume, J.L., Lambiotte, R., Lefebvre, E.: Fast unfolding of communities in large networks. J. Stat. Mech. Theory Exp. **2008**(10), P10008 (2008)
8. Fortunato, S., Barthélemy, M.: Resolution limit in community detection. PNAS **104**(1), 36–41 (2007)
9. Coscia, M., Rossetti, G., Giannotti, F., Pedreschi, D.: Demon: a local-first discovery method for overlapping communities. In: Agarwal, D., Pei, J. (eds.) KDD, Q.Y. 0001, pp. 615–623. ACM (2012)
10. Rosvall, M., Bergstrom, C.T.: Maps of random walks on complex networks reveal community structure. Proc. Natl. Acad. Sci. **105**(4), 1118–1123 (2008)
11. Rousseeuw, P.J.: Silhouettes: a graphical aid to the interpretation and validation of cluster analysis. J. Comput. Appl. Math. **20**, 53–65 (1987)

---

[2]http://www.sobigdata.eu.

Estimating Peer-Influence Effects Under Homophily: Randomized Treatments and Insights

Niloy Biswas and Edoardo M. Airoldi

Abstract When doing causal inference on networks, there is interference among the units. In a social network setting, such interference among individuals is known as peer-influence. Estimating the causal effect of peer-influence under the presence of homophily presents various challenges. In this paper, we present results quantifying the error incurred from ignoring homophily when estimating peer-influence on networks. We then present randomized treatment strategies on networks which can help disentangle homophily from the estimation of peer-influence.

1 Introduction

When doing causal inference there is often *interference* among the units of interest. Interference is when the response to treatment of a unit is affected by the treatments assigned to its neighbors. In a social network setting, where a unit corresponds to an individual and an individual's neighborhood corresponds to their peers, such interference among individuals is known as peer-influence. With the increased usage of social media and availability of network data, understanding the casual effects of peer-influence has garnered much interest. The research area yields a wide range of applications. For example, in advertising Bakshy [5] examined the impact of friends' product affiliation on advertisements via randomized experiments on Facebook users; Aral [3] used randomized experiments on Facebook to examine how firms can design social media marketing campaigns to create peer-influence. In politics, Bond [6] assessed voting behavior results from randomized experiments on Facebook (where political mobilization messages were delivered to Facebook users via a randomized control trial during the 2010 US congressional elections) to find that effect of peer-influence on voting turnout was greater than the effect of the direct messages themselves.

N. Biswas (✉) · E. M. Airoldi
Department of Statistics, Harvard University, Cambridge, USA
e-mail: niloy_biswas@g.harvard.edu

E. M. Airoldi
e-mail: airoldi@fas.harvard.edu

© Springer International Publishing AG 2018
S. Cornelius et al. (eds.), *Complex Networks IX*, Springer Proceedings
in Complexity, https://doi.org/10.1007/978-3-319-73198-8_28

323

There is recent work on methodology for estimation of causal peer-influence effects (e.g., Toulis and Kao [10], Athey et al. [4]). However, identifying and estimating peer-influence under the presence of homophily have long remained a challenging problem [2, 8, 9]. This is the problem of identifying to what extent the response of an individual is attributable to the treatments given to its neighbors (peer-influence) or attributable to a latent, intrinsic similarity between peers (homophily). This paper makes several contributions toward tackling this problem. In particular, we: (i) Introduce a framework for modeling peer-influence and homophily; (ii) under different models of peer-influence and homophily, quantify the error incurred from ignoring homophily in the estimation of the peer-influence effect; (iii) under a stochastic block model framework for the network, devise randomized treatment strategies which can help disentangle latent homophily from the estimation of peer-influence. Our randomized treatment strategies can also be applied in a more general setting, for general inference of network features in the presence of latent homophily.

2 Peer-Influence Under Homophily: Results and Inference Strategies

2.1 A General Framework for Modeling Peer-Influence and Homophily

Peer-influence is used to denote when the response of one individual is affected by the treatments assigned to its neighbors (e.g., friends in a social network). For individual i, this can be represented by $\mathbf{peer}((Z_j)_{j\in\mathcal{N}_i})$, where $(Z_j)_{j\in\mathcal{N}_i}$ are the treatments assigned to the neighbors of i and $\mathbf{peer}(\cdot)$ is a function taking values in the space of responses.

Homophily represents the latent, intrinsic similarity between close individuals in a network. For $j = 1, \ldots, N$, let X_j be independent and identically distributed random variables corresponding to the latent variable associated with individual j in the network. Then, for individual i, homophily can be represented by $\mathbf{hom}((X_j)_{j\in\mathcal{N}_i})$, where $(X_j)_{j\in\mathcal{N}_i}$ are the latent variables in the neighborhood of i and $\mathbf{hom}(\cdot)$ is a function taking values in the space of responses.

We now introduce the general framework used in our analysis of peer-influence and homophily. Suppose we are interested in the responses of N units in a network. This is represented by the random variables Y_i for $i = 1, \ldots, n$. The response Y_i of the ith unit depends on its treatment, peer-influence, and latent homophily. Our full model is given by:

$$Y_i(Z_i = 0, (Z_j)_{j\in\mathcal{N}_i}, (X_j)_{j\in\mathcal{N}_i}) = \alpha + \beta_0 \mathbf{peer}((Z_j)_{j\in\mathcal{N}_i}) + h_0 \mathbf{hom}((X_j)_{j\in\mathcal{N}_i}) + \epsilon_i(0, \sigma_Y^2) \tag{1}$$

$$Y_i(Z_i = 1, (Z_j)_{j \in \mathcal{N}_i}, (X_j)_{j \in \mathcal{N}_i}) = \tau + Y_i(Z_i = 0, (Z_j)_{j \in \mathcal{N}_i}, (X_j)_{j \in \mathcal{N}_i})$$
$$+ \beta_1 \mathbf{peer}((Z_j)_{j \in \mathcal{N}_i}) + h_1 \mathbf{hom}((X_j)_{j \in \mathcal{N}_i}) \quad (2)$$

$\epsilon_i(0, \sigma_Y^2)$ for $i = 1, \ldots, N$ are the noise terms in the network, indepedent and identically distributed according to an unknown distribution with zero mean and variance σ_Y^2. β_0, β_1 are the unknown peer-influence parameters, and h_0, h_1 are the unknown homophily parameters. Latent effects due to homophily in the model are represented by indepedent and identically distributed random variables $(X_i)_{i=1}^N$ with mean 1 and variance σ_X^2. $\mathbf{Z}$ are the assigned treatments.

Under different models of the peer-influence $\mathbf{peer}(\cdot)$ and homophily $\mathbf{hom}(\cdot)$, we will focus on estimating peer-influence and homophily parameters β_0 and h_0, respectively, assuming the variances are known. Note that our analysis here is focussed on inference concerning the untreated individuals (1 above), but all the methodology can be easily applied to the treated individuals in the network (2 above). In our analysis, we consider the significance of the following factors in the inference of peer-influence under the presence of homophily, their consequences for the design of experiments:

1. Modeling of peer-influence: as a binary ($\mathbf{peer}((Z_j)_{j \in \mathcal{N}_i}) = \mathbf{1}_{\sum_{j \in \mathcal{N}_i} Z_j > 0}$) or a linear ($\mathbf{peer}((Z_j)_{j \in \mathcal{N}_i}) = \sum_{j \in \mathcal{N}_i} Z_j$) effect.
2. Modeling of homophily: as an unnormalized ($\mathbf{hom}((X_j)_{j \in \mathcal{N}_i}) = \sum_{j \in \mathcal{N}_i} X_j$) or normalized ($\mathbf{hom}((X_j)_{j \in \mathcal{N}_i}) = \sum_{j \in \mathcal{N}_i} X_j / |\mathcal{N}_i|$) latent factor. Unnormalized homophily corresponds to when dense regions of the network have a stronger homophily effect compared to more sparse regions. Normalized homophily corresponds to when the homophily effect is not affected by the density of different regions in the network.
3. Choice of peer-influence estimate: as a difference of means estimate (for binary peer-influence) or as the average of stratified estimates (for linear peer-influence).
4. Allocation of treatments: fixed optimal treatment allocation or randomized treatment.

In this short paper, we only discuss results and strategies for the case of binary peer-influence under unnormalized homophily. Discussion of the other cases is included in the appendix.

Binary peer-influence effect with unnormalized homophily. Consider the binary peer-influence model with unnormalized homophily. For the untreated individuals, we have

$$Y_i(Z_i = 0, (Z_j)_{j \in \mathcal{N}_i}) = \alpha + \beta_0 \mathbf{1}_{\sum_{j \in \mathcal{N}_i} Z_j > 0} + h_0 \sum_{j \in \mathcal{N}_i} X_j + \epsilon_i(0, \sigma_Y^2) \quad (3)$$

where $\epsilon_i(0, \sigma_Y^2)$ are independent and identically distributed with zero mean and σ_Y^2 variance.

Consider estimating the peer-influence parameter β_o using a difference in means estimator. Partition the set of untreated individuals into sets $M_0^{(0)} := \{i : Z_i = 0, \sum_{j \in \mathcal{N}_i} Z_j = 0\}$ (the set of untreated individuals with no treated neighbors) and $M_0^{(1)} := \{i : Z_i = 0, \sum_{j \in \mathcal{N}_i} Z_j > 0\}$ (the set of untreated individuals with at least one treated neighbors). Then, the difference in means estimator for β_0 is given by:

$$\hat{\beta}_0 = \underset{i \in M_0^{(1)}}{avg\ Y_i} - \underset{i \in M_0^{(0)}}{avg\ Y_i} \tag{4}$$

Under the negligence of latent homophily in the model, this difference of means estimator for peer-influence would appear unbiased. However, the presence of latent homophily actually interferes and introduces bias to the estimation of peer-influence, as highlighted in Theorem 1 below.

Theorem 1 *Consider the difference in means estimator $\hat{\beta}_0$ for binary peer-influence effect β_0. Under the presence of unnormalized homophily in our model (3), the mean squared error of $\hat{\beta}_0$ (conditional on the treatment $\mathbf{Z}$) is:*

$$\mathbb{E}[(\hat{\beta}_0 - \beta_0)^2 | \mathbf{Z}] = \left(h_0 \left(\underset{i \in M_0^{(1)}}{avg}\ |\mathcal{N}_i| - \underset{i \in M_0^{(0)}}{avg}\ |\mathcal{N}_i| \right) \right)^2$$
$$+ h_0^2 \sigma_X^2 \left(\underset{i,j \in M_0^{(0)}}{avg}\ |\mathcal{N}_i \cap \mathcal{N}_j| + \underset{i,j \in M_0^{(1)}}{avg}\ |\mathcal{N}_i \cap \mathcal{N}_j| - 2 \underset{i \in M_0^{(0)}, j \in M_0^{(1)}}{avg}\ |\mathcal{N}_i \cap \mathcal{N}_j| \right)$$
$$+ \sigma_Y^2 \left(\frac{1}{|M_0^{(0)}|} + \frac{1}{|M_0^{(1)}|} \right) \tag{5}$$

We can interpret (5) to understand the optimal treatment allocation with respect to minimizing the bias and variance. For binary peer-influence effect with unnormalized homophily, the bias of $\hat{\beta}_0$ is minimized through an assignment of treatments $\mathbf{Z}$ which manages to balance the average homophily effect (corresponding to average vertex degrees) between individuals in $M_0^{(1)}$ and $M_0^{(0)}$. Under such balanced treatment assignment, unbiasedness is achieved when

$$\underset{i \in M_0^{(1)}}{avg}\ |\mathcal{N}_i| = \underset{i \in M_1^{(1)}}{avg}\ |\mathcal{N}_i| = \underset{i \in M_0^{(0)} \cup M_0^{(1)}}{avg}\ |\mathcal{N}_i|,$$

where $M_0^{(0)} \cup M_0^{(1)}$ is the set of all (untreated) individuals. For binary peer-influence effect with unnormalized homophily, the variance of $\hat{\beta}_0$ is minimized through treatments $\mathbf{Z}$ which:

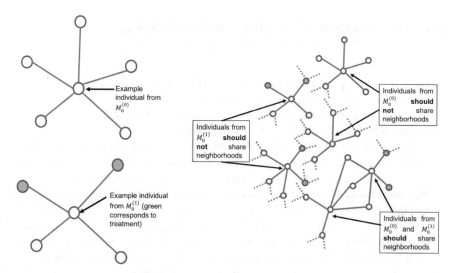

Fig. 1 Optimal treatment allocation

1. Ensure $|M_0^{(1)}| = |M_0^{(0)}|$, such that there is balance between the number of individuals which are affected and not affected by peer-influence (in set $M_0^{(1)}$ and $M_0^{(0)}$, respectively).

2. Ensure that the individuals in $M_0^{(1)}$ are mixed with individuals in $M_0^{(0)}$ as well as possible. In particular, this corresponds to choosing $\mathbf{Z}$ such that elements in $M_0^{(0)}$ have minimal shared neighborhoods between themselves (minimizing $\underset{i,j \in M_0^{(0)}}{avg} |\mathcal{N}_i \cap \mathcal{N}_j|$), elements in $M_0^{(1)}$ have minimal shared neighborhoods between themselves (minimizing $\underset{i,j \in M_0^{(1)}}{avg} |\mathcal{N}_i \cap \mathcal{N}_j|$), and shared neighborhoods between elements of $M_0^{(0)}$ and $M_0^{(1)}$ are maximal, respectively (maximizing $\underset{i \in M_0^{(0)}, j \in M_0^{(1)}}{avg} |\mathcal{N}_i \cap \mathcal{N}_j|$). This is illustrated through Fig. 1.

Having developed conceptual insights into what treatment assignments are optimal for inferring peer-influence in the presence of homophily, we can further extend our analysis to consider the cases of randomized treatment. In particular, by considering randomized treatment under a stochastic block model framework (e.g., see Holland [7], Airoldi [1]), we can take advantage of symmetries and exchangeability to gain insight into the optimal design of randomized experiments under such framework. This is explored in Sect. 2.2 where randomized treatment designs which disentangle homophily from the estimation of peer-influence are considered.

2.2 Disentangling Homophily from Estimation of Peer-Influence: Randomized Treatment Strategies

We now propose a general strategy for reducing bias in the inference of binary peer-influence under the presence of homophily. It is applicable to weighted, directed graphs which are clustered. Furthermore, our strategy does not assume any model for homophily $\mathbf{hom}(\cdot)$, which can remain unknown.

Suppose we have a graph G of N vertices which is clustered into r clusters. Assume that given clustering of the graph captures the covariates of the individuals in the network, such that individuals with same or similar covariates are members of the same cluster. Under such a clustering, we fit a corresponding stochastic block model onto the network of N individuals in r communities. Note that by fitting such a stochastic block model, we are implicitly assuming that individuals in the same cluster are exchangeable (hence the need to have a good cluster for this assumption to be justified). Denote the communities of the fitted stochastic block model by the sets $B_1, \ldots, B_r$, which are of respective sizes $A_1, \ldots, A_r$ (where $A_1 + \cdots + A_r = N$). Let $\mathbf{P}$ be the $r \times r$ adjacency probability matrix between the r communities. Values $A_1, \ldots, A_r$ directly are obtained from the cluster sizes, and the entries of the matrix $\mathbf{P}$ can be estimated using MLEs (e.g., in the unweighted graph case, we can choose $[\mathbf{P}]_{i,j} = \frac{\text{number of edges from cluster i to j}}{|A_i||A_j|}$). Within each community B_s, different individuals are affected by different levels of peer-influence. For example, in the binary peer-influence case, untreated individuals are either in set $M_0^{(0)}$ (no treated neighbors—not affected by peer-influence) or in set $M_0^{(1)}$ (at least one treated neighbors—affected by peer-influence); in the linear peer-influence case, untreated individuals in $M_0^{(k)}$ are affected by the k-levels of peer-influence.

When estimating peer-influence, the bias due to homophily arises from imbalances in the homophily effect between the sets of individuals with different levels of peer-influence. This motivates the key idea in our design of randomized treatments to remove bias from homophily: We want to design experiments such that in every community B_s, an equal number of individuals are affected and not affected by peer-influence. By the construction of the cluster, every community B_s has a similar effect due to latent homophily. Therefore by designing randomized experiments which ensure that every such cluster B_s has an equal (expected) number of individuals with different levels of peer-influence, we reduce the bias in the estimation of peer-influence arising from latent homophily. For randomized treatments where individuals in cluster s are treated independently with probability θ_s, our strategy described leads to constrained optimization problems for θ_s. This can then be solved to obtain optimal θ_s^{opt} values as required for reducing bias in the estimation of peer-influence under the presence of latent, unknown homophily. We now highlight our strategy in detail for the estimation of binary peer-influence under the presence of homophily (linear peer-influence case in appendix).

An algorithm for inference of binary peer-influence. For a weighted, directed graph G of N vertices which is clustered into r clusters, consider a corresponding stochastic block model of N individuals in r communities. Denote the communities of the SBM (clusters of G) by the sets $B_1, \ldots, B_r$, which are of respective sizes $A_1, \ldots, A_r$ (where $A_1 + \cdots + A_r = N$). Let $\mathbf{P}$ be the $r \times r$ adjacency probability matrix between the r communities. We assign treatments independently to individuals such that individuals in B_s are treated with probability θ_s for $s = 1, \ldots, r$. We want to choose θ_s with the aim of reducing bias, such that homophily does not interfere with the estimation of peer-influence. Note that the general bias of the binary peer-influence estimator $\hat{\beta}_0$ is

$$\underset{i \in M_0^{(1)}}{avg} \; \mathbb{E}_X[\mathbf{hom}((X)_{j \in \mathcal{N}_i})] - \underset{i \in M_0^{(0)}}{avg} \; \mathbb{E}_X[\mathbf{hom}((X)_{j \in \mathcal{N}_i})].$$

This highlights that the bias in our estimation arises from an imbalance in the average homophily effect between the sets $M_0^{(1)}$ and $M_0^{(0)}$ (the individuals who are and are not affected by peer-influence, respectively). This observation motivates the key idea in our design of randomized treatments to remove bias from homophily: We want to design experiments such that in every cluster B_s, an equal number of individuals are affected and not affected by peer-influence. For randomized treatment assignment, this means we want

$$\forall s = 1, \ldots, r, \quad \mathbb{E}[|M_0^{(1)} \cap B_s|] = \mathbb{E}[|M_0^{(0)} \cap B_s|]. \tag{6}$$

Let us now derive a result about $M_0^{(0)}$ and $M_0^{(1)}$ under our framework to proceed further with (6).

Proposition 1 *Consider a stochastic block model (SBM) of N individuals in r communities. Denote the communities of the SBM by the sets $B_1, \ldots, B_r$, which are of respective sizes $A_1, \ldots, A_r$ (where $A_1 + \cdots + A_r = N$). Let $\mathbf{P}$ be the $r \times r$ adjacency probability matrix between the r communities. We assign treatments independently to individuals such that individuals in B_s are treated with probability θ_s for $s = 1, \ldots, r$. Under such setup, let $M_0^{(0)}$ denote the set of untreated individuals which have no treated neighbors and let $M_0^{(1)}$ denote the set of untreated individuals which have at least one treated neighbor. For ease of notation, let $\{s \in M_0^{(0)}\}$, $\{s \in M_0^{(1)}\}$ denote the event that a fixed vertex in community s is in the set $M_0^{(0)}$, $M_0^{(1)}$, respectively. Then,*

$$\mathbb{P}(s \in M_0^{(0)}) = (1 - \theta_s) \prod_{v=1}^{r} (1 - P_{s,v} \theta_v)^{A_v - I_{v=s}}, \; and \tag{7}$$

$$\mathbb{P}(s \in M_0^{(1)}) = (1 - \theta_s) \left(1 - \prod_{v=1}^{r} (1 - P_{s,v} \theta_v)^{A_v - I_{v=s}} \right). \tag{8}$$

Using Proposition (1), we can now directly derive an algorithm to reduce the effect of homophily during inference. Let s denote any vertex in the graph which is in community B_s. Note that $\mathbb{E}[|M_0^{(0)} \cap B_s|] = A_s \mathbb{P}(s \in M_0^{(0)})$ and $\mathbb{E}[|M_0^{(1)} \cap B_s|] = A_s \mathbb{P}(s \in M_0^{(1)}) = A_s(1 - \mathbb{P}(s \in M_0^{(0)}))$, as all untreated individuals are in either $M_0^{(0)}$ or $M_0^{(1)}$. This gives,

$$\mathbb{E}[|M_0^{(1)} \cap B_s|] = \mathbb{E}[|M_0^{(0)} \cap B_s|] \iff \mathbb{P}(s \in M_0^{(0)}) = \frac{1}{2}\mathbb{P}(Z_s = 0)$$

$$\iff (1 - \theta_s) \prod_{v=1}^{r} (1 - P_{s,v}\theta_v)^{A_v - \mathbf{1}_{v=s}} = \frac{1}{2}(1 - \theta_s)$$

$$\iff \sum_{v=1}^{r} (A_v - \mathbf{1}_{v=s})log(1 - P_{s,v}\theta_v) + log(2) = 0$$

For $|P_{s,v}| \approx 0$, $log(1 - P_{s,v}\theta_v) \approx -P_{s,v}\theta_v$. This allows us to approximate the optimal θ_s values by simply solving a set of linear equations. Our algorithm is given below.

Algorithm 1: Randomized treatment design for more accurate inference of peer-influence

---

1 function optimal_treatment_values $(\mathbf{A}(G), \mathbf{B}(G))$;

    Input : Adjacency matrix $\mathbf{A}$ and clustering $\mathbf{B}$ (with r clusters) of some graph G

    Output: Treatment probabilities $\theta \in [0, 1]^r$ for Bernoulli assignment on each cluster

2 Fit an SBM, giving an adjacency matrix $\mathbf{P}$ for clusters $B_1, \ldots, B_r$ of sizes $A_1, \ldots, A_r$.

3 Choose treatment probabilities for the clusters $\theta \in [0, 1]^r$ as the solution to:

$$\begin{pmatrix} P_{1,1}(A_1 - 1) & P_{1,2}A_2 & \cdots & \cdots & P_{1,r}A_r \\ P_{2,1}A_1 & P_{2,2}(A_2 - 1) & \cdots & \cdots & P_{2,r}A_r \\ \vdots & \vdots & \vdots & \cdots & \vdots \\ P_{r-1,1}A_1 & P_{r-1,2}A_2 & \cdots & P_{r-1,r-1}(A_{r-1} - 1) & P_{r-1,r}A_r \\ P_{r,1}A_1 & P_{r,2}A_2 & \cdots & P_{r,r-1}A_{r-1} & P_{r,r}(A_r - 1) \end{pmatrix} \begin{pmatrix} \theta_1 \\ \theta_2 \\ \vdots \\ \theta_{r-1} \\ \theta_r \end{pmatrix} = log(2) \begin{pmatrix} 1 \\ 1 \\ \vdots \\ 1 \\ 1 \end{pmatrix} \tag{9}$$

In practice, if (9) does not have a solution in $[0, 1]^r$, we can solve the constrained optimization problem of minimizing

$$\left\| \begin{pmatrix} P_{1,1}(A_1 - 1) & P_{1,2}A_2 & \cdots & \cdots & P_{1,r}A_r \\ P_{2,1}A_1 & P_{2,2}(A_2 - 1) & \cdots & \cdots & P_{2,r}A_r \\ \vdots & \vdots & \vdots & \cdots & \vdots \\ P_{r-1,1}A_1 & P_{r-1,2}A_2 & \cdots & P_{r-1,r-1}(A_{r-1} - 1) & P_{r-1,r}A_r \\ P_{r,1}A_1 & P_{r,2}A_2 & \cdots & P_{r,r-1}A_{r-1} & P_{r,r}(A_r - 1) \end{pmatrix} \begin{pmatrix} \theta_1 \\ \theta_2 \\ \vdots \\ \theta_{r-1} \\ \theta_r \end{pmatrix} - log(2) \begin{pmatrix} 1 \\ 1 \\ \vdots \\ 1 \\ 1 \end{pmatrix} \right\| \tag{10}$$

for $\theta \in [0, 1]^r$ and for some chosen norm $\| \ \|$ on $\mathbb{R}^d$ (e.g., L^2). Note that under the optimal treatment probabilities θ^{opt} obtained from (9), the total expected number of treated individuals is $\sum_{s=1}^r A_s \theta_s^{opt}$. In practice, often it is desirable to control the expected number of individuals treated under randomized treatment. This is can be done under our framework by considering the constrained optimization problem of minimizing the norm in (9) subject to $\theta \in [0, 1]^r$ and $\sum_{s=1}^r A_s \theta_s = Nx$, where x is our chosen percentage of individuals treated.

Analysis of treatment strategies via simulations. We highlight the performance of our randomized treatment strategy compared to alternatives via numerical results from Monte Carlo simulations under a stochastic block model. We consider the bias and mean squared error of our optimal randomized treatment compared to other common randomized treatment strategies. Unsuccessful treatment occurs when either one of the sets $M_0^{(0)}$ or $M_0^{(1)}$ is empty, and the difference in means estimator for binary peer-influence (4) is ill-defined.

We consider the unnormalized sum of latent variables $(\mathbf{hom}((X)_{j \in \mathcal{N}_i}) = \sum_{j \in \mathcal{N}_i} X_j$ for X_j i.i.d. latent random variables with mean 1, variance σ_X) as our homophily function. The baseline simulation model (with binary peer-influence, unnormalized homophily) here is:

$$Y_i(Z_i = 0, (Z_j)_{j \in \mathcal{N}_i}, (X_j)_{j \in \mathcal{N}_i}) = \alpha + \beta_0 1_{\sum_{j \in \mathcal{N}_i} Z_j > 0} + h_0 \sum_{j \in \mathcal{N}_i} X_j + \epsilon_i(0, \sigma_Y^2)$$

$$Y_i(Z_i = 1, (Z_j)_{j \in \mathcal{N}_i}, (X_j)_{j \in \mathcal{N}_i}) = \tau + Y_i(Z_i = 0, (Z_j)_{j \in \mathcal{N}_i}, (X_j)_{j \in \mathcal{N}_i}) + \beta_1 1_{\sum_{j \in \mathcal{N}_i} Z_j > 0} + h_1 \sum_{j \in \mathcal{N}_i} X_j$$

for $\alpha = 3, \beta_0 = 0.1, h_0 = 1, \tau = 0.2, \beta_1 = 0.05, h_1 = 0.5, \sigma_Y = 1.5^2, \sigma_X = 1^2$.

SBM Graph Simulation. The following figures display the bias and variance of the difference in means estimator (y-axis) against the controlled expected percentage of treated individuals (x-axis) for our strategy in comparison with other common randomized treatment strategies. The figures highlight that our randomized treatment strategy leads to improved estimation of peer-influence. We are working with a directed SBM of 1530 vertices and 7 clusters with

$$(A_1, A_2, A_3, A_4, A_5, A_6, A_7) = (600, 340, 200, 150, 100, 90, 50), \ \mathbf{P} = \begin{pmatrix} 50 & 10 & 20 & 5 & 5 & 15 & 3 \\ 10 & 30 & 5 & 15 & 15 & 10 & 5 \\ 10 & 5 & 40 & 5 & 10 & 13 & 12 \\ 4 & 5 & 10 & 25 & 15 & 14 & 12 \\ 14 & 15 & 10 & 5 & 20 & 10 & 10 \\ 13 & 14 & 5 & 2 & 10 & 35 & 15 \\ 10 & 14 & 14 & 5 & 5 & 10 & 45 \end{pmatrix} \Big/ 1530.$$

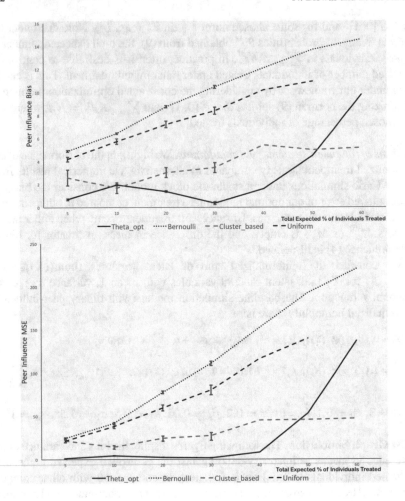

2.3 Concluding Remarks

When doing causal inference in networks, neglecting latent homophily can lead to inaccurate inference of peer-influence. In this paper, we have introduced a general framework for modeling peer-influence and homophily, quantified the error incurred from ignoring homophily, and devised randomised treatment strategies which allow the estimation of peer-influence in the presence of homophily. Simulations highlight our method's performance relative to other randomized treatment strategies. This work is a preliminary insight into a forthcoming project. Our future extensions will involve further statistical analysis and theoretical guarantees on the performance of the randomized treatment strategies, and results from experimentation on large real-world social networks.

A   Appendices

A.1   Peer-Influence Under Homophily: Results and Inference Strategies

Binary peer-influence effect with normalized homophily: Consider now binary peer-influence effect with normalized homophily. For the untreated individuals, we have

$$Y_i(Z_i = 0, (Z_j)_{j \in \mathcal{N}_i}) = \alpha + \beta_0 \mathbf{1}_{\sum_{j \in \mathcal{N}_i} Z_j > 0} + h_0 \sum_{j \in \mathcal{N}_i} \frac{X_j}{|\mathcal{N}_i|} + \epsilon_i(0, \sigma_Y^2) \quad (11)$$

where $\epsilon_i(0, \sigma_Y^2)$ are idependent and identically distributed with zero mean and σ_Y^2 variance.

As before, consider estimating the peer-influence parameter β_o using a difference in means estimator. Partition the set of untreated individuals into sets $M_0^{(0)} := \{i : Z_i = 0, \sum_{j \in \mathcal{N}_i} Z_j = 0\}$ (the set of untreated individuals with no treated neighbors) and $M_0^{(1)} := \{i : Z_i = 0, \sum_{j \in \mathcal{N}_i} Z_j > 0\}$ (the set of untreated individuals with at least one treated neighbors). Then, the difference in means estimator for β_0 is given by:

$$\hat{\beta}_0 = \underset{i \in M_0^{(1)}}{avg\ Y_i} - \underset{i \in M_0^{(0)}}{avg\ Y_i} \quad (12)$$

Unlike in the case with unnormalized homophily, the difference of means estimator for peer-influence remains unbiased in the presence of normalized homophily. This is further highlighted in Theorem 2 below. Furthermore, for most sparse and dense models for the underlying graph, Theorem 2 can be used to show that $\hat{\beta}_0$ is a consistent estimator of peer-influence under normalized homophily.

Theorem 2 *Consider the difference in means estimator $\hat{\beta}_0$ for binary peer-influence effect β_0. Under the presence of normalized homophily in our model (11), the mean squared error of $\hat{\beta}_0$ (conditional on the treatment $\mathbf{Z}$) is:*

$$\mathbb{E}[(\hat{\beta}_0 - \beta_0)^2 | \mathbf{Z}] =$$

$$h_0^2 \sigma_X^2 \left(\underset{i,j \in M_0^{(0)}}{avg} \frac{|\mathcal{N}_i \cap \mathcal{N}_j|}{|\mathcal{N}_i||\mathcal{N}_j|} + \underset{i,j \in M_0^{(1)}}{avg} \frac{|\mathcal{N}_i \cap \mathcal{N}_j|}{|\mathcal{N}_i||\mathcal{N}_j|} - 2 \underset{i \in M_0^{(0)}, j \in M_0^{(1)}}{avg} \frac{|\mathcal{N}_i \cap \mathcal{N}_j|}{|\mathcal{N}_i||\mathcal{N}_j|} \right) + \sigma_Y^2 \left(\frac{1}{|M_0^{(0)}|} + \frac{1}{|M_0^{(1)}|} \right)$$

$$(13)$$

Linear peer-influence effect with unnormalized homophily: We now consider modeling peer-influence as a linear function of the number of treated neighbors

peer$((Z_j)_{j \in \mathcal{N}_i}) = \sum_{j \in \mathcal{N}_i} Z_j$. For the untreated individuals under unnormalized homophily, this gives:

$$Y_i(Z_i = 0, (Z_j)_{j \in \mathcal{N}_i}) = \alpha + \beta_0 \sum_{j \in \mathcal{N}_i} Z_j + h_0 \sum_{j \in \mathcal{N}_i} X_j + \epsilon_i(0, \sigma_Y^2) \qquad (14)$$

where $\epsilon_i(0, \sigma_Y^2)$ are idependent and identically distributed with zero mean and σ_Y^2 variance.

Consider estimating the peer-influence parameter β_0. Generalizing our methodology from the binary peer-influence case, we now develop a stratified estimator for β_0. Let

$$M_0^{(k)} := \{i : Z_i = 0, \sum_{j \in \mathcal{N}_i} Z_j = k\}$$

be the set of untreated individuals which have k treated neighbors. Then, an average of difference in means estimator for peer-influence is:

$$\hat{\beta}_0 = \frac{\sum_k \hat{\beta}_0^{(k)}}{\sum_k 1} \; for \; \hat{\beta}_0^{(k)} = \frac{1}{k}\left(\frac{\sum_{i \in M_0^{(k)}} Y_i}{|M_0^{(k)}|} - \frac{\sum_{i \in M_0^{(0)}} Y_i}{|M_0^{(0)}|}\right) = \frac{1}{k}\left(\underset{i \in M_0^{(k)}}{avg\,} Y_i - \underset{i \in M_0^{(0)}}{avg\,} Y_i\right).$$
$$(15)$$

where we average over all k such that $|M_0^{(k)}| > 0$ (so that $\hat{\beta}_0^{(k)}$ is well-defined). Note that here we are averaging over the class of estimators $\hat{\beta}_0^{(k)}$ under the assumption of linear peer-influence. In the case of nonlinearity, we can also consider each $\hat{\beta}_0^{(k)}$ separately to understand the kth-level peer-influence effect in the network.

The presence of latent unnormalized homophily interferes and introduces bias to the estimation of linear peer-influence, as highlighted in Theorem 3 below.

Theorem 3 *Consider the estimator $\hat{\beta}_0$ for linear peer-influence effect β_0. Under the presence of unnormalized homophily in our model (3), the mean squared error of $\hat{\beta}_0$ (conditional on the treatment $\mathbf{Z}$) is:*

$$\mathbb{E}[(\hat{\beta}_0 - \beta_0)^2 | \mathbf{Z}] =$$

$$\left(\frac{h_0}{\sum_{k>0} 1} \sum_{k>0} \frac{1}{k}\left(\underset{i \in M_0^{(k)}}{avg\,} |\mathcal{N}_i| - \underset{i \in M_0^{(0)}}{avg\,} |\mathcal{N}_i|\right)\right)^2$$

$$+ \frac{1}{(\sum_{k>0} 1)^2} \sum_{k,l>0} \frac{1}{kl}\left[h_0^2 \sigma_X^2\left(\underset{i \in M_0^{(k)}, j \in M_0^{(l)}}{avg\,} |\mathcal{N}_i \cap \mathcal{N}_j| + \underset{i,j \in M_0^{(0)}}{avg\,} |\mathcal{N}_i \cap \mathcal{N}_j| - 2\underset{i \in M_0^{(0)}, j \in M_0^{(k)}}{avg\,} |\mathcal{N}_i \cap \mathcal{N}_j|\right)\right.$$

$$\left. + \sigma_Y^2\left(\frac{1}{|M_0^{(0)}|} + \frac{I_{k=l}}{|M_0^{(k)}|}\right)\right]$$
$$(16)$$

Equation (16) highlights that unbiasedness estimation via optimal treatment allocation may be difficult computationally, as now we need to ensure balance across all the strata $(M_0^{(k)})_{k \geq 0}$. This motivates an alternative approach of unbiased estimation.

Linear peer-influence effect with normalized homophily: For the peer-influence effect on untreated individuals under normalized homophily, we obtain:

$$Y_i(Z_i = 0, (Z_j)_{j \in \mathcal{N}_i}) = \alpha + \beta_0 \sum_{j \in \mathcal{N}_i} Z_j + h_0 \sum_{j \in \mathcal{N}_i} \frac{X_j}{|\mathcal{N}_i|} + \epsilon_i(0, \sigma_Y^2) \qquad (17)$$

where $\epsilon_i(0, \sigma_Y^2)$ are idependent and identically distributed with zero mean and σ_Y^2 variance.

To estimate the peer-influence parameter β_0, the same stratified estimator as in the linear peer-influence with unnormalized homophily case can be applied:

$$\hat{\beta}_0 = \frac{\sum_k \hat{\beta}_0^{(k)}}{\sum_k 1} \ for \ \hat{\beta}_0^{(k)} = \frac{1}{k} \left(\frac{\sum_{i \in M_0^{(k)}} Y_i}{|M_0^{(k)}|} - \frac{\sum_{i \in M_0^{(0)}} Y_i}{|M_0^{(0)}|} \right) = \frac{1}{k} \left(\underset{i \in M_0^{(k)}}{avg} \ Y_i - \underset{i \in M_0^{(0)}}{avg} \ Y_i \right).$$

$$(18)$$

where $M_0^{(k)} := \{i : Z_i = 0, \sum_{j \in \mathcal{N}_i} Z_j = k\}$ and we are averaging over all k such that $|M_0^{(k)}| > 0$.

In the presence of normalized homophily, $\hat{\beta}_0$ remains an unbiased estimator of peer-influence. This is highlighted in Theorem 4 below.

Theorem 4 *Consider the estimator $\hat{\beta}_0$ for linear peer-influence effect β_0. Under the presence of normalized homophily in our model (11), $\hat{\beta}_0$ is unbiased and the mean squared error of $\hat{\beta}_0$ (conditional on the treatment $\mathbf{Z}$) is:*

$$\mathbb{E}[(\hat{\beta}_0 - \beta_0)^2 | \mathbf{Z}] =$$

$$\frac{1}{(\sum_{k>0} 1)^2} \sum_{k,l>0} \frac{1}{kl} \left[h_0^2 \sigma_X^2 \left(\underset{i \in M_0^{(k)}, j \in M_0^{(l)}}{avg} \frac{|\mathcal{N}_i \cap \mathcal{N}_j|}{|\mathcal{N}_i||\mathcal{N}_j|} + \underset{i,j \in M_0^{(0)}}{avg} \frac{|\mathcal{N}_i \cap \mathcal{N}_j|}{|\mathcal{N}_i||\mathcal{N}_j|} - 2 \underset{i \in M_0^{(0)}, j \in M_0^{(k)}}{avg} \frac{|\mathcal{N}_i \cap \mathcal{N}_j|}{|\mathcal{N}_i||\mathcal{N}_j|} \right) \right.$$

$$\left. + \sigma_Y^2 \left(\frac{1}{|M_0^{(0)}|} + \frac{\mathbf{1}_{k=l}}{|M_0^{(k)}|} \right) \right]$$

$$(19)$$

The difference of means estimator for linear peer-influence remains unbiased in the presence of normalized homophily. Furthermore, for most sparse and dense models for the underlying graph, Theorem 2 can be used to show that $\hat{\beta}_0$ is a consistent estimator of linear peer-influence under normalized homophily.

A.2   Disentangling Homophily from Estimation of Peer-Influence: Randomized Treatment Strategies

An algorithm for inference of linear peer-influence. We now use our general framework to design randomized treatments for the inference of linear peer-influence effects under homophily. We proceed to find the optimal treatment probabilities θ_s for $s = 1, \ldots, r$ under a stochastic block model with r communities as before.

Let $M_0^{(k)}$ denote the set of untreated individuals which have k neighbors (note that we are abusing notation here: Now, $M_0^{(1)}$ represents untreated individuals which have exactly 1 neighbor, rather than at least 1 neighbor as before in the binary peer-influence case). First, we derive a proposition about $M_0^{(k)}$ under our framework.

Proposition 2 *Consider a stochastic block model (SBM) of N individuals in r communities. Denote the communities of the SBM by the sets $B_1, \ldots, B_r$, which are of respective sizes $A_1, \ldots, A_r$ (where $A_1 + \cdots + A_r = N$). Let P be the $r \times r$ adjacency probability matrix between the r communities. We assign treatments independently to individuals such that individuals in B_s are treated with probability θ_s for $s = 1, \ldots, r$. Under such setup, let $M_0^{(k)}$ denote the set of untreated individuals which have k treated neighbors. For ease of notation, let $\{s \in M_0^{(k)}\}$ denote the event that a fixed vertex in community s is in the set $M_0^{(k)}$. Then,*

$$\mathbb{P}(s \in M_0^{(k)}) = (1 - \theta_s) \sum_{\substack{t_1, \ldots, t_r: \\ \forall v=1, \ldots, r \ 0 \le t_v \le A_v - I_{\{v=s\}}, \\ t_1 + \cdots + t_r = k}} \left(\prod_{v=1}^{r} Bin(t_v; A_v - I_{\{v=s\}}, \theta_v P_{s,v}) \right)$$

$$(20)$$

where $Bin(t_v; A_v - I_{\{v=s\}}, \theta_v P_{s,v}) = \binom{A_v - I_{\{v=s\}}}{t_v} (\theta_v P_{s,v})^{t_v} (1 - \theta_v P_{s,v})^{A_v - I_{\{v=s\}} - t_v}$.

The main idea behind the homophily disentangling strategy is to ensure that in every community B_s on our stochastic block model, there are equal (expected) numbers of individuals being affected by different levels of peer-influence. In the case of linear peer-influence, this means choosing treatment values such that inside every community s, each individual has an equal probability of being in sets $M_0^{(k)}$ for different peer-influence levels k. Under a stochastic block model, values of k range from 0 to $N - 1$ (as one individual can have at most $N - 1$ treated neighbors). However, in practice, we can choose to consider $k = 0, 1, \ldots, K$ where K is the maximum degree of the actual observed network. Therefore, through an optimal assignment of treatments, we wish to satisfy

$$\forall s = 1, \ldots, r, \quad \mathbb{P}(s \in M_0^{(0)}) = \mathbb{P}(s \in M_0^{(1)}) = \cdots = \mathbb{P}(s \in M_0^{(K-1)}) = \mathbb{P}(s \in M_0^{(K)}),$$

where expressions for each $\mathbb{P}(s \in M_0^{(k)})$ as functions of θ_s for $s = 1, \ldots, r$ are obtained from Proposition 2 above. This gives Kr conditions to satisfy for r variables $\theta_s \in [0, 1]$ (for $s = 1, \ldots, r$), so we can approach this as a constrained optimization problem as considered in the binary peer-influence case before.

B   Tables of Main Results

B.1   Analytical Results

Model	Peer-Influence Estimate	Results
Binary peer-influence with unnormalized homophily: $\textbf{peer}((Z_j)_{j\in\mathcal{N}_i}) = 1\sum_{j\in\mathcal{N}_i} Z_j > 0$ $\textbf{hom}((X_j)_{j\in\mathcal{N}_i}) = \sum_{j\in\mathcal{N}_i} X_j$	$\hat{\beta}_0 = \underset{i\in M_0^{(1)}}{avg}\, Y_i - \underset{i\in M_0^{(0)}}{avg}\, Y_i$ for $M_0^{(0)} := \{i: Z_i = 0, \sum_{j\in\mathcal{N}_i} Z_j = 0\}$ and $M_0^{(1)} := \{i: Z_i = 0, \sum_{j\in\mathcal{N}_i} Z_j > 0\}$	Bias of $\hat{\beta}_0$ (conditional on **Z**): $h_0\left(\underset{i\in M_0^{(1)}}{avg}\lvert\mathcal{N}_i\rvert - \underset{i\in M_0^{(0)}}{avg}\lvert\mathcal{N}_i\rvert\right)$. Variance of $\hat{\beta}_0$ (conditional on **Z**): $h_0^2\sigma_X^2\left(\underset{i,j\in M_0^{(0)}}{avg}\lvert\mathcal{N}_i\cap\mathcal{N}_j\rvert + \underset{i,j\in M_0^{(1)}}{avg}\lvert\mathcal{N}_i\cap\mathcal{N}_j\rvert - 2\underset{\substack{i\in M_0^{(0)}\\ j\in M_0^{(1)}}}{avg}\lvert\mathcal{N}_i\cap\mathcal{N}_j\rvert\right) + \sigma_Y^2\left(\frac{1}{\lvert M_0^{(0)}\rvert} + \frac{1}{\lvert M_0^{(1)}\rvert}\right)$
Binary peer-influence with normalized homophily: $\textbf{peer}((Z_j)_{j\in\mathcal{N}_i}) = 1\sum_{j\in\mathcal{N}_i} Z_j > 0$ $\textbf{hom}((X_j)_{j\in\mathcal{N}_i}) = \sum_{j\in\mathcal{N}_i} \frac{X_j}{\lvert\mathcal{N}_i\rvert}$	$\hat{\beta}_0 = \underset{i\in M_0^{(1)}}{avg}\, Y_i - \underset{i\in M_0^{(0)}}{avg}\, Y_i$ for $M_0^{(0)} := \{i: Z_i = 0, \sum_{j\in\mathcal{N}_i} Z_j = 0\}$ and $M_0^{(1)} := \{i: Z_i = 0, \sum_{j\in\mathcal{N}_i} Z_j > 0\}$	Bias of $\hat{\beta}_0$ (conditional on **Z**): **0.** Variance of $\hat{\beta}_0$ (conditional on **Z**): $h_0^2\sigma_X^2\left(\underset{i,j\in M_0^{(0)}}{avg}\frac{\lvert\mathcal{N}_i\cap\mathcal{N}_j\rvert}{\lvert\mathcal{N}_i\rvert\lvert\mathcal{N}_j\rvert} + \underset{i,j\in M_0^{(1)}}{avg}\frac{\lvert\mathcal{N}_i\cap\mathcal{N}_j\rvert}{\lvert\mathcal{N}_i\rvert\lvert\mathcal{N}_j\rvert} - 2\underset{\substack{i\in M_0^{(0)}\\ j\in M_0^{(1)}}}{avg}\frac{\lvert\mathcal{N}_i\cap\mathcal{N}_j\rvert}{\lvert\mathcal{N}_i\rvert\lvert\mathcal{N}_j\rvert}\right) + \sigma_Y^2\left(\frac{1}{\lvert M_0^{(0)}\rvert} + \frac{1}{\lvert M_0^{(1)}\rvert}\right)$
Linear peer-influence with unnormalized homophily: $\textbf{peer}((Z_j)_{j\in\mathcal{N}_i}) = \sum_{j\in\mathcal{N}_i} Z_j$ $\textbf{hom}((X_j)_{j\in\mathcal{N}_i}) = \sum_{j\in\mathcal{N}_i} X_j$	$\hat{\beta}_0 = \dfrac{\sum_k \hat{\beta}_0^{(k)}}{\sum_k 1}$ for $\hat{\beta}_0^{(k)} := \frac{1}{k}\left(\underset{i\in M_0^{(k)}}{avg}\, Y_i - \underset{i\in M_0^{(0)}}{avg}\, Y_i\right)$, $M_0^{(k)} := \{i: Z_i = 0, \sum_{j\in\mathcal{N}_i} Z_j = k\}$.	Bias of $\hat{\beta}_0$ (conditional on **Z**): $\dfrac{h_0}{\sum_{k>0}1}\sum_{k>0}\frac{1}{k}\left(\underset{i\in M_0^{(k)}}{avg}\lvert\mathcal{N}_i\rvert - \underset{i\in M_0^{(0)}}{avg}\lvert\mathcal{N}_i\rvert\right)$ Variance of $\hat{\beta}_0$ (conditional on **Z**): $\dfrac{1}{(\sum_{k>0}1)^2}\sum_{k,l>0}\frac{1}{kl}\left[h_0^2\sigma_X^2\left(\underset{i,j\in M_0^{(k)}}{avg}\lvert\mathcal{N}_i\cap\mathcal{N}_j\rvert + \underset{i,j\in M_0^{(0)}}{avg}\lvert\mathcal{N}_i\cap\mathcal{N}_j\rvert\right.\right.$ $\left.\left. - 2\underset{\substack{i\in M_0^{(0)}\\ j\in M_0^{(k)}}}{avg}\lvert\mathcal{N}_i\cap\mathcal{N}_j\rvert\right) + \sigma_Y^2\left(\frac{1}{\lvert M_0^{(0)}\rvert} + \frac{1_{k=l}}{\lvert M_0^{(k)}\rvert}\right)\right]$
Linear peer-influence with normalized homophily: $\textbf{peer}((Z_j)_{j\in\mathcal{N}_i}) = \sum_{j\in\mathcal{N}_i} Z_j$ $\textbf{hom}((X_j)_{j\in\mathcal{N}_i}) = \sum_{j\in\mathcal{N}_i} \frac{X_j}{\lvert\mathcal{N}_i\rvert}$	$\hat{\beta}_0 = \dfrac{\sum_k \hat{\beta}_0^{(k)}}{\sum_k 1}$ for $\hat{\beta}_0^{(k)} := \frac{1}{k}\left(\underset{i\in M_0^{(k)}}{avg}\, Y_i - \underset{i\in M_0^{(0)}}{avg}\, Y_i\right)$, $M_0^{(k)} := \{i: Z_i = 0, \sum_{j\in\mathcal{N}_i} Z_j = k\}$.	Bias of $\hat{\beta}_0$ (conditional on **Z**): 0 Variance of $\hat{\beta}_0$ (conditional on **Z**): $\dfrac{1}{(\sum_{k>0}1)^2}\sum_{k,l>0}\frac{1}{kl}\left[h_0^2\sigma_X^2\left(\underset{i,j\in M_0^{(k)}}{avg}\frac{\lvert\mathcal{N}_i\cap\mathcal{N}_j\rvert}{\lvert\mathcal{N}_i\rvert\lvert\mathcal{N}_j\rvert} + \underset{i,j\in M_0^{(0)}}{avg}\frac{\lvert\mathcal{N}_i\cap\mathcal{N}_j\rvert}{\lvert\mathcal{N}_i\rvert\lvert\mathcal{N}_j\rvert}\right.\right.$ $\left.\left. - 2\underset{\substack{i\in M_0^{(0)}\\ j\in M_0^{(k)}}}{avg}\frac{\lvert\mathcal{N}_i\cap\mathcal{N}_j\rvert}{\lvert\mathcal{N}_i\rvert\lvert\mathcal{N}_j\rvert}\right) + \sigma_Y^2\left(\frac{1}{\lvert M_0^{(0)}\rvert} + \frac{1_{k=l}}{\lvert M_0^{(k)}\rvert}\right)\right]$

B.2   *Randomized Treatment Strategies to Disentangle Homophily*

Model	Peer-Influence Estimate	Randomised treatment strategy
Binary peer-influence, $$\mathbf{peer}((Z_j)_{j\in N_i}) = \mathbf{1}_{\sum_{j\in N_i} Z_j > 0},$$ with an *unknown* homophily function $\mathbf{hom}((X_j)_{j\in N_i})$.	$$\hat{\beta}_0 = \underset{i\in M_0^{(1)}}{avg} Y_i - \underset{i\in M_0^{(0)}}{avg} Y_i$$ for $M_0^{(0)} := \{i: Z_i = 0, \sum_{j\in N_i} Z_j = 0\}$ (the set of untreated individuals with no treated neighbours) and $M_0^{(1)} := \{i: Z_i = 0, \sum_{j\in N_i} Z_j > 0\}$ (the set of untreated individuals with at least one treated neighbour).	For graph G of N vertices which is clustered into r clusters, consider a corresponding Stochastic Block Model of N individuals in r communities $B_1,...,B_r$. We assign treatments independently such that individuals in B_s are treated with probability θ_s for $s = 1,...r$. We choose θ_s such that homophily does not interfere with the estimation of binary peer-influence. Let $\{s \in M_0^{(0)}\}$, $\{s \in M_0^{(1)}\}$ denote the event that a fixed vertex in community s is in the set $M_0^{(0)}$, $M_0^{(1)}$ respectively. In our optimal assignment of treatments we wish to satisfy $\mathbb{P}(s \in M_0^{(0)}) = \mathbb{P}(s \in M_0^{(1)})$, where $\mathbb{P}(s \in M_0^{(0)}) = (1 - \theta_s)\prod_{v=1}^r (1 - P_{s,v}\theta_v)^{A_v - 1_{v=s}}$.. This can be approximated by solving the contrained optimisation problem of minimising: $$\left\| \begin{pmatrix} P_{1,1}(A_1 - 1) & P_{1,2}A_2 & \cdots & & P_{1,r}A_r \\ P_{2,1}A_1 & P_{2,2}(A_2 - 1) & \cdots & & P_{2,r}A_r \\ \vdots & \vdots & & & \vdots \\ P_{r-1,1}A_1 & P_{r-1,2}A_2 & \cdots & P_{r-1,r-1}(A_{r-1} - 1) & P_{r-1,r}A_r \\ P_{r,1}A_1 & P_{r,2}A_2 & \cdots & P_{r,r-1}A_{r-1} & P_{r,r}(A_r - 1) \end{pmatrix} \begin{pmatrix} \theta_1 \\ \theta_2 \\ \vdots \\ \theta_{r-1} \\ \theta_r \end{pmatrix} - \log(2) \begin{pmatrix} 1 \\ 1 \\ \vdots \\ 1 \\ 1 \end{pmatrix} \right\|$$ for $\theta \in [0,1]^r$ and some chosen norm $\|\ \|$ on $\mathbb{R}^d$ (e.g. L^2).
Linear peer-influence, $$\mathbf{peer}((Z_j)_{j\in N_i}) = \sum_{j\in N_i} Z_j,$$ with an *unknown* homophily function $\mathbf{hom}((X_j)_{j\in N_i})$.	$$\hat{\beta}_0 = \frac{\sum_k \hat{\beta}_0^{(k)}}{\sum_k 1}$$ for $$\hat{\beta}_0^{(k)} := \frac{1}{k}\left(\underset{i\in M_0^{(k)}}{avg} Y_i - \underset{i\in M_0^{(0)}}{avg} Y_i \right),$$ $M_0^{(k)} := \{i: Z_i = 0, \sum_{j\in N_i} Z_j = k\}$ (the set of untreated individuals which have k treated neighbours).	For graph G of N vertices which is clustered into r clusters, consider a corresponding Stochastic Block Model of N individuals in r communities $B_1,...,B_r$. We assign treatments independently such that individuals in B_s are treated with probability θ_s for $s = 1,...r$. We choose θ_s such that homophily does not interfere with the estimation of binary peer-influence. Let $\{s \in M_0^{(k)}\}$ denote the event that a fixed vertex in community s is in the set $M_0^{(k)}$. In our optimal assignment of treatments we wish to satisfy $$\forall s = 1,...r, \quad \mathbb{P}(s \in M_0^{(1)}) = \mathbb{P}(s \in M_0^{(2)}) = ... = \mathbb{P}(s \in M_0^{(K-1)}) = \mathbb{P}(s \in M_0^{(K)}),$$ where $$\mathbb{P}(s \in M_0^{(k)}) = (1 - \theta_s) \sum_{\substack{t_1,...,t_r \\ \forall v=1,...r,\ 0 \le t_v \le A_v - 1_{v=s} \\ t_1 + ... + t_r = k}} \left(\prod_{v=1}^r \mathbf{Bin}(t_v; A_v - 1_{v=s}, \theta_v P_{s,v}) \right)$$ for $\mathbf{Bin}(t_v; A_v - 1_{v=s}, \theta_v P_{s,v}) = \binom{A_v - 1_{v=s}}{t_v}(\theta_v P_{s,v})^{t_v}(1 - \theta_v P_{s,v})^{A_v - 1_{v=s} - t_v}$. We approach this as a constrained optimisation problem as in the binary peer-influence case.

C   Proofs

C.1   Proof of Theorem 1 (See p. xxx)

Theorem 1 *Consider the difference in means estimator $\hat{\beta}_0$ for binary peer-influence effect β_0. Under the presence of unnormalized homophily in our model (3), the mean squared error of $\hat{\beta}_0$ (conditional on the treatment $\mathbf{Z}$) is:*

$$\mathbb{E}[(\hat{\beta}_0 - \beta_0)^2 | \mathbf{Z}] = \left(h_0 \left(\underset{i \in M_0^{(1)}}{avg} |\mathcal{N}_i| - \underset{i \in M_0^{(0)}}{avg} |\mathcal{N}_i| \right) \right)^2$$

$$+ h_0^2 \sigma_X^2 \left(\underset{i,j \in M_0^{(0)}}{avg} |\mathcal{N}_i \cap \mathcal{N}_j| + \underset{i,j \in M_0^{(1)}}{avg} |\mathcal{N}_i \cap \mathcal{N}_j| - 2 \underset{i \in M_0^{(0)}, j \in M_0^{(1)}}{avg} |\mathcal{N}_i \cap \mathcal{N}_j| \right) + \sigma_Y^2 \left(\frac{1}{|M_0^{(0)}|} + \frac{1}{|M_0^{(1)}|} \right)$$

$$(5)$$

Proof Recall the definition of the difference in means estimator for binary peer-influence (4).

$$\hat{\beta}_0 = \underset{i \in M_0^{(1)}}{avg} Y_i - \underset{i \in M_0^{(0)}}{avg} Y_i$$

where $M_0^{(0)} := \{i : Z_i = 0, \sum_{j \in \mathcal{N}_i} Z_j = 0\}$ (the set of untreated individuals with no treated neighbors) and $M_0^{(1)} := \{i : Z_i = 0, \sum_{j \in \mathcal{N}_i} Z_j > 0\}$ (the set of untreated individuals with at least one treated neighbors). The response variables $(Y_i)_{i=1,\ldots,N}$ are defined by:

$$Y_i(Z_i = 0, (Z_j)_{j \in \mathcal{N}_i}) = \alpha + \beta_0 \mathbf{1}_{\sum_{j \in \mathcal{N}_i} Z_j > 0} + h_0 \sum_{j \in \mathcal{N}_i} X_j + \epsilon_i(0, \sigma_Y^2)$$

$$Y_i(Z_i = 1, (Z_j)_{j \in \mathcal{N}_i}, (X_j)_{j \in \mathcal{N}_i}) = \tau + Y_i(Z_i = 0, (Z_j)_{j \in \mathcal{N}_i}, (X_j)_{j \in \mathcal{N}_i}) + \beta_1 \mathbf{1}_{\sum_{j \in \mathcal{N}_i} Z_j > 0} + h_1 \sum_{j \in \mathcal{N}_i} X_j$$

$\epsilon_i(0, \sigma_Y^2)$ for $i = 1, \ldots, N$ are the noise terms in the network, indepedent and identically distributed with zero mean and variance σ_Y^2. Note that the sets $M_0^{(0)}$ and $M_0^{(1)}$ are $\mathbf{Z}$ measurable and that latent homophily variables $\mathbf{X} = (X_j)_{j=1,\ldots,N}$ are independent of $\mathbf{Z} = (Z_j)_{j=1,\ldots,N}$. Therefore,

$$\mathbb{E}[\hat{\beta}_0 | \mathbf{Z}] = \frac{\sum_{i \in M_0^{(1)}} \mathbb{E}[Y_i | \mathbf{Z}]}{|M_0^{(1)}|} - \frac{\sum_{i \in M_0^{(0)}} \mathbb{E}[Y_i | \mathbf{Z}]}{|M_0^{(0)}|}$$

$$= \frac{\sum_{i \in M_0^{(1)}} \mathbb{E}[\beta_0 + \sum_{j \in \mathcal{N}_i} X_j]}{|M_0^{(1)}|} - \frac{\sum_{i \in M_0^{(0)}} \mathbb{E}[\sum_{j \in \mathcal{N}_i} X_j]}{|M_0^{(0)}|}$$

$$= \beta_0 + \frac{\sum_{i \in M_0^{(1)}} h_0 |\mathcal{N}_i|}{|M_0^{(1)}|} - \frac{\sum_{i \in M_0^{(0)}} h_0 |\mathcal{N}_i|}{|M_0^{(0)}|}$$

$$= \beta_0 + h_0 \left(\underset{i \in M_0^{(1)}}{avg} |\mathcal{N}_i| - \underset{i \in M_0^{(1)}}{avg} |\mathcal{N}_i| \right).$$

This gives the bias of $\hat{\beta}_0$: $\mathbb{E}\left[\hat{\beta}_0 - \beta_0 | \mathbf{Z} \right] = h_0 \left(\underset{i \in M_0^{(1)}}{avg} |\mathcal{N}_i| - \underset{i \in M_0^{(1)}}{avg} |\mathcal{N}_i| \right)$. Similarly,

$$var[\hat{\beta}_0 | \mathbf{Z}] = var\left(\frac{\sum_{i \in M_0^{(1)}} Y_i}{|M_0^{(1)}|} - \frac{\sum_{j \in M_0^{(0)}} Y_j}{|M_0^{(0)}|} \,\bigg|\, \mathbf{Z} \right)$$

$$= \frac{var(\sum_{i \in M_0^{(1)}} Y_i \mid \mathbf{Z})}{|M_0^{(1)}|^2} + \frac{var(\sum_{j \in M_0^{(0)}} Y_j \mid \mathbf{Z})}{|M_0^{(0)}|^2} - \frac{2cov(\sum_{i \in M_0^{(1)}} Y_i, \sum_{j \in M_0^{(0)}} Y_j \mid \mathbf{Z})}{|M_0^{(0)}||M_0^{(1)}|}$$

$$= \frac{\sum_{i \in M_0^{(1)}} \sum_{k \in M_0^{(1)}} cov(Y_i, Y_k \mid \mathbf{Z})}{|M_0^{(1)}|^2} + \frac{\sum_{j \in M_0^{(0)}} \sum_{l \in M_0^{(0)}} cov(Y_j, Y_l \mid \mathbf{Z})}{|M_0^{(0)}|^2}$$

$$- \frac{2 \sum_{i \in M_0^{(1)}} \sum_{j \in M_0^{(0)}} cov(Y_i, Y_j \mid \mathbf{Z})}{|M_0^{(0)}||M_0^{(1)}|}$$

$$= \underset{i,k \in M_0^{(1)}}{avg} cov(Y_i, Y_k \mid \mathbf{Z}) + \underset{j,l \in M_0^{(0)}}{avg} cov(Y_i, Y_k \mid \mathbf{Z}) - 2 \underset{i \in M_0^{(0)}, j \in M_0^{(1)}}{avg} cov(Y_i, Y_j \mid \mathbf{Z}).$$

For $i \in M_0^{(1)}$ and $k \in M_0^{(1)}$, by the law of total covariance and as $\mathbf{X}$ are i.i.d.,

$$cov(Y_i, Y_k \mid \mathbf{Z}) = \mathbb{E}[cov(Y_i, Y_k \mid \mathbf{X}, \mathbf{Z}) \mid \mathbf{Z}] + cov\left(\mathbb{E}[Y_i | \mathbf{X}, \mathbf{Z}], \mathbb{E}[Y_k | \mathbf{X}, \mathbf{Z}] \mid \mathbf{Z} \right)$$

$$= \sigma_Y^2 \mathbb{1}_{\{i=k\}} + cov\left(\alpha + \beta_0 + h_0 \sum_{a \in \mathcal{N}_i} X_a, \alpha + \beta_0 + h_0 \sum_{b \in \mathcal{N}_k} X_b \right)$$

$$= \sigma_Y^2 \mathbb{1}_{\{i=k\}} + h_0^2 cov\left(\sum_{a \in \mathcal{N}_i} X_a, \sum_{b \in \mathcal{N}_k} X_b \right)$$

$$= \sigma_Y^2 \mathbb{1}_{\{i=k\}} + h_0^2 \sigma_X^2 |\mathcal{N}_i \cap \mathcal{N}_j|$$

Similarly for $j \in M_0^{(0)}$ and $l \in M_0^{(0)}$,

$$cov(Y_j, Y_l \mid \mathbf{Z}) = \sigma_Y^2 \mathbb{1}_{\{j=l\}} + cov\left(\alpha + h_0 \sum_{a \in \mathcal{N}_j} X_a, \alpha + h_0 \sum_{b \in \mathcal{N}_l} X_b \,\bigg|\, \mathbf{Z} \right)$$

$$= \sigma_Y^2 \mathbb{1}_{\{j=l\}} + h_0^2 \sigma_X^2 |\mathcal{N}_j \cap \mathcal{N}_l|$$

For $i \in M_0^{(1)}$ and $j \in M_0^{(0)}$, by the law of total covariance and as $\mathbf{X}$ are i.i.d.,

$$cov(Y_i, Y_j \mid \mathbf{Z}) = \mathbb{E}[cov(Y_i, Y_j \mid \mathbf{X}, \mathbf{Z}) \mid \mathbf{Z}] + cov\left(\mathbb{E}[Y_i \mid \mathbf{X}, \mathbf{Z}], \mathbb{E}[Y_k \mid \mathbf{X}, \mathbf{Z}] \mid \mathbf{Z}\right)$$

$$= 0 + cov\left(\alpha + \beta_0 + h_0 \sum_{a \in \mathcal{N}_i} X_a, \alpha + h_0 \sum_{b \in \mathcal{N}_k} X_b\right)$$

$$= h_0^2\, cov\left(\sum_{a \in \mathcal{N}_i} X_a, \sum_{b \in \mathcal{N}_k} X_b\right)$$

$$= h_0^2 \sigma_X^2 |\mathcal{N}_i \cap \mathcal{N}_j|.$$

Therefore,

$$var[\hat{\beta}_0 \mid \mathbf{Z}] = \operatorname*{avg}_{i,k \in M_0^{(1)}} cov(Y_i, Y_k \mid \mathbf{Z}) + \operatorname*{avg}_{j,l \in M_0^{(0)}} cov(Y_i, Y_k \mid \mathbf{Z}) - 2 \operatorname*{avg}_{i \in M_0^{(0)}, j \in M_0^{(1)}} cov(Y_i, Y_j \mid \mathbf{Z})$$

$$= \operatorname*{avg}_{i,k \in M_0^{(1)}}\left(\sigma_Y^2 \mathbb{1}_{\{i=k\}} + h_0^2 \sigma_X^2 |\mathcal{N}_i \cap \mathcal{N}_k|\right) + \operatorname*{avg}_{j,l \in M_0^{(0)}}\left(\sigma_Y^2 \mathbb{1}_{\{j=l\}} + h_0^2 \sigma_X^2 |\mathcal{N}_j \cap \mathcal{N}_l|\right)$$

$$- 2 \operatorname*{avg}_{i \in M_0^{(0)}, j \in M_0^{(1)}}\left(h_0^2 \sigma_X^2 |\mathcal{N}_j \cap \mathcal{N}_l|\right)$$

$$= h_0^2 \sigma_X^2 \left(\operatorname*{avg}_{i,j \in M_0^{(0)}} |\mathcal{N}_i \cap \mathcal{N}_j| + \operatorname*{avg}_{i,j \in M_0^{(1)}} |\mathcal{N}_i \cap \mathcal{N}_j| - 2 \operatorname*{avg}_{i \in M_0^{(0)}, j \in M_0^{(1)}} |\mathcal{N}_i \cap \mathcal{N}_j|\right)$$

$$+ \sigma_Y^2\left(\frac{1}{|M_0^{(0)}|} + \frac{1}{|M_0^{(1)}|}\right)$$

Now we can recall the bias–variance decomposition of the MSE to obtain

$$\mathbb{E}[(\hat{\beta}_0 - \beta_0)^2 \mid \mathbf{Z}] = \left(\mathbb{E}[\hat{\beta}_0 - \beta_0 \mid \mathbf{Z}]\right)^2 + var[\hat{\beta}_0 \mid \mathbf{Z}]$$

$$= \left(h_0\left(\operatorname*{avg}_{i \in M_0^{(1)}} |\mathcal{N}_i| - \operatorname*{avg}_{i \in M_0^{(0)}} |\mathcal{N}_i|\right)\right)^2$$

$$+ h_0^2 \sigma_X^2 \left(\operatorname*{avg}_{i,j \in M_0^{(0)}} |\mathcal{N}_i \cap \mathcal{N}_j| + \operatorname*{avg}_{i,j \in M_0^{(1)}} |\mathcal{N}_i \cap \mathcal{N}_j| - 2 \operatorname*{avg}_{i \in M_0^{(0)}, j \in M_0^{(1)}} |\mathcal{N}_i \cap \mathcal{N}_j|\right)$$

$$+ \sigma_Y^2\left(\frac{1}{|M_0^{(0)}|} + \frac{1}{|M_0^{(1)}|}\right)$$

as required. $\qquad\qquad\square$

C.2 Proof of Theorem 3 (See p. xxx)

Theorem 3 *Consider the difference in means estimator $\hat{\beta}_0$ for binary peer-influence effect β_0. Under the presence of unnormalized homophily in our model (3), the mean* *squared error of $\hat{\beta}_0$ (conditional on the treatment $\mathbf{Z}$) is:*

$$\mathbb{E}[(\hat{\beta}_0 - \beta_0)^2|\mathbf{Z}] = \left(h_0\left(\underset{i\in M_0^{(1)}}{avg}|\mathcal{N}_i| - \underset{i\in M_0^{(0)}}{avg}|\mathcal{N}_i|\right)\right)^2$$

$$+ h_0^2\sigma_X^2\left(\underset{i,j\in M_0^{(0)}}{avg}|\mathcal{N}_i\cap\mathcal{N}_j| + \underset{i,j\in M_0^{(1)}}{avg}|\mathcal{N}_i\cap\mathcal{N}_j| - 2\underset{i\in M_0^{(0)},j\in M_0^{(1)}}{avg}|\mathcal{N}_i\cap\mathcal{N}_j|\right)$$

$$+ \sigma_Y^2\left(\frac{1}{|M_0^{(0)}|} + \frac{1}{|M_0^{(1)}|}\right) \tag{16}$$

Proof We proceed as similar to the binary peer-influence estimator case. Recall the definition of the estimator for linear peer-influence (21):

$$\hat{\beta}_0 = \frac{\sum_k \hat{\beta}_0^{(k)}}{\sum_k 1} \ for \ \hat{\beta}_0^{(k)} = \frac{1}{k}\left(\frac{\sum_{i\in M_0^{(k)}}Y_i}{|M_0^{(k)}|} - \frac{\sum_{i\in M_0^{(0)}}Y_i}{|M_0^{(0)}|}\right) = \frac{1}{k}\left(\underset{i\in M_0^{(k)}}{avg}Y_i - \underset{i\in M_0^{(0)}}{avg}Y_i\right), \tag{21}$$

where $M_0^{(k)} := \{i : Z_i = 0, \sum_{j\in\mathcal{N}_i} Z_j = k\}$ (the set of untreated individuals with k treated neighbors). The response variables $(Y_i)_{i=1,\dots,N}$ are defined by:

$$Y_i(Z_i = 0, (Z_j)_{j\in\mathcal{N}_i}) = \alpha + \beta_0 \sum_{j\in\mathcal{N}_i} Z_j + h_0 \sum_{j\in\mathcal{N}_i} X_j + \epsilon_i(0, \sigma_Y^2)$$

$$Y_i(Z_i = 1, (Z_j)_{j\in\mathcal{N}_i}, (X_j)_{j\in\mathcal{N}_i}) = \tau + Y_i(Z_i = 0, (Z_j)_{j\in\mathcal{N}_i}, (X_j)_{j\in\mathcal{N}_i}) + \beta_1 \sum_{j\in\mathcal{N}_i} Z_j + h_1 \sum_{j\in\mathcal{N}_i} X_j$$

$\epsilon_i(0, \sigma_Y^2)$ for $i = 1, \dots, N$ are the noise terms in the network, indepedent and identically distributed with zero mean and variance σ_Y^2. Note that sets $M_0^{(k)}$ are $\mathbf{Z}$ measurable and that latent homophily variables $\mathbf{X} = (X_j)_{j=1,\dots,N}$ are independent of $\mathbf{Z} = (Z_j)_{j=1,\dots,N}$. Therefore,

$$\mathbb{E}[\hat{\beta}_0^{(k)}|\mathbf{Z}] = \frac{1}{k}\left(\frac{\sum_{i\in M_0^{(k)}}\mathbb{E}[Y_i|\mathbf{Z}]}{|M_0^{(k)}|} - \frac{\sum_{i\in M_0^{(0)}}\mathbb{E}[Y_i|\mathbf{Z}]}{|M_0^{(0)}|}\right)$$

$$= \frac{1}{k}\left(\frac{\sum_{i\in M_0^{(k)}}\mathbb{E}[k\beta_0 + \sum_{j\in\mathcal{N}_i}X_j|\mathbf{Z}]}{|M_0^{(k)}|} - \frac{\sum_{i\in M_0^{(0)}}\mathbb{E}[\sum_{j\in\mathcal{N}_i}X_j|\mathbf{Z}]}{|M_0^{(0)}|}\right)$$

$$= \beta_0 + \frac{1}{k}\left(\frac{\sum_{i\in M_0^{(k)}}\mathbb{E}[\sum_{j\in\mathcal{N}_i}X_j|\mathbf{Z}]}{|M_0^{(k)}|} - \frac{\sum_{i\in M_0^{(0)}}\mathbb{E}[\sum_{j\in\mathcal{N}_i}X_j|\mathbf{Z}]}{|M_0^{(0)}|}\right)$$

$$= \beta_0 + \frac{1}{k}\left(\frac{\sum_{i\in M_0^{(k)}}h_0|\mathcal{N}_i|}{|M_0^{(k)}|} - \frac{\sum_{i\in M_0^{(0)}}h_0|\mathcal{N}_i|}{|M_0^{(0)}|}\right)$$

$$= \beta_0 + \frac{h_0}{k} \left(\underset{i \in M_0^{(k)}}{avg} |\mathcal{N}_i| - \underset{i \in M_0^{(1)}}{avg} |\mathcal{N}_i| \right),$$

which gives the bias of the estimator $\hat{\beta}_0 = \frac{\sum_k \hat{\beta}_0^{(k)}}{\sum_k 1}$ to be:

$$\mathbb{E}[\hat{\beta}_0 - \beta_0 | \mathbf{Z}] = \frac{h_0}{\sum_{k>0} 1} \sum_{k>0} \frac{1}{k} \left(\underset{i \in M_0^{(k)}}{avg} |\mathcal{N}_i| - \underset{i \in M_0^{(0)}}{avg} |\mathcal{N}_i| \right).$$

Similarly, $var[\hat{\beta}_0 | \mathbf{Z}] = \frac{1}{(\sum_k 1)^2} \sum_{k>0} \sum_{l>0} cov(\hat{\beta}_0^{(k)}, \hat{\beta}_0^{(l)})$, where

$$cov(\hat{\beta}_0^{(k)}, \hat{\beta}_0^{(l)} | \mathbf{Z}) = \frac{1}{kl} cov \left(\frac{\sum_{i \in M_0^{(k)}} Y_i}{|M_0^{(k)}|} - \frac{\sum_{j \in M_0^{(0)}} Y_j}{|M_0^{(0)}|}, \frac{\sum_{i \in M_0^{(l)}} Y_i}{|M_0^{(l)}|} - \frac{\sum_{j \in M_0^{(0)}} Y_j}{|M_0^{(0)}|} \Big| \mathbf{Z} \right)$$

$$= \frac{1}{kl} \left(\frac{\sum_{i \in M_0^{(k)}, j \in M_0^{(l)}} cov(Y_i, Y_j | \mathbf{Z})}{|M_0^{(k)}| |M_0^{(l)}|} + \frac{\sum_{i \in M_0^{(0)}, j \in M_0^{(0)}} cov(Y_i, Y_j | \mathbf{Z})}{|M_0^{(0)}|^2} \right.$$

$$\left. - \frac{\sum_{i \in M_0^{(k)}, j \in M_0^{(0)}} cov(Y_i, Y_j | \mathbf{Z})}{|M_0^{(k)}| |M_0^{(0)}|} - \frac{\sum_{i \in M_0^{(0)}, j \in M_0^{(l)}} cov(Y_i, Y_j | \mathbf{Z})}{|M_0^{(0)}| |M_0^{(l)}|} \right).$$

For $i \in M_0^{(k)}$ and $j \in M_0^{(l)}$, by the law of total covariance and as $\mathbf{X}$ are i.i.d.,

$$cov(Y_i, Y_j | \mathbf{Z}) = \mathbb{E}[cov(Y_i, Y_j | \mathbf{X}, \mathbf{Z}) | \mathbf{Z}] + cov \left(\mathbb{E}[Y_i | \mathbf{X}, \mathbf{Z}], \mathbb{E}[Y_j | \mathbf{X}, \mathbf{Z}] \Big| \mathbf{Z} \right)$$

$$= \sigma_Y^2 \mathbb{1}_{\{i=j\}} + cov \left(\alpha + k\beta_0 + h_0 \sum_{a \in \mathcal{N}_i} X_a, \alpha + l\beta_0 + h_0 \sum_{b \in \mathcal{N}_j} X_b \right)$$

$$= \sigma_Y^2 \mathbb{1}_{\{i=j\}} + h_0^2 cov \left(\sum_{a \in \mathcal{N}_i} X_a, \sum_{b \in \mathcal{N}_j} X_b \right)$$

$$= \sigma_Y^2 \mathbb{1}_{\{i=j\}} + h_0^2 \sigma_X^2 |\mathcal{N}_i \cap \mathcal{N}_j|.$$

This gives

$$cov(\hat{\beta}_0^{(k)}, \hat{\beta}_0^{(l)} | \mathbf{Z}) = \frac{1}{kl} \left(\frac{\sum_{i \in M_0^{(k)}, j \in M_0^{(l)}} cov(Y_i, Y_j | \mathbf{Z})}{|M_0^{(k)}| |M_0^{(l)}|} + \frac{\sum_{i \in M_0^{(0)}, j \in M_0^{(0)}} cov(Y_i, Y_j | \mathbf{Z})}{|M_0^{(0)}|^2} \right.$$

$$\left. - \frac{\sum_{i \in M_0^{(k)}, j \in M_0^{(0)}} cov(Y_i, Y_j | \mathbf{Z})}{|M_0^{(k)}| |M_0^{(0)}|} - \frac{\sum_{i \in M_0^{(0)}, j \in M_0^{(l)}} cov(Y_i, Y_j | \mathbf{Z})}{|M_0^{(0)}| |M_0^{(l)}|} \right)$$

$$= \frac{1}{kl} \left(\frac{\sum_{i \in M_0^{(k)}, j \in M_0^{(l)}} \sigma_Y^2 \mathbb{1}_{\{i=j\}} + h_0^2 \sigma_X^2 |\mathcal{N}_i \cap \mathcal{N}_j|}{|M_0^{(k)}| |M_0^{(l)}|} \right.$$

$$+ \frac{\sum_{i \in M_0^{(0)}, j \in M_0^{(0)}} \sigma_Y^2 \mathbb{1}_{\{i=j\}} + h_0^2 \sigma_X^2 |\mathcal{N}_i \cap \mathcal{N}_j|}{|M_0^{(0)}|^2}$$

$$\left. - \frac{\sum_{i \in M_0^{(k)}, j \in M_0^{(0)}} h_0^2 \sigma_X^2 |\mathcal{N}_i \cap \mathcal{N}_j|}{|M_0^{(k)}| |M_0^{(0)}|} - \frac{\sum_{i \in M_0^{(0)}, j \in M_0^{(l)}} h_0^2 \sigma_X^2 |\mathcal{N}_i \cap \mathcal{N}_j|}{|M_0^{(0)}| |M_0^{(l)}|} \right)$$

$$
= \frac{1}{kl}\left(\sigma_Y^2 \frac{\mathbf{1}_{\{l=k\}}}{|M^{(k)}|} + h_0^2\sigma_X^2 \frac{\sum_{i\in M_0^{(k)}, j\in M_0^{(l)}} |\mathcal{N}_i \cap \mathcal{N}_j|}{|M_0^{(k)}||M_0^{(l)}|} \right.
$$

$$
+ \sigma_Y^2 \frac{1}{|M^{(0)}|} + h_0^2\sigma_X^2 \frac{\sum_{i\in M_0^{(0)}, j\in M_0^{(0)}} |\mathcal{N}_i \cap \mathcal{N}_j|}{|M_0^{(0)}||M_0^{(0)}|}
$$

$$
\left. - h_0^2\sigma_X^2 \frac{\sum_{i\in M_0^{(k)}, j\in M_0^{(0)}} |\mathcal{N}_i \cap \mathcal{N}_j|}{|M_0^{(k)}||M_0^{(0)}|} - h_0^2\sigma_X^2 \frac{\sum_{i\in M_0^{(0)}, j\in M_0^{(l)}} |\mathcal{N}_i \cap \mathcal{N}_j|}{|M_0^{(0)}||M_0^{(l)}|} \right)
$$

$$
= \frac{1}{kl}\left[h_0^2\sigma_X^2 \left(\operatorname*{avg}_{i\in M_0^{(k)}, j\in M_0^{(l)}} |\mathcal{N}_i \cap \mathcal{N}_j| + \operatorname*{avg}_{i,j\in M_0^{(0)}} |\mathcal{N}_i \cap \mathcal{N}_j| - \operatorname*{avg}_{i\in M_0^{(k)}, j\in M_0^{(0)}} |\mathcal{N}_i \cap \mathcal{N}_j| \right.\right.
$$

$$
\left.\left. - \operatorname*{avg}_{i\in M_0^{(0)}, j\in M_0^{(l)}} |\mathcal{N}_i \cap \mathcal{N}_j| \right) + \sigma_Y^2 \left(\frac{1}{|M_0^{(0)}|} + \frac{\mathbf{1}_{k=l}}{|M_0^{(k)}|} \right) \right].
$$

Now, we can recall the bias–variance decomposition of the MSE to obtain

$$
\mathbb{E}[(\hat{\beta}_0 - \beta_0)^2 | \mathbf{Z}] = \left(\mathbb{E}[\hat{\beta}_0 - \beta_0 | \mathbf{Z}] \right)^2 + var[\hat{\beta}_0 | \mathbf{Z}]
$$

$$
= \left(\mathbb{E}[\hat{\beta}_0 - \beta_0 | \mathbf{Z}] \right)^2 + \frac{1}{(\sum_k 1)^2} \sum_{k>0}\sum_{l>0} cov(\hat{\beta}_0^{(k)}, \hat{\beta}_0^{(l)})
$$

$$
= \left(\frac{h_0}{\sum_{k>0} 1} \sum_{k>0} \frac{1}{k} \left(\operatorname*{avg}_{i\in M_0^{(k)}} |\mathcal{N}_i| - \operatorname*{avg}_{i\in M_0^{(0)}} |\mathcal{N}_i| \right) \right)^2
$$

$$
+ \frac{1}{(\sum_{k>0} 1)^2} \sum_{k,l>0} \frac{1}{kl} \left[h_0^2\sigma_X^2 \left(\operatorname*{avg}_{i\in M_0^{(k)}, j\in M_0^{(l)}} |\mathcal{N}_i \cap \mathcal{N}_j| + \operatorname*{avg}_{i,j\in M_0^{(0)}} |\mathcal{N}_i \cap \mathcal{N}_j| \right.\right.
$$

$$
\left.\left. - \operatorname*{avg}_{i\in M_0^{(0)}, j\in M_0^{(k)}} |\mathcal{N}_i \cap \mathcal{N}_j| \right) - \operatorname*{avg}_{i\in M_0^{(l)}, j\in M_0^{(0)}} |\mathcal{N}_i \cap \mathcal{N}_j| + \sigma_Y^2 \left(\frac{1}{|M_0^{(0)}|} + \frac{\mathbf{1}_{k=l}}{|M_0^{(k)}|} \right) \right]
$$

$$
= \left(\frac{h_0}{\sum_{k>0} 1} \sum_{k>0} \frac{1}{k} \left(\operatorname*{avg}_{i\in M_0^{(k)}} |\mathcal{N}_i| - \operatorname*{avg}_{i\in M_0^{(0)}} |\mathcal{N}_i| \right) \right)^2
$$

$$
+ \frac{1}{(\sum_{k>0} 1)^2} \sum_{k,l>0} \frac{1}{kl} \left[h_0^2\sigma_X^2 \left(\operatorname*{avg}_{i\in M_0^{(k)}, j\in M_0^{(l)}} |\mathcal{N}_i \cap \mathcal{N}_j| + \operatorname*{avg}_{i,j\in M_0^{(0)}} |\mathcal{N}_i \cap \mathcal{N}_j| \right.\right.
$$

$$
\left.\left. - 2 \operatorname*{avg}_{i\in M_0^{(0)}, j\in M_0^{(k)}} |\mathcal{N}_i \cap \mathcal{N}_j| \right) + \sigma_Y^2 \left(\frac{1}{|M_0^{(0)}|} + \frac{\mathbf{1}_{k=l}}{|M_0^{(k)}|} \right) \right]
$$

as required. $\square$

C.3 Proof of Theorem 1 (See p. xxx)

Proposition 1 *Consider a stochastic block model (SBM) of N individuals in r communities. Denote the communities of the SBM by the sets $B_1, \dots, B_r$, which are of respective sizes $A_1, \dots, A_r$ (where $A_1 + \dots + A_r = N$). Let P be the $r \times r$ adjacency probability matrix between the r communities. We assign treatments independently to individuals such that individuals in B_s are treated with probability θ_s for $s = 1, \dots, r$. Under such setup, let $M_0^{(0)}$ denote the set of untreated individuals* which have no treated neighbors and let $M_0^{(1)}$ denote the set of untreated individuals which have at least one treated neighbor. For ease of notation, let $\{s \in M_0^{(0)}\}$, $\{s \in M_0^{(1)}\}$ denote the event that a fixed vertex in community s is in the sets $M_0^{(0)}$, $M_0^{(1)}$ respectively. Then,

$$\mathbb{P}(s \in M_0^{(0)}) = (1 - \theta_s) \prod_{v=1}^{r} (1 - P_{s,v}\theta_v)^{A_v - 1_{v=s}}, \text{ and} \tag{7}$$

$$\mathbb{P}(s \in M_0^{(1)}) = (1 - \theta_s)\left(1 - \prod_{v=1}^{r}(1 - P_{s,v}\theta_v)^{A_v - 1_{v=s}}\right). \tag{8}$$

Proof Note that each vertex in the graph is assigned treatment independently and that under the stochastic block model the events of any pair of vertices being adjacent are independent. Therefore,

$$\mathbb{P}(s \text{ has } 0 \text{ treated neighbors} \mid s \text{ is untreated}) = \mathbb{P}(s \text{ has } 0 \text{ treated neighbors})$$

for all k and $s = 1, \ldots, r$. This gives

$$\begin{aligned}
\mathbb{P}(s \in M_0^{(0)}) &= \mathbb{P}(s \text{ is untreated})\mathbb{P}(s \text{ has } 0 \text{ treated neighbors}) \\
&= (1 - \theta_s)\mathbb{P}(s \text{ has } 0 \text{ treated neighbors}) \\
&= (1 - \theta_s)\mathbb{P}\left(\bigcap_{v=1}^{r}\{s \text{ has } 0 \text{ treated neighbors in } B_v\}\right) \\
&= (1 - \theta_s)\prod_{v=1}^{r}\mathbb{P}(s \text{ has } 0 \text{ treated neighbors in } B_v) \\
&= (1 - \theta_s)\prod_{v=1}^{r}(1 - P_{s,v}\theta_v)^{A_v - 1_{v=s}}.
\end{aligned}$$

where the $A_v - 1_{v=s}$ arises from noting that s can have at most $A_s - 1$ neighbors in B_s (it cannot connect to itself). Note that sets $M_0^{(0)}$ and $M_0^{(1)}$ partition the set of untreated individuals. Therefore,

$$\begin{aligned}
\mathbb{P}(s \in M_0^{(1)}) &= \mathbb{P}(Z_s = 0) - \mathbb{P}(s \in M_0^{(0)}) \\
&= (1 - \theta_s) - (1 - \theta_s)\prod_{v=1}^{r}(1 - P_{s,v}\theta_v)^{A_v - 1_{v=s}} \\
&= (1 - \theta_s)\left(1 - \prod_{v=1}^{r}(1 - P_{s,v}\theta_v)^{A_v - 1_{v=s}}\right).
\end{aligned}$$

□

C.4 Proof of Proposition 2 (See p. xxx)

Proposition 2 *Consider a stochastic block model (SBM) of N individuals in r communities. Denote the communities of the SBM by the sets $B_1, \ldots, B_r$, which are of respective sizes $A_1, \ldots, A_r$ (where $A_1 + \cdots + A_r = N$). Let $\boldsymbol{P}$ be the $r \times r$ adjacency probability matrix between the r communities. We assign treatments independently to individuals such that individuals in B_s are treated with probability θ_s for $s = 1, \ldots, r$. Under such setup, let $M_0^{(k)}$ denote the set of untreated individuals which have k treated neighbors. For ease of notation, let $\{s \in M_0^{(k)}\}$ denote the event that a fixed vertex in community s is in the set $M_0^{(k)}$. Then,*

$$\mathbb{P}(s \in M_0^{(k)}) = (1 - \theta_s) \sum_{\substack{t_1, \ldots, t_r: \\ \forall v=1, \ldots, r \ 0 \le t_v \le A_v - \mathbf{1}_{\{v=s\}}, \\ t_1 + \cdots + t_r = k}} \left(\prod_{v=1}^{r} \boldsymbol{Bin}(t_v; A_v - \boldsymbol{1}_{\{v=s\}}, \theta_v P_{s,v}) \right)$$

(20)

where $\boldsymbol{Bin}(t_v; A_v - \boldsymbol{1}_{\{v=s\}}, \theta_v P_{s,v}) = \binom{A_v - \boldsymbol{1}_{\{v=s\}}}{t_v}(\theta_v P_{s,v})^{t_v}(1 - \theta_v P_{s,v})^{A_v - \boldsymbol{1}_{\{v=s\}} - t_v}$.

Proof Note that each vertex in the graph is assigned treatment indpendently. Therefore,

$$\mathbb{P}(s \text{ has k treated neighbors} \mid s \text{ is untreated}) = \mathbb{P}(s \text{ has k treated neighbors})$$

for all k and $s = 1, \ldots, r$. This gives

$$\mathbb{P}(s \in M_0^{(k)}) = \mathbb{P}(s \text{ is untreated})\mathbb{P}(s \text{ has k treated neighbors})$$
$$= (1 - \theta_s)\mathbb{P}(s \text{ has k treated neighbors})$$

$$= (1 - \theta_s) \sum_{\substack{t_1, \ldots, t_r: \\ \forall v=1, \ldots, r \ 0 \le t_v \le A_v - \mathbf{1}_{\{v=s\}}, \\ t_1 + \cdots + t_r = k}} \mathbb{P}\left(\bigcap_{v=1}^{r} \{s \text{ has } t_v \text{ treated neighbors in } B_v\} \right)$$

$$= (1 - \theta_s) \sum_{\substack{t_1, \ldots, t_r: \\ \forall v=1, \ldots, r \ 0 \le t_v \le A_v - \mathbf{1}_{\{v=s\}}, \\ t_1 + \cdots + t_r = k}} \left(\prod_{v=1}^{r} \mathbb{P}(s \text{ has } t_v \text{ treated neighbors in } B_v) \right).$$

We now wish to evaluate $\mathbb{P}(s \text{ has } t_k \text{ treated neighbors in } B_v)$. Let n_v be the number of neighbors s (denoting a fixed individual in community B_s) has in B_v. Under a stochastic block model setup,

$$n_v \sim Bin(A_v - \boldsymbol{1}_{v=s}, P_{s,v})$$

$$t_v | n_v \sim Bin(n_v, \theta_v)$$

where the $A_v - \mathbf{1}_{v=s}$ arises from noting that s can have at most $A_s - 1$ neighbors in B_s (it cannot connect to itself). We want the unconditional distribution of t_v. Recall that moment generating function of $X \sim Bin(N, p)$ is $\mathbb{E}(z^X) = ((1 - p) + pz)^N$. Therefore,

$$\mathbb{E}[z^{t_v}] = \mathbb{E}[\mathbb{E}[z^{t_v}|n_v]] = \mathbb{E}\left[\left((1 - \theta_s) + \theta_s z\right)^{n_v}\right] = \left((1 - P_{s,v}) + P_{s,v}\left((1 - \theta_s) + \theta_s z\right)\right)^{A_v - \mathbf{1}_{v=s}}$$

$$= \left((1 - \theta_s P_{s,v}) + P_{s,v}\theta_s z\right)^{A_v},$$

giving $t_v \sim Bin(A_v - \mathbf{1}_{v=s}, \theta_s P_{s,v})$. This gives

$$\mathbb{P}(s \text{ has } t_v \text{ treated neighbors in } B_v) = Bin(t_v; A_v - \mathbf{1}_{\{v=s\}}, \theta_v P_{s,v})$$

from which (20) directly follows. $\qquad\square$

References

1. E.M. Airoldi, D.M. Blei, S.E. Fienberg, E.P. Xing, Mixed membership stochastic block- models. J. Mach. Learn. Res. (2008)
2. J.D. Angrist, The perils of peer effects. Labour Econ. (2014)
3. S. Aral, D. Walker, Creating social contagion through viral product design: a randomized trial of peer influence in networks. Manag. Sci. (2011)
4. S. Athey, D. Eckles, G.W. Imbens, Exact P-values for network interference. J. Am. Stat. Assoc. (2016)
5. E. Bakshy, D. Eckles, R. Yan, I. Rosenn, Social influence in social advertising: Evidence from field experiments, in *Proceedings of the 13th ACM Conference on Electronic Commerce*, 2012
6. R.M. Bond, C.J. Fariss, J.J. Jones, A.D.I. Kramer, C. Marlow, J.E. Settle, J.H. Fowler, A 61-million-person experiment in social influence and political mobilization. Nature (2012)
7. P. Holland, K. Laskey, S. Leinhardt, Stochastic block models: first steps. Soc. Netw. (1983)
8. C.F. Manski, Identification of endogenous social effects: the reflection problem. Rev. Econ. Stud. (1993)
9. C.R. Shalizi, A.C. Thomas, Homophily and contagion are generically confounded in observational social network studies. Sociol. Methods Res. (2011)
10. P. Toulis, E. Kao, Estimation of causal peer influence effects. J. Mach. Learn. Res. (2013)

Author Index

© Springer International Publishing AG 2018
S. Cornelius et al. (eds.), *Complex Networks IX*, Springer Proceedings
in Complexity, https://doi.org/10.1007/978-3-319-73198-8

Printed in the United States
By Bookmasters